ANALYSIS AND BEYOND

An Introduction with Examples and Exercises

ANALYSIS AND BEYOND

An Introduction with Examples and Exercises

Shigeru Kanemitsu
Kyushu Institute of Technology, Japan

Takako Kuzumaki
Gifu University, Japan

Jianya Liu
Shandong University, China

World Scientific

NEW JERSEY • LONDON • SINGAPORE • BEIJING • SHANGHAI • HONG KONG • TAIPEI • CHENNAI • TOKYO

Published by

World Scientific Publishing Co. Pte. Ltd.
5 Toh Tuck Link, Singapore 596224
USA office: 27 Warren Street, Suite 401-402, Hackensack, NJ 07601
UK office: 57 Shelton Street, Covent Garden, London WC2H 9HE

Library of Congress Cataloging-in-Publication Data
Names: Kanemitsu, Shigeru, author. | Kuzumaki, Takako, 1960– author. |
Liu, Jianya (Mathematics professor), author.
Title: Analysis and beyond : an introduction with examples and exercises /
Shigeru Kanemitsu, Takako Kuzumaki, Jianya Liu.
Description: Hackensack, New Jersey : World Scientific., [2021] |
Includes bibliographical references and index.
Identifiers: LCCN 2020044531 | ISBN 9789811224485 (hardcover) |
ISBN 9789811224492 (ebook for institutions) | ISBN 9789811224508 (ebook for individuals)
Subjects: LCSH: Mathematical analysis. | Calculus.
Classification: LCC QA300 .K354 2021 | DDC 515--dc23
LC record available at https://lccn.loc.gov/2020044531

British Library Cataloguing-in-Publication Data
A catalogue record for this book is available from the British Library.

For any available supplementary material, please visit
https://www.worldscientific.com/worldscibooks/10.1142/11943#t=suppl

Printed in Singapore

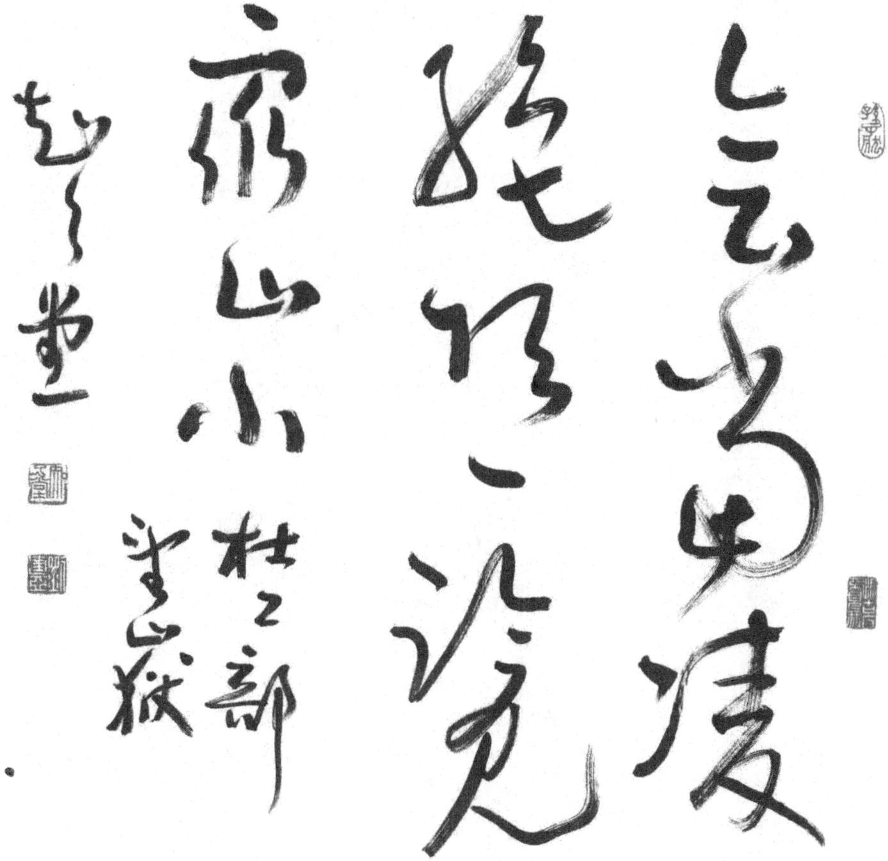
會當凌絕頂一覽眾山小
杜工部 望嶽

Preface

The aim of the book is to familiarize the reader who has just entered the university with sound basics of analysis to such an extent that the reader can go on to any areas of mathematics and related disciplines, i.e. the book serves as a smooth analytic continuation to advanced material. To explain how smooth, we recall an episode of D. Hilbert. When Mr. and Mrs. Hilbert invited their guests, Mrs. Hilbert noticed her husband's tie was spotted and she told him to go upstairs to change his tie. But after the arrival of their guests, Hilbert did not come down. One of the young guests was asked by Mrs. Hilbert to check what is going on with him. When he went upstairs to the bedroom, he found Hilbert sound asleep. The reason was that for him to untie his necktie is **analytically continued** to going to bed and sleeping. We hope this much smoothness will also be achieved for the readers towards the advanced material in this book.

We quote the following which is suggestive of the role of analysis.
"The beautiful ineffectual dreamer who comes to grief against hard facts. One always feels that Goethe's judgements are so true. True in the larger analysis," J. Joyce, Ulysses, p. 111, Penguin Books, 1986. We recall Goethe's words, "Geometry is frozen music" and in the middle age, geometry and mathematics are rather synonymous.

In one sentence, the book can be described as an elucidation of the **reversible process of differentiation and integration** by way of two featured principles: the **chain rule** and its inverse, **change of variable** and the **Leibniz rule** and its inverse, **integration by parts**. The chain rule or **differentiation of composite functions** is ubiquitous since almost all (a.a.) functions are composite functions of (elementary) functions and the change of variable method is ubiquitous as its reverse process. The Leibniz rule or **differentiation of the product of two functions** is essential since it makes differentiation non-linear and the method of integration by parts as ubiquitous as its reverse process.

In many mediocre textbooks, these two techniques for integration are frequently used for finding anti-derivatives and many people view them as computational tools. Indeed, above all things, to try to find anti-derivatives by an ingenious method would be rather a waste of time just like trying to find a factorization of a polynomial by hand. For it is known by Liouville's theorem that only a very small class of integrals can be expressed in terms of elementary functions. We make a rather smart use of them by elucidating the principles in every possible way as follows.

Integration by parts is used in unifying the sum and integral in Riemann-Stieltjes integration and at the far end in developing the theory of distributions in Chapter 15. Integration by parts for the Stieltjes integral with step function entails the partial summation which in turn includes the telescoping series technique ((2.8) below) as its special case. At the very basic level, the sum formula for a geometric progression (Theorem 1.2) is also a special case of the telescoping series.

The method of change of variable is used in defining integrals over a domain which are described by parameters including the inner product of differential forms. In line and surface integrals, the form of integration is not the same as with the change of variable, but as a manifestation of the local-global principle, which we may simply think of them as that.

process: diff.	inverse process: int.
chain rule = diff. compos. ftns	change of variable
Leibniz rule = diff. product of ftns	integration by parts
local	global

Table 1. Inverse processes

Analysis is based on calculus and the very terminology "calculus" which arises from the ancient Latin word *calcul* meaning counting (stone) coins, suggests that it is the computational study of infinitesimal quantities, i.e. treatment of various limits. The essence of analysis lies in the local-global principle, i.e. differentiation consists of the speculation that there is a rule that governs infinitesimal change and integration lies in the speculation that collecting these local data gives rise to the whole: Genotype determines phenotype, to borrow a passage from molecular biology. Our two main principles are in conformity with this core of analysis, i.e. while keeping linearity by the chain rule, we further warp from it by Leibniz rule. For treating the delicate limit process, some classical books are of great reference, e.g. [Whittaker and Watson (1927)].

As one of the most basic subjects, calculus has the role of abridging the gap between high school mathematics and conceptual material taught in universities. There are some very comprehensive and standard books including [Hille (1972)] [Dieudonné (1969)], [Rudin (1966)], [Takagi (1938, 2010)] which look however written for those who have already acquired the very basics to learn more about analysis. In an ordinary course for one semester, it would be difficult for the student to digest such books. The present book is meant for filling a gap between computation-oriented textbooks and more research-oriented books such as [Apostol (1957)], [Buck (1965)] [Rudin (1953)], [Widder (1989)], thus the title "analysis and beyond". To achieve this, there must be some new ways of arranging and presenting the material. Apparently, if we spend too much space for one variable calculus, the goal will never be reachable. As a general rule, we adopt some novel treatments of the material and make use of several new tools. These simplify many results which would require more time and effort otherwise. Some other features are in order which distinguish this book from others.

Firstly we unify as much notion as possible in one paradigm or in a hierarchy. One typical feature is the full use of two main techniques in integration as mentioned above.

Another method often used is the clearing of denominators. By this, differentiation is thrown into a more lucid setting enabling the transition to the several variables case smooth. This is used coupled with the method of partial fraction expansion. To find the partial fraction expansion, we employ complex variables and clear the denominator to apply the method of undetermined coefficients in the same way as we apply it to determine the Taylor coefficients. By this we can cover both finite difference equations and linear differential equations. The method is sound since it is assured by Liouville's theorem [Chakraborty *et al.* (2016)] in complex analysis. Speaking of complex analysis, we shall make good use of the book [Chakraborty *et al.* (2016), Corollary 1.3, p. 28] which is a twin sister of the present one since many results in real analysis have the complete form in complex analysis.

Secondly, we intentionally use a lot of operators so that the reader will be familiar with their use in more advanced subjects. In addition to several differentiation-related operators including the derivative and its inverse, gradient, difference and shift operators are also introduced. For example, the well-known telescoping series technique is viewed as the product of operators $\Delta^{-1}\Delta = I$ as in (2.9) below. As mentioned above, this is a special case of partial summation.

Thirdly we provide some advanced subjects in the book whenever they give a better perspective on the material. For example, Lemma 11.2 serves not only as mnemonics for the change of variable formula for multiple integrals but also for a precursor of homological algebra in § 14.1. We state rudiments of topological spaces in Chapter 16 which complement and elucidate the axioms of real number system §2.2. For further details on topological spaces, cf. [Engelking (1977)], [Kelley (1955)] etc.

Last but not least, we would like to express our hearty thanks to Professor Jay Mehta for his devoted and efficient help in preparing the final version of the book. We thank Mr. Y. Kamata for preparing some figures. As always, the book has been completed thanks to kind and constant encouragement by Ms. Kwong Lai Fun to whom we are very grateful.

The cover picture shows Mt. Tai in summer.
The calligraphy is from *Gazing at Mt. Tai* by Du Fu.

釋文：
會當凌絕頂，一覽眾山小。
杜工部望嶽，知了堂。

印章：
孫子狀；恨古人不見我；知了堂；劉建亞。

The first line is interpreted as:
Try to ascend the mountain's crest
It dwarfs all peaks under our foot.

Then followed by:
The poet Du gazing at Mt. Tai—title of the poem
Zhi-Liao-Tang—pen name of Jianya Liu
Finally there are stamped two seals.

S. Kanemitsu, T. Kuzumaki, and J.-Y. Liu

Contents

Chapter 1

Preliminaries

1.1 Mathematical induction

We denote the set of natural numbers by $\mathbb{N}$:

$$\mathbb{N} = \{1, 2, 3, \cdots\}$$

in which we assume the principle of mathematical induction (complete induction).

"Mathematical induction" (sometimes simply "**induction**") refers to the principle of proving a proposition with regard to all natural numbers $\geq n_0 \in \mathbb{N}$ by proving first that

(i) it is true for $n_0 \in \mathbb{N}$, and then that
(ii) if we assume its validity for $(n_0 \leq) \, n \in \mathbf{N}$, then it is valid also for $n+1$.

Instead of (ii) we may adopt either

(ii′) if we assume the proposition is valid for $n_0, n_0+1, \cdots, n$, then it is also valid for $n+1$,

Proposition 1.1. *Suppose X is a subset of $\mathbb{N}$. Then if* (i) $1 \in X$ *and* (ii) *if* $1, \cdots, n-1 \in X$, *then* $n \in X$ *holds, then* $X = \mathbb{N}$.

This implies the important principle.

Theorem 1.1. *Suppose X is a non-empty subset of $\mathbb{N}$. Then X has the minimal element* $\min X$.

Proof. Suppose there does not exist $\min X$. Then $1 \notin X$. If $1, \cdots, n-1 \notin X$, then $n \notin X$ and mathematical induction implies that $X = \varnothing$, a contradiction. □

Exercise 1. Prove the following binomial theorem by mathematical induction.

Binomial theorem. Provided that the product of a and b are commutative, we have

$$(a+b)^n = \sum_{r=0}^{n} \binom{n}{r} a^{n-r} b^r, \tag{1.1}$$

where $\binom{n}{r} = {}_nC_r = \frac{n!}{r!(n-r)!} = \frac{n(n-1)\cdots(n-r+1)}{r!}$ is called the **binomial coefficient** expressing the number of combinations of choosing r objects from n objects.

$n! = 1 \cdot 2 \cdot \cdots \cdot n$ is called the **factorial** of n and $0! = 1$. The symbol is used probably because of its extremely large size

$$n! = \Gamma(n+1) \sim \sqrt{2\pi n}\left(\frac{n}{e}\right)^n, \tag{1.2}$$

where e indicates the base to the natural logarithm to be introduced in Remark 12.1. The formula is referred to as the Stirling formula. [The identity

$$\binom{z}{r} = \frac{z\,(z-1)\cdots(z-r+1)}{r!}, \quad \binom{z}{0} = 1 \tag{1.3}$$

holds for all $z \in \mathbb{C}$ when viewed as the definition of a polynomial of degree r for $0 \le r \in \mathbb{Z}$.]

Remark 1.1. In the above solution, what is essential is the identity

$$\binom{n-1}{r} + \binom{n-1}{r-1} = \binom{n}{r} \tag{1.4}$$

which is the basis of **umbral calculus** as well as the familiar mnemonics, the Pascal triangle. This can be proved computationally but the following interpretation is much more instructive. In choosing r objects from n objects, one specifies one particular object x, say and distinguish the choice by whether it is contained or not. The number of combinations not containing x is $\binom{n-1}{r}$ since one has to choose r objects while the combination containing x is $\binom{n-1}{r-1}$ since one needs to choose the remaining $r-1$ objects.

Examples of umbral calculus will be given in Exercise 7 and §13.7.

Example 1.1. The special case of the binomial theorem

$$2^n = \sum_{r=0}^{n} \binom{n}{r} \tag{1.5}$$

counts the number of all subsets of a set with n elements. For if $X = \{x_1, \cdots, x_n\}$, then the left-hand side checks if x_1 belongs to the subset Y or not. To each case there are two possibilities of x_2, and so on. Hence there are 2^n subsets. On the other hand, the right-hand side counts the number of subsets having 0, 1, $\cdots, n$ elements. Suggested by this, we denote the family of all subsets of a set X by 2^X (or sometimes $\mathcal{P}(X)$), called the **power set**.

Exercise 2. Prove that

$$\lim_{h\to 0} \frac{(x+h)^n - x^n}{h} = nx^{n-1}.$$

The binomial theorem appears in a few places below including Example 8.5.

Exercise 3. Let the gcd (the greatest common divisor, g.c.d.) of $m, n \in \mathbb{N}$ be $d \in \mathbb{N}$. Prove that there exist $x, y \in \mathbb{Z}$ such that $mx + ny = d$.

Solution. Let $\mathfrak{i} = \{mx + ny | x, y \in \mathbb{Z}\}$ and $\mathfrak{i}_+ = \mathfrak{i} \cap \mathbb{N}$. Then since $m, n \in \mathfrak{i}_+ \subset \mathbb{N}$, $\mathfrak{i}_+$ is non-empty and there exists $d_0 = \min \mathfrak{i}_+ = mx_0 + ny_0$. Hence $d | d_0$. For any member $mx + ny \in \mathfrak{i}_+$ by Euclidean division (Theorem 4.5), $mx + ny = d_0 q + r$, $0 \le r < d_0$, then r must be 0. For otherwise, $1 \le r = mx + ny - d_0 q \in \mathfrak{i}_+$, $r < d_0$, a contradiction. Hence $d_0 | m$ and $d_0 | n$, so that $d_0 | d$. Hence $d = d_0 = mx_0 + ny_0$.

Remark 1.2. In the above Exercise we may suppose $\mathfrak{i}$ is an additive subgroup which is closed under multiplication by any element of $a \in \mathbb{Z}$. Then since $\pm n \in \mathfrak{i}$, it is a non-empty subset of $\mathbb{Z}$ and the argument follows along the same lines and we conclude that any member of $\mathfrak{i}$ is a multiple of $d = \min \mathfrak{i}_+$, so that $\mathfrak{i} = d\mathbb{Z}$, i.e. it is a principal ideal: Euclidean domain is a PID.

Definition 1.1. For the meaning of $\mathbb{R}^n$ we refer to §1.3 below. In linear algebra, we usually denote an n-dimensional real vector $\boldsymbol{x}$ as a **column vector**

$$\boldsymbol{x} = \begin{pmatrix} x_1 \\ x_2 \\ \vdots \\ x_n \end{pmatrix} \in \mathbb{R}^n. \tag{1.6}$$

In what follows we often use the notation by transposing it into row vector to save space:

$$\boldsymbol{x} = {}^t(x_1, \cdots, x_n) \in \mathbb{R}^n \tag{1.7}$$

t indicating the transpose.

The following exercise will be continued on to Exercise 145.

Exercise 4. Define the length (norm) $|\boldsymbol{x}|(\geq 0)$ of a vector $\boldsymbol{x}$ by

$$|\boldsymbol{x}|^2 = \sum_{j=1}^{n} x_j^2. \tag{1.8}$$

For $\boldsymbol{x}_1, \boldsymbol{x}_2, \cdots, \boldsymbol{x}_m \in \mathbb{R}^n$ prove the **triangular inequality** by induction: $\left|\sum_{k=1}^{m} \boldsymbol{x}_k\right| \leq \sum_{k=1}^{m} |\boldsymbol{x}_k|$. Also, locate the case where the equality holds true.

Solution. The case of two vectors is the key. For $\boldsymbol{x} = {}^t(x_1, \cdots, x_n), \boldsymbol{y} = {}^t(y_1, \cdots, y_n) \in \mathbb{R}^n$, we introduce the standard **inner product** (**scalar product**):

$$(\boldsymbol{x}, \boldsymbol{y}) = \boldsymbol{x} \cdot \boldsymbol{y} = \sum_{j=1}^{n} x_j y_j \in \mathbb{R}. \tag{1.9}$$

In the sequel we use both symbols and terms interchangeably. We may say the inner product $\boldsymbol{x} \cdot \boldsymbol{y}$ or the scalar product $(\boldsymbol{x}, \boldsymbol{y})$ since in literature both are used. Then $|\boldsymbol{x}|^2 = (\boldsymbol{x}, \boldsymbol{x})$ and we have the **Lagrange formula**

$$|\boldsymbol{x}|^2 |\boldsymbol{y}|^2 = (\boldsymbol{x}, \boldsymbol{y})^2 + \frac{1}{2} \sum_{1 \leq i, j \leq n} (x_i y_j - x_j y_i)^2, \tag{1.10}$$

whence we conclude the important **Cauchy-Schwarz inequality**

$$(\boldsymbol{x}, \boldsymbol{y})^2 \leq |\boldsymbol{x}|^2 |\boldsymbol{y}|^2. \tag{1.11}$$

The equality holds when $\boldsymbol{x} // \boldsymbol{y}$. Cf. Exercise 62.
Another proof depends on the quadratic inequality (supposing $\boldsymbol{x} \neq \boldsymbol{o}$)

$$0 \leq (t\boldsymbol{x} + \boldsymbol{y}, t\boldsymbol{x} + \boldsymbol{y}) = |\boldsymbol{x}|^2 t^2 + 2(\boldsymbol{x}, \boldsymbol{y})t + |\boldsymbol{y}|^2. \tag{1.12}$$

Exercise 5. Define the distance between two vectors $\boldsymbol{x}, \boldsymbol{y} \in \mathbb{R}^n$ by

$$d(\boldsymbol{x}, \boldsymbol{y}) = |\boldsymbol{x} - \boldsymbol{y}|. \tag{1.13}$$

Prove that $d : \mathbb{R}^n \times \mathbb{R}^n \to \mathbb{R}_{\geq 0}$ satisfies the conditions for the distance function:
(i) $d(\boldsymbol{x}, \boldsymbol{y}) \geq 0$ and $d(\boldsymbol{x}, \boldsymbol{y}) = 0$ if and only if $\boldsymbol{x} = \boldsymbol{y}$.
(ii) $d(\boldsymbol{x}, \boldsymbol{y}) = d(\boldsymbol{y}, \boldsymbol{x})$.
(iii) (the triangular inequality) $d(\boldsymbol{x}, \boldsymbol{z}) \leq d(\boldsymbol{x}, \boldsymbol{y}) + d(\boldsymbol{y}, \boldsymbol{z})$.

Exercise 6. Prove the arithmetic-geometric mean inequality by induction: For $a_i > 0$, $1 \le i \le k$ we have

$$\frac{A_k}{k} \ge (a_1 \cdots a_k)^{\frac{1}{k}}, \tag{1.14}$$

where $A_k = \sum_1^k a_i$.

Solution. We note the equality

$$\frac{A_{k+1}}{k+1} = \frac{A_k + a_{k+1}}{k+1} = \frac{A_k}{k} + \frac{1}{k+1}\left(a_{k+1} - \frac{A_k}{k}\right), \tag{1.15}$$

and

$$\left(\frac{A_{k+1}}{k+1}\right)^{k+1} = \sum_{m=0}^{k+1} \binom{k+1}{m} \left(\frac{A_k}{k}\right)^{k-m+1} \left(\frac{1}{k+1}\right)^m \left(a_{k+1} - \frac{A_k}{k}\right)^m. \tag{1.16}$$

The summation of $2s$-term and $(2s+1)$-term is not negative as following

$$\begin{aligned}
&\binom{k+1}{2s}\left(\frac{A_k}{k}\right)^{k-2s+1}\left(\frac{1}{k+1}\right)^{2s}\left(a_{k+1} - \frac{A_k}{k}\right)^{2s} \\
&+ \binom{k+1}{2s+1}\left(\frac{A_k}{k}\right)^{k-2s}\left(\frac{1}{k+1}\right)^{2s+1}\left(a_{k+1} - \frac{A_k}{k}\right)^{2s+1} \\
&= \binom{k+1}{2s}\left(\frac{A_k}{k}\right)^{k-2s}\left(\frac{1}{k+1}\right)^{2s}\left(a_{k+1} - \frac{A_k}{k}\right)^{2s} \\
&\qquad \times \frac{k(k-2s+1)a_{k+1} + 2s(k+2)A_k}{k(2s+1)(k+1)} \ge 0.
\end{aligned}$$

Hence

$$\begin{aligned}
\left(\frac{A_{k+1}}{k+1}\right)^{k+1} &\ge \left(\frac{A_k}{k}\right)^{k+1} + \left(\frac{A_k}{k}\right)^k \left(a_{k+1} - \frac{A_k}{k}\right) \\
&= \left(\frac{A_k}{k}\right)^k a_{k+1} \ge a_1 \cdots a_k \cdot a_{k+1}.
\end{aligned} \tag{1.17}$$

Exercise 7. Suppose f and g are n times differentiable. Prove **Leibniz's formula**

$$(fg)^{(n)} = \sum_{r=0}^{n} \binom{n}{r} f^{(n-r)} g^{(r)}. \tag{1.18}$$

Remark 1.3. The case $n = 1$ is the most important differentiation rule for the product of two differentiable functions. Differentiation is by definition a linear process but it accommodates the product, making its applicability to non-linear problems.

$$(fg)' = f'g + fg'. \tag{1.19}$$

Introducing the differential operator $D = \frac{\mathrm{d}}{\mathrm{d}x}$ and its inverse operator, the anti-derivative, $D^{-1} = \int$, we derive the formula for integration by parts from (1.19):

$$D^{-1}f'g = fg - D^{-1}fg'. \tag{1.20}$$

Cf. (10.18).

Exercise 8. Prove **de Moivre's formula**

$$(\cos\theta + i\sin\theta)^n = \cos n\theta + i\sin n\theta, \quad n \in \mathbb{Z}, \tag{1.21}$$

where i, the imaginary unit, will be introduced in §1.6. For the moment, we just use the rule $i^2 = -1$. Cf. Example 1.3.

1.2 Sum formula for a geometric sequence

Theorem 1.2. *The* **sum formula for a geometric progression**

$$(1-r)\sum_{k=0}^{n-1} r^k = 1 - r^n, \quad \sum_{k=0}^{n-1} r^k = \frac{1-r^n}{1-r} \quad (r \neq 1), \tag{1.22}$$

and the factorization formula

$$(a-b)\sum_{k=0}^{n-1} a^{n-k-1}b^k = a^n - b^n, \tag{1.23}$$

are equivalent.

Proof follows under the substitution $r = \frac{b}{a}$. The proof of (1.22) follows from (2.8) since the LHS is $\sum_{k=0}^{n-1}(r^k - r^{k+1})$. Hence the sum formula for a geometric progression is also a special case of integration by parts. Cf. Remark 10.2.

Exercise 9. Solve the quintic equation $z^5 = 1$ in $\mathbb{R}$ and in $\mathbb{C}$.

Solution. We have the decomposition

$$z^5 - 1 = (z-1)f(z), \quad f(z) = z^4 + z^3 + z^2 + z + 1. \tag{1.24}$$

If $0 \neq x \in \mathbb{R}$, then $f(x) = x^2\left(X^2 + X - 1\right)$ with $X = x + \frac{1}{x}$. Since $X^2 + X - 1 = \left(X + \frac{1}{2}\right)^2 - \frac{5}{4}$ and $X = x + \frac{1}{x} \geq 2$ if $x > 0$ and $-X =$

$-x - \frac{1}{x} \geq 2$ for $x < 0$, it follows that $|X + \frac{1}{2}| \geq \min\{\frac{5}{2}, \frac{3}{2}\}$, whence $X^2 + X - 1 = \frac{9}{4} - \frac{5}{4} > 0$. Hence $z = 1$ is the only solution in $\mathbb{R}$. In $\mathbb{C}$, $X^2 + X - 1 = 0$ is solvable and $X = \frac{-1 \pm \sqrt{5}}{2}$. Substituting this in $X = z + \frac{1}{z}$ and solving in z, we obtain

$$z = \frac{-1+\sqrt{5} \pm \sqrt{10+2\sqrt{5}}i}{4}, \frac{-1-\sqrt{5} \pm \sqrt{10-2\sqrt{5}}i}{4}.$$

Cf. Example 1.7.

Exercise 10. Prove that

$$\lim_{b \to x} \frac{b^n - x^n}{b - x} = nx^{n-1}.$$

Exercise 11. Prove that for $|r| < 1$ we have

$$\lim_{n \to \infty} r^n = 0, \tag{1.25}$$

whence that

$$\lim_{n \to \infty} \sum_{k=0}^{n-1} r^k = \frac{1}{1-r}, \tag{1.26}$$

i.e. the **geometric series** is convergent for $|r| < 1$:

$$\sum_{n=0}^{\infty} r^n = \frac{1}{1-r}. \tag{1.27}$$

Solution. Since $\frac{1}{|r|} > 1$, we may write $\frac{1}{|r|} = 1 + c$, where $c > 0$. Hence

$$|r|^n = (1+c)^{-n} \leq \frac{1}{1+cn} = O\left(\frac{1}{n}\right) \to 0 \tag{1.28}$$

as $n \to \infty$ by Lemma 2.2.

The geometric series will appear in a few places below, in Theorem 3.6, g-adic expansion in Chapter 9, etc. For the treatment of the divergent geometric series with $r = -1$, cf. (12.4) and subsequent argument.

Exercise 12. Suppose $r \neq 1$ and $r^q = 1$ for $2 \leq q \in \mathbb{N}$. Then prove that

$$\sum_{n=1}^{q} nr^n = \frac{qr}{r-1} \tag{1.29}$$

and that

$$\sum_{n=1}^{q-1} n(r^n - r^{-n}) = q\frac{r^{1/2} + r^{-1/2}}{r^{1/2} - r^{-1/2}}. \tag{1.30}$$

1.3 Cartesian product

The book [Davis and Hersh (1986)] begins with the passage

"In November 1619, René Descartes, a twenty-three-year-old Frenchman, dreamed of a world unified by mathematics, a world in which all intellectual matters could be dealt with rationality by logical computation. 18 years since then he wrote the most famous book 'Discourse'."

In [Kotre (1995), p. 25] the author refers to the "great Cartesian Theater," meaning apparently the notion of reason as described by "Cogito ergo sum."

In this section we are mainly concerned with his mathematically most important invention of coordinates, *Cartesian product*. This is so fundamental that in engineering disciplines there is misunderstanding that it is so trivial. However, we note that the mathematically trivial fact to the effect that the matrices, which are again thought of very common and trivial, are just coordinates and are regarded as embedded in the Cartesian product is not well conceived by engineers. This can be perceived when one introduces the distance (norm) of two matrices. The range of applicability of Cartesian products is so wide and diverse that we will glimpse just part of it by concrete examples. As is often the case, whenever one needs to construct a real entity which corresponds to an abstract system, one appeals to the Cartesian product (direct product), or the coordinates (sequences). The notion is one of the two great mathematical inventions of R. Descartes who introduced coordinates into the study of geometry, whence the name "Cartesian". We recall that as opposed to the Cartesian product, the free product is the most general construction from a given family of sets. It is indeed a dual concept of the direct product in case of groups. Cf. [Kitajima and Kanemitsu (2012)].

Also **given a set with a certain property, one uses its subsets (as generators), Cartesian product, equivalence classes to construct a new entity having a similar property**.

The Cartesian product or the direct product of two sets X, Y is the set of all ordered pairs (x, y), where an ordered pair means that two such pairs (x, y) and (x', y') are equal if and only if $x = x'$ and $y = y'$:

$$X \times Y = \{(x, y) | x \in X, y \in Y\}. \tag{1.31}$$

Exercise 13. (Kuratowski-Wiener) Prove that an ordered pair (x, y) may be expressed as $\{x, \{x, y\}\}$.

Similarly, from n sets $X_1, \cdots, X_n$ we define the Cartesian product $X_1 \times$

$\cdots \times X_n$ as the set of all n-tuples $(x_1, \cdots, x_n)$ with $x_j \in X_j\,(1 \le j \le n)$:

$$X_1 \times \cdots \times X_n = \{(x_1, \cdots, x_n) | x_j \in X_j\,(1 \le j \le n)\}, \tag{1.32}$$

where two n-tuples are equal if and only if each corresponding entries are equal. When they are equal, $X_1 = \cdots = X_n = X$, we write

$$X^n = X \times \cdots \times X = \{(x_1, \cdots, x_n) | x_j \in X\}. \tag{1.33}$$

There is an enormous amount of examples. E.g. in Exercise 4 we have encountered the n-dimensional Euclidean space $\mathbb{R}^n$. There we used the column vector expression since this appears in linear algebra as a result of defining the product of a matrix and the vector: $A\boldsymbol{x}$ and for this the column vector expression is convenient. If one uses the row vector expression, then the product will be $\boldsymbol{x}^t A$ with the transposed matrix ${}^tA$.

We now state the most general definition.

Definition 1.2. Let Λ be an arbitrary index set and let X_λ $(\lambda \in \Lambda)$ be a family of sets indexed by Λ. Let a be a function from Λ to X_λ $(\lambda \in \Lambda)$. Then we denote the function as

$$a = (a_\lambda) = (a_\lambda)_{\lambda \in \Lambda} \tag{1.34}$$

and we often refer to this as coordinates or a sequence in case when Λ is a countable set. We denote the set of all such functions a by

$$\prod X_\lambda = \prod_{\lambda \in \Lambda} X_\lambda = \{(a_\lambda)_{\lambda \in \Lambda}\} \tag{1.35}$$

and refer to it as the **Cartesian product** (or direct product) of X_λ $(\lambda \in \Lambda)$. By definition, two functions with the same domain (of definition) coincide if and only if each of their values coincide, which means that two sequences (a_λ) and (b_λ) coincide if and only if $a_\lambda = b_\lambda$ for each $\lambda \in \Lambda$. In the case where X_λ $(\lambda \in \Lambda)$ are some algebraic systems with identity e_λ, then we denote the subset of $\prod X_\lambda$ consisting of those all but finite number of whose entries are the identities by $\sum X_\lambda$ (or sometimes $\oplus X_\lambda$) and refer to it as the **direct sum**.

Example 1.2.

(i) If in (1.35), all the X_λ's are Abelian groups with unity (very often denoted 0_λ), then the Cartesian product (1.35) becomes an Abelian group with respect to *componentwise addition*:

$$(a_\lambda) + (b_\lambda) = (a_\lambda + b_\lambda) \text{ for each } \lambda \in \Lambda \tag{1.36}$$

with the identity (0_λ).

(ii) If in (1.35), all the X_λ's are rings, then the Cartesian product (1.35) becomes a ring with respect to *componentwise* addition and multiplication:

$$(a_\lambda) + (b_\lambda) = (a_\lambda + b_\lambda),\ (a_\lambda)(b_\lambda) = (a_\lambda b_\lambda),\ \text{for each } \lambda \in \Lambda. \tag{1.37}$$

Definition 1.3. Let R be a commutative ring with unity 1. A sequence (coordinate) γ is a function from the set of non-negative integers $\mathbb{Z}_{\geq 0} = \mathbb{N} \cup \{0\}$ (or any countable set, e.g. $\mathbb{N}$) into a ring R. Hence it can be expressed as

$$\gamma = (c_0, c_1, \cdots) \in R^\infty = \prod R = R \times R \times \cdots, \tag{1.38}$$

where $\gamma(n) = c_n$ and $n \in \mathbb{N} \cup \{0\}$ is called the index set. Among the elements of R^∞ there are defined natural operations—componentwise addition and multiplication: For $\alpha = (a_0, a_1, \cdots)$, $\beta = (b_0, b_1, \cdots) \in R^\infty$,

$$\begin{aligned} \alpha + \beta &= (a_0 + b_0, a_1 + b_1, \cdots), \\ \alpha\beta &= (a_0 b_0, a_0 b_1 + a_1 b_0, \cdots, c_j, \cdots), \end{aligned} \tag{1.39}$$

where

$$c_j = \sum_{k+\ell=j} a_k b_\ell \tag{1.40}$$

is the **Cauchy product**. The scalar product by elements of R may be defined by

$$c\alpha = (ca_0, ca_1, \cdots). \tag{1.41}$$

With these operations R^∞ forms an R-module as well as a ring. By the definition of the equality in R^∞, we may express γ in the form of a power series (**formal power series**)

$$\gamma = c_0 + c_1 X + c_2 X^2 + \cdots \tag{1.42}$$

with the understanding that two formal power series coincide if and only if all the corresponding coefficients are equal. Here the symbol X is called an **indeterminate** and X^j just shows the $(j+1)$st position of the coefficient c_j in the sequence (1.38). The set of all the formal power series over R is denoted by

$$R[[X]] = \{c_0 + c_1 X + c_2 X^2 + \cdots | c_j \in R\} \tag{1.43}$$

on which there is defined the componentwise addition and the new operation, the Cauchy product as multiplication: For two formal power series $\alpha(X) = a_0 + a_1 X + a_2 X^2 + \cdots$, we define correspondingly to (1.39)

$$\begin{aligned} \alpha(X) + \beta(X) &= a_0 + b_0 + (a_1 + b_1)X + \cdots, \\ \alpha(X)\beta(X) &= a_0 b_0 + (a_0 b_1 + a_1 b_0)X + \cdots + c_j X^j + \cdots, \end{aligned} \tag{1.44}$$

which explains the reason why we introduce the Cauchy product (1.40). The scalar product by elements of R may be defined as

$$c\alpha(X) = ca_0 + ca_1X + \cdots . \tag{1.45}$$

With these operations $R[[X]]$ forms an R-module as well as a ring (i.e. an algebra), called the **formal power series ring**.

A **polynomial** γ is a terminating formal power series, i.e. a sequence with all but finite number of coordinates being 0. Therefore there exists the maximal index $n \in \mathbb{N} \cup \{0\}$ such that $c_n \neq 0$, $c_j = 0$, $j > n+1$:

$$\gamma = (c_0, c_1, \cdots, c_n, 0, 0, \cdots), \tag{1.46}$$

which, correspondingly to (1.42), may be expressed as

$$\gamma = c_0 + c_1X + c_2X^2 + \cdots + c_nX^n \tag{1.47}$$

called a polynomial of **degree** n (denoted: $\deg\gamma = n$). c_n is called the leading coefficient. We assume that $\deg\gamma = -\infty$ for $\gamma = (0, 0, \cdots, 0)$. Correspondingly to (1.43), we denote the set of all polynomials over R by $R[X]$:

$$R[X] = \{c_0 + c_1X + c_2X^2 + \cdots + c_nX^n | c_j \in R\}. \tag{1.48}$$

Correspondingly to (1.44) and (1.45), we may define the operations on $R[X]$: For $\alpha(X) = a_0 + a_1X + \cdots + a_mX^m$, $\beta(X) = b_0 + b_1X + \cdots + b_nX^n$

$$\alpha(X) + \beta(X) = a_0 + b_0 + (a_1 + b_1)X + \cdots + (a_\ell + b_\ell)X^\ell, \ell = \max\{m, n\}, \tag{1.49}$$

$$\alpha(X)\beta(X) = a_0b_0 + (a_0b_1 + a_1b_0)X + \cdots + a_mb_nX^{m+n},$$

under which $R[X]$ forms a ring—the polynomial ring over R. The scalar product by elements of R may be defined

$$c\alpha(X) = ca_0 + ca_1X + \cdots + ca_mX^m. \tag{1.50}$$

With these operations $R[X]$ forms an R-algebra referred to as the **polynomial ring** over R.

Theorem 1.3. *For any polynomials $f, g \in R[X]$ we have*

$$\deg(fg) = \deg(f) + \deg(g). \tag{1.51}$$

We view R as embedded in $R[X]$ by the injection

$$f : R \to R[X]; \quad f(c) = c \tag{1.52}$$

and we usually identify R with its image under this injective isomorphism and regard $R \subset R[X]$. Writing X for $1X$, we see that the power X^j in $R[X]$

gives the monomial $1X^j$ and cX^j is the product of c and $1X^j$ in $R[X]$. In this way, the polynomial in (1.47) may be viewed as constructed from c_j's and X by the operations in $R[X]$. This interpretation is the same as the construction of a group ring $R[G]$, where the semi-group G is the set of all powers of X.

Exercise 14. Prove that the mapping in (1.52) is an injective ring homomorphism.

1.4 Algebraic structure

Definition 1.4. Let K be a non-empty set. A binary operation $*$ is a mapping from the direct product to K: $* : K \times K \to K$ which maps each ordered pair (a, b) to the 'product' $a * b \in K$. A set K is called a (commutative) **field** if there are defined four (binary) operations in arithmetic—addition, subtraction, multiplication and division (save for one by 0) and these can always be conducted within K.

More precisely, we suppose the following. There are two binary operations, called addition and multiplication, $+$ and $\cdot$ defined in K.
(I) K is an **Abelian group** with respect to addition. I.e.
(0) if $a, b \in K$, then $a + b \in K$.
(i) (associative law). For $a, b, c \in K$,

$$(a + b) + c = a + (b + c).$$

(ii) (existence of identity (unity) element). There is the zero element 0 such that for any $a \in K$

$$a + 0 = 0 + a = a.$$

(iii) (existence of inverse elements). For each a there is the additive inverse $-a$ such that

$$a + (-a) = -a + a = 0.$$

(iv) (commutative law). For $a, b \in K$,

$$a + b = b + a.$$

(II) $K^\times = K - \{0\}$ is a multiplicative Abelian group, i.e.
(0) if $a, b \in K$, then $a \cdot b \in K$ (for simplicity we write ab for $a \cdot b$).
(i) (associative law). For $a, b, c \in K$,

$$(ab)c = a(bc).$$

(ii) (existence of identity (unit) element). There is the identity element 1 such that for any $a \in K$

$$a1 = 1a = a.$$

(iii) (existence of inverse elements). For $a \neq 0$, there is the multiplicative inverse a^{-1} such that

$$aa^{-1} = a^{-1}a = 1. \tag{1.53}$$

(iv) (commutative law). For $a, b \in K$,

$$ab = ba.$$

(v) (distributive law). For $a, b, c \in K$,

$$a(b + c) = ab + ac, \quad (a + b)c = ac + bc.$$

Remark 1.4. (i) If K satisfies (I), (0)–(iii), K is called a (non-Abelian) **group**. If K satisfies (I) and (II) (0)–(i) and (v), we say that K is a (non-commutative) **ring**. If it satisfies multiplicative commutativity in (iv), it is called a commutative ring.

We define subtraction as the inverse operation of addition:

$$a - b = a + (-b),$$

and division as one of multiplication:

$$a \div b = ab^{-1}$$

the principle of Yin and Yang.

Therefore there are essentially two operations in a field, addition and multiplication, and (v) determines the relation between $+$ and $\cdot$. There are other types of distribution laws. (i) is also essential in the sense that they claim the result of two ways of operations, a and b first and then c, and b and c first and then a, is the same. Without this rule, there would occur an excessive complication. Indeed, if we denote the number of ways of putting parentheses in the product of n elements by $f(n)$, then we can find the explicit formula for it: (13.70). Additive commutativity is always assumed in a field or more generally in a ring.

Let V be an additive Abelian group and let F be a field which serves as $K^{\times}$ in (II).

(II)′ F acts on V, i.e.
(0)+(v) if $a \in F$, $u, v \in V$ then $av \in V$ and $a(u+v) = au + av$ holds true
(i)+(v) For $a, b \in F$, $u \in V$, $(a+b)v = av + bv$ and $(ab)v = a(bv)$ hold true
(ii) For $v \in V$ we have $1v = v$

V is then called a **vector space** (or a linear space) over F. It is possible to construct the direct product and direct sum of vector spaces as in §1.3. The n-dimensional Euclidean space in Definition 1.1 is a typical example. By the basic unit vectors (1.86), it may expressed as

$$\mathbb{R}^n = \mathbb{R}\boldsymbol{e}_1 \oplus \cdots \oplus \mathbb{R}\boldsymbol{e}_n. \tag{1.54}$$

Exercise 15. Prove that there is only one identity element.

Exercise 16. (Uniqueness of the inverse) Prove that for each a, there is a unique inverse element.

Solution. If there is another element a' satisfying the same equality as (1.53):

$$aa' = a'a = 1.$$

Then $a' = a'1 = a'aa^{-1} = 1a^{-1} = a^{-1}$.

Exercise 17. Prove that if a and b have inverses, then

$$(ab)^{-1} = b^{-1}a^{-1}. \tag{1.55}$$

Solution. Since $b^{-1}a^{-1}ab = abb^{-1}a^{-1} = 1$, we may appeal to Exercise 16 to conclude (1.55).

Exercise 18. Use the above rule, especially (v) to deduce that for all $a \in K$, $a0 = 0a = 0$.

Solution. Putting $b = c = 0$ in the distributive law in Definition 1.4, we get $a0 = a0 + a0$. Adding $-a0$ to both sides proves the assertion.

Exercise 19. Explain the reason why you cannot divide by 0.

Solution. Suppose division by 0 is possible in K in the sense of Remark 1.4, i.e. suppose that the inverse element 0^{-1} exists in K. Then $00^{-1} = 0^{-1}0 = 1$. However, by Exercise 18, $0^{-1}0 = 0$, whence we conclude that $1 = 0$. Hence for any $a \in K$, we have $a = a1 = a0 = 0$, i.e. $K = \{0\}$, a singleton, which is usually ruled out from consideration. Recently there is a new theory being developed of absolute mathematics over this one-element field.

1.5 Equivalence relation and classes

In the set of natural numbers, only addition and multiplication are always possible. Therefore we extend the range of $\mathbb{N}$ to construct the set $\mathbb{Z}$ of (rational) integers:

$$\mathbb{Z} = \{0, \pm 1, \pm 2, \cdots\}$$

in which addition and subtraction (and multiplication) are possible and $\mathbb{Z}$ forms a ring. A **ring** is nearly a field with division not always possible. The ring $\mathbb{Z}$ of integers is of particular type being a Euclidean domain, i.e. a domain in which the Euclidean division is possible.

The notion of equivalence classes and a quotient space is one of the most fundamental ones in modern mathematics and in daily life matters as well.

Definition 1.5. A (binary) relation $\sim$ defined on a set $X \neq \varnothing$ is called an **equivalence relation** if it satisfies
(i) (reflexive law) $x \sim x$.
(ii) (symmetric law) If $x \sim y$, then $y \sim x$.
(iii) (transitive law) If $x \sim y$ and $y \sim z$, then $x \sim z$.
If $x \sim y$, then x is said to be equivalent to y or conversely in view of (ii). The set C_x of all elements y equivalent to x is called an **equivalence class** containing x:

$$C_x = \{y \in X \mid y \sim x\}. \tag{1.56}$$

The following theorem provides us with a **quotient space** or a classification $X/\sim$ of the set X into disjoint union of mutually inequivalent classes.

Theorem 1.4. *The following statements are equivalent.*
(i) $x \sim y$
(ii) $C_x = C_y$
(iii) $C_x \cap C_y \neq \varnothing$.

Let Λ be a **complete set of representatives** of X with respect to the equivalence relation $\sim$ (that is, the set of all inequivalent elements of X w.r.t. $\sim$). Then $X/\sim = \cup_{\lambda \in \Lambda} C_\lambda$ is a quotient space of X.

Many concrete examples of quotient spaces will be given in what follows. Here we state a few most familiar examples.

- Two vectors (directed line segments in the plane or in the space) are equivalent if they can be overlapped via translation. A complete set of representatives may be given as the set of all directed line segments with initial points at the origin, which may be denoted by the coordinates of the terminating point.

- Two planar triangles are equivalent (congruent) if they can be overlapped via translation, rotation and reflection (the Euclidean group). Thus we may consider only one triangle as one representing the class of all congruent triangles.

- Two integers a, b are said to be congruent modulo a given natural number m written: $a \equiv b \bmod m$ if there is an integer q such that $a - b = mq$. This example will repeatedly appear.

- Isomorphic images constitute equivalence classes. We often identify an algebraic sub-system with its isomorphic image in a bigger system and call it an **embedding**.

On these examples light may be shed from a more advanced standpoint, i.e. as the group action. In the second example, the group acting is the **Euclidean group** and in the first, it is the subgroup of translations. In the third it is the subgroup consisting of all multiples of m.

Exercise 20. In $\mathbb{N} \times \mathbb{N}$ we define the relation

$$(n, m) \sim (n', m') \iff n + m' = n' + m. \tag{1.57}$$

Prove that this is an equivalence relation. In $X = \mathbb{N} \times \mathbb{N}/ \sim = \{C_{n,m}\}$ define the addition by

$$C_{(n,m)} + C_{(n',m')} = C_{(n+n',m+m')}. \tag{1.58}$$

Then X is an Abelian group. Elements of X have the form $C_{(n+m,n)} \leftrightarrow m$, $C_{(n,n)} \leftrightarrow 0$ of $C_{(n,n+m)} \leftrightarrow -m$. Define the product by $C_{(n,m)}C_{(n',m')} = C_{(nn'+mm',nm'+mn')}$. We have

$$\begin{aligned} &C_{(n_1+m_1,n_1)}C_{(n_2+m_2,n_2)} = C_{(n_3+m_1m_2,n_3)}, \\ &C_{(n_1+m_1,n_1)}C_{(n_2,n_2+m_2)} = C_{(n_3,n_3+m_1m_2)}, \\ &C_{(n_1,n_1+m_1)}C_{(n_2,n_2+m_2)} = C_{(n_3+m_1m_2,n_3)}, C_{(n,n)}C_{(n',m')} = C_{(n,n)}, \end{aligned} \tag{1.59}$$

in X. Then prove that X is a ring in the sense of Definition 1.4, which we interpret as the ring of rational integers $\mathbb{Z}$.

Exercise 21. We define the relation $\sim$ in $\mathbb{Z} \times (\mathbb{Z} - \{0\})$ as follows. Two elements $(a_1, b_1) \sim (a_2, b_2)$ if $a_1 b_2 - a_2 b_1 = 0$. Prove that this is an equivalence relation and describe equivalence classes. We denote the equivalence class containing (a_1, b_1) by a_1/b_1 or $\frac{a_1}{b_1}$. Then, defining addition and multiplication by $a_1/b_1 + a_2/b_2 = (a_1 b_2 + a_2 b_1)/(b_1 b_2)$, $(a_1/b_1)(a_2/b_2) = (a_1 a_2)/(b_1 b_2)$ prove that $\big(\mathbb{Z} \times (\mathbb{Z} - \{0\})\big)/ \sim$ forms a field — the **rational number field** — $\mathbb{Q}$ in the sense of Definition 1.4.

Exercise 22. Generalizing the case of integers modulo an integer m, prove that the following relation in an Abelian group G is an equivalence relation, where H is a subgroup of G, i.e. it is a subset of G forming a group with respect to the operation in G.

$$a \sim b \Longleftrightarrow a - b \in H. \tag{1.60}$$

The quotient space G/H is called a quotient group. This construction of course applies to the case of a vector space.

1.6 Construction of complex numbers

From §1.3 we know that the totality of 2-dimensional real vectors $\mathbb{R}^2$ form a **vector space** under the addition (translation) and scalar multiplication. In addition to the scalar product ((1.9)) for $\boldsymbol{z} = {}^t(x, y)$, $\boldsymbol{z}' = {}^t(x', y')$:

$$\boldsymbol{z} \cdot \boldsymbol{z}' = xx' + yy' \in \mathbb{R},$$

there is defined the **vector product**:

$$\boldsymbol{z} \times \boldsymbol{z}' = \begin{vmatrix} x & x' \\ y & y' \end{vmatrix} = xy' - x'y \in \mathbb{R},$$

where the middle term indicates the determinant.

Example 1.3. (The first construction) We introduce the vector $\bar{\boldsymbol{z}} = \begin{pmatrix} x \\ -y \end{pmatrix}$, which is a reflection of $\boldsymbol{z}$ with respect to the x-axis, and combine these two multiplications in the following $*$-operation due to Gauss:

$$\boldsymbol{z} * \boldsymbol{z}' = \begin{pmatrix} \bar{\boldsymbol{z}} \cdot \boldsymbol{z}' \\ \bar{\boldsymbol{z}} \times \boldsymbol{z}' \end{pmatrix} = \begin{pmatrix} xx' - yy' \\ xy' + x'y \end{pmatrix}.$$

E.g. if we label $\begin{pmatrix} 0 \\ 1 \end{pmatrix}$ by i, then we have

$$i^2 = i * i = \begin{pmatrix} 0 \\ 1 \end{pmatrix} * \begin{pmatrix} 0 \\ 1 \end{pmatrix} = \begin{pmatrix} -1 \\ 0 \end{pmatrix}, \quad i^2 = -1.$$

In view of

$$\boldsymbol{z} = \begin{pmatrix} x \\ y \end{pmatrix} = x \begin{pmatrix} 1 \\ 0 \end{pmatrix} + y \begin{pmatrix} 0 \\ 1 \end{pmatrix} = x\boldsymbol{e}_1 + yi,$$

where $\boldsymbol{e}_1 = \begin{pmatrix} 1 \\ 0 \end{pmatrix}$, we think of the vector $\boldsymbol{z}$ as a "number", a **complex number**, $z = x + iy$:

$$\mathbb{R}^2 \ni \boldsymbol{z} = \begin{pmatrix} x \\ y \end{pmatrix} \longleftrightarrow z = x + iy \in \mathbb{C}. \tag{1.61}$$

Since we just denote the basis vector $\boldsymbol{e}_2 = \begin{pmatrix} 0 \\ 1 \end{pmatrix}$ as i, the above correspondence is bijective and moreover, it turns out that the star product is the same as ordinary multiplication of numbers, with i^2 replaced by -1:

$$zz' = z \cdot z' = (x + iy)(x' + iy') = xx' - yy' + i(xy' + x'y) \longleftrightarrow \boldsymbol{z} * \boldsymbol{z}'.$$

We can easily prove that the system $(\mathbb{R}^2, +, *)$ forms a field, which we denote by $\mathbb{C} = (\mathbb{C}, +, \cdot)$ and refer to it as the **field of complex numbers**. In electrical engineering, it is customary to denote i by j because i is kept for current.

Proof. The only point is the division and here is a proof. $\boldsymbol{z} = \begin{pmatrix} x \\ y \end{pmatrix} = \boldsymbol{o} = \begin{pmatrix} 0 \\ 0 \end{pmatrix}$ if and only if $\begin{pmatrix} x \\ y \end{pmatrix} = \begin{pmatrix} 0 \\ 0 \end{pmatrix}$, i.e. if and only if $\boldsymbol{z} = \boldsymbol{o}$, i.e. if and only if $|\boldsymbol{z}| = \sqrt{x^2 + y^2} = 0$. Hence for each $\boldsymbol{z} \neq \boldsymbol{o}$, there exists the inverse element $\boldsymbol{z}^{-1} = \frac{1}{x^2 + y^2} \begin{pmatrix} x \\ -y \end{pmatrix} = \frac{1}{|\boldsymbol{z}|^2} \bar{\boldsymbol{z}}$, where $\bar{\boldsymbol{z}} = \begin{pmatrix} x \\ -y \end{pmatrix}$, so that $\boldsymbol{z}^{-1} * \boldsymbol{z} = \frac{1}{|\boldsymbol{z}|^2} \bar{\boldsymbol{z}} * \boldsymbol{z} = \begin{pmatrix} 1 \\ 0 \end{pmatrix}$. □

In the notation $z = x + iy$ $(x, y \in \mathbb{R})$, we call x [y] the real [imaginary] part of z, denoted $\operatorname{Re} z$ [$\operatorname{Im} z$]. Call the corresponding complex number to the vector $\bar{\boldsymbol{z}} = \begin{pmatrix} x \\ -y \end{pmatrix}$ the (complex) **conjugate**, denoted $\bar{z} = x - iy$. Recalling the length (norm, absolute value) of a vector in (1.8), we introduce the **absolute value** of the complex number $z = x + iy$ by

$$|z| = |x + iy| = |\boldsymbol{z}| = \sqrt{x^2 + y^2}.$$

Then we have

$$z\bar{z} = |z|^2, \quad \operatorname{Re} z = \frac{z + \bar{z}}{2}, \quad \operatorname{Im} z = \frac{z - \bar{z}}{2i} \tag{1.62}$$

and

$$|\operatorname{Re} z|, |\operatorname{Im} z| \le |z|. \tag{1.63}$$

We may of course define the distance between two numbers z, z' by

$$d(z, z') = |z - z'| = \sqrt{(x - x')^2 + (y - y')^2}.$$

For this to be a distance function it must satisfy the triangular inequality, which will be shown in Exercise 23, (iii).

Then $(\mathbb{C}, d)$ is a metric space which is **complete** because $\mathbb{R}^2$ is so. These are the simplest examples of **Hilbert spaces**. Hence we may develop analysis on it, **complex analysis**.

Example 1.4. (The second construction) Consider the 2-dimensional subspace of the 4-dimensional real vector space

$$M = \{\boldsymbol{z} = (x, -y, y, x) | x, y \in \mathbb{R}\} \subset \mathbb{R}^4$$

and introduce the componentwise addition (translation) and the new multiplication $\times$ for $\boldsymbol{z}_j = (x_j, -y_j, y_j, x_j), j = 1, 2$

$$\boldsymbol{z}_1 \times \boldsymbol{z}_2 = (x_1x_2 - y_1y_2, -(x_1y_2 + x_2y_1), x_1y_2 + x_2y_1, x_1x_2 - y_1y_2). \tag{1.64}$$

Then $(M, +, \times)$ forms a field isomorphic to $\mathbb{C}$. We use *only the Abelian group structure of* $\mathbb{R}^4$ *to show that* M *is an Abelian group.* To show that the set $M - \{\boldsymbol{o}\}$ is a multiplicative group, we cannot use any property of $\mathbb{R}^4$. The inverse of $\boldsymbol{z} = (x, -y, y, x) \ne \boldsymbol{o}$ is $\boldsymbol{z}^{-1} = \frac{1}{x^2+y^2}(x, y, -y, x)$. $\sqrt{x^2 + y^2}$ is not the norm of $\boldsymbol{z} = (x, -y, y, x)$ as viewed as an element of $\mathbb{R}^4$ because the norm in that space is $\sqrt{x^2 + y^2}$. It is indeed, the determinant of the matrix $Z = \begin{pmatrix} x & -y \\ y & x \end{pmatrix}$ corresponding to $\boldsymbol{z} = (x, -y, y, x)$. Indeed,

$$M = \mathbb{R}^t(1, 0, 0, 1) \oplus \mathbb{R}^t(0, -1, 1, 0) \tag{1.65}$$

and this is almost the same as Example 1.3.

Exercise 24 below gives a matrix version of this example. Comparing these two, we are led to the identification of a matrix as a Cartesian coordinate of its entries. Then as is done in Exercise 49, the norm of a matrix $A = (a_{ij})$ is to be $\|A\|^2 = \sum_{i.j} |a_{ij}|^2$.

Let $\mathbb{R}[X]$ denote the ring of all polynomials with real coefficients, where the polynomial ring is nothing other than the direct sum $\sum \mathbb{R}$ of infinitely many copies of $\mathbb{R}$, which is a subset of the Cartesian product.

Example 1.5. (The third construction) Let j denote a root (in the algebraic closure of $\mathbb{R}$ **provided that it exists**) of the irreducible polynomial

X^2+1 over $\mathbb{R}$, irreducible because for any real number α, $\alpha^2+1>0$ and X^2+1 cannot be decomposed into a product of linear factors. The adjoint $\mathbb{R}(j)=\{a+bj|a,b\in\mathbb{R}\}$ is a field, which is seen to be isomorphic to $\mathbb{C}$. However, without assuming the existence of an algebraic closure (which we are constructing), we cannot form the smallest subfield that contains both $\mathbb{R}$ and j and only the factor ring in Examples 1.6 works. Once $\mathbb{C}$ is formed by one of three ways, one can form a splitting field like $\mathbb{Q}(\sqrt{2})$, $\mathbb{Q}(\zeta_5)$, where ζ_5 is a root of the polynomial $f(z)$ in (1.24).

Example 1.6. (The fourth construction) Let $\mathbb{R}[X]$ denote the ring of all polynomials with real coefficients, where the polynomial ring is nothing other than the direct sum $\sum\mathbb{R}$ (see §1.3).

The factor ring $\mathbb{R}[X]/(X^2+1)$ forms a field which can be conveniently expressed as the adjoint $\mathbb{R}(j)=\{a+bj|a,b\in\mathbb{R}\}$ in Example 1.5.

Since

$$\mathbb{R}[X]/(X^2+1)=\{a+bX \bmod (X^2+1)|a,b\in\mathbb{R}\},$$

we may prove that the mapping $a+bX\to a+bj$ is a field isomorphism under the condition that $\mathbb{R}(j)$ exists. As remarked in Example 1.5, the factor ring exists and provides the third construction.

Exercise 23. (i) Prove the **multiplication formula** for the absolute value

$$|z_1z_2|=|z_1||z_2| \tag{1.66}$$

by the 2-dimensional Lagrange formula

$$(x_1^2+y_1^2)(x_2^2+y_2^2)=(x_1x_2-y_1y_2)^2+(x_1y_2+x_2y_1)^2 \tag{1.67}$$

(which was used by Lagrange in the proof of his Two Squares Theorem to the effect that the product of two integers in the form of a sum of two squares is again a sum of two squares).
(ii) Prove $\overline{z_1z_2}\ \bar{z}_1\bar{z}_2$. Prove (1.66) by this and (1.62).
(iii) Prove the **triangular inequalities**

$$||z_1|-|z_2||\le|z_1\pm z_2|\le|z_1|+|z_2| \tag{1.68}$$

and generalize (1.68) to

$$\left|\sum_{k=1}^n z_k\right|\le\sum_{k=1}^n|z_k|,\quad z_k\in\mathbb{C},\ 1\le k\le n.$$

Solution. We give two different proofs of (iii). The first proof rests on (1.62), (1.63) and (1.67):

$$\begin{aligned}|z_1 + z_2|^2 &= (z_1 + z_2)\overline{(z_1 + z_2)} = z_1\overline{z_1} + 2\,\mathrm{Re}\, z_1\overline{z_2} + z_2\overline{z_2}\\ &= |z_1|^2 + |z_2|^2 + 2\,\mathrm{Re}\, z_1\overline{z_2} \le |z_1|^2 + |z_2|^2 + 2|z_1\overline{z_2}|\\ &\le |z_1|^2 + |z_2|^2 + 2|z_1||\overline{z_2}| = (|z_1| + |z_2|)^2\end{aligned}$$

whence the second inequality of (1.68) follows. The first inequality of (1.68) follows from it by subtraction.

The second proof follows from the Cauchy-Schwarz inequality (1.11):

$$|\boldsymbol{z}_1 \cdot \boldsymbol{z}_2| \le |\boldsymbol{z}_1||\boldsymbol{z}_2|. \tag{1.69}$$

Indeed,

$$\begin{aligned}|z_1 + z_2|^2 &= |\boldsymbol{z}_1 + \boldsymbol{z}_2|^2 = (\boldsymbol{z}_1 + \boldsymbol{z}_2)\cdot(\boldsymbol{z}_1 + \boldsymbol{z}_2) = |\boldsymbol{z}_1|^2 + |\boldsymbol{z}_2|^2 + 2\boldsymbol{z}_1\cdot\boldsymbol{z}_2\\ &\le |\boldsymbol{z}_1|^2 + |\boldsymbol{z}_2|^2 + 2|\boldsymbol{z}_1||\boldsymbol{z}_2| = (|\boldsymbol{z}_1| + |\boldsymbol{z}_2|)^2 = (|z_1| + |z_2|)^2.\end{aligned}$$

Note that (1.69) is a consequence of (1.67), whence that it is the Lagrange identity (1.67) that underlies the triangular inequality and that (1.68) verifies that the distance function introduced above is indeed a distance function.

Exercise 24. This exercise is a matrix version of Example 1.4 and you will learn how simple it is to work with matrices than with Cartesian coordinates.

Prove that the set of all matrices of the form

$$Z = \begin{pmatrix} x & -y \\ y & x \end{pmatrix} \tag{1.70}$$

with $x, y \in \mathbb{R}$ forms a field with respect to ordinary addition and multiplication of matrices, and that it is isomorphic to the complex number field $\mathbb{C}$. Then check that (1.67) can be interpreted as the multiplicativity of the determinants: $|Z_1||Z_2| = |Z_1Z_2|$.

In addition to the rectangular coordinates $\boldsymbol{z} = \begin{pmatrix} x \\ y \end{pmatrix}$ the **polar coordinates** $\begin{pmatrix} r \\ \theta \end{pmatrix}$ are of great use. If $\begin{pmatrix} r \\ \theta \end{pmatrix}$ is the polar coordinate of the point $\boldsymbol{z} = \begin{pmatrix} x \\ y \end{pmatrix} \ne \mathbf{o}$ then

$$\begin{cases} x = r\cos\theta, \\ y = r\sin\theta, \end{cases} \tag{1.71}$$

and we have the *vector-valued function*

$$\boldsymbol{x} = \begin{pmatrix} x \\ y \end{pmatrix} = \boldsymbol{\phi}(\boldsymbol{t}) = \begin{pmatrix} x(\boldsymbol{t}) \\ y(\boldsymbol{t}) \end{pmatrix} = \begin{pmatrix} r\cos\theta \\ r\sin\theta \end{pmatrix}, \qquad \boldsymbol{t} = \begin{pmatrix} r \\ \theta \end{pmatrix}. \tag{1.72}$$

For the corresponding complex number $z = x + iy$, this reads

$$z = r(\cos\theta + i\sin\theta) = re^{i\theta}, \tag{1.73}$$

where the last equality is (8.123). This is called the **polar form** of $z \neq 0$ and $r = |z|$ is the absolute value of z defined above and θ, denoted by $\arg z$ (or amp z), is called the **argument** of z. At the origin $z = 0$, the absolute value is 0 and the argument is indefinite. Fixing a value θ_0 of $\arg z$, the argument is a multi-valued function expressed as

$$\arg z = \theta_0 + 2\pi n \quad (n \in \mathbb{Z}).$$

To handle the multi-valuedness of $\arg z$, we often restrict its range to one cycle $[0, 2\pi)$ or $(-\pi, \pi]$, in which case we may treat it as a single-valued function.

To find the polar form of $z = x + iy \neq 0$, we first calculate $r = |z| = \sqrt{x^2 + y^2}$. Then factor out r to get $z = r\left(\frac{x}{r} + i\frac{y}{r}\right)$.

Then find θ such that

$$\cos\theta = \frac{x}{r}, \quad \sin\theta = \frac{y}{r}.$$

Substituting this, we get (1.73). Since $\tan\theta = \frac{y}{x}$, we may define $\arg z$ as $\arctan\frac{y}{x}$.

Exercise 25. (i) For $z_1, z_2 \neq 0$, prove that

$$\arg(z_1 z_2) = \arg z_1 + \arg z_2$$

and that

$$\arg\left(\frac{z_1}{z_2}\right) = \arg z_1 - \arg z_2.$$

Interpret the results in the light of the exponential law (8.112).
(ii) Deduce from (i) that for $0 \neq z \in \mathbb{C}$ and $\theta \in \mathbb{R}$, $(\cos\theta + i\sin\theta)z$ is the point on the circle $r = |z|$ rotated in the positive direction by θ. What about $(\cos\theta - i\sin\theta)z$?
(iii) Interpret (ii) from the point of view of Euler's identity (8.117). Cf. Remark 8.2.

Example 1.7. We find the fifth root of 2 in $\mathbb{C}$. We are to find z such that $z^5 = 2$. Raising the polar form $z = r(\cos\theta + i\sin\theta)$ of z to the fifth power by (1.73), we obtain

$$z^5 = r^5(\cos 5\theta + i\sin 5\theta),$$

i.e. the polar coordinate is $\begin{pmatrix} r^5 \\ 5\theta \end{pmatrix}$. On the other hand, the polar form of 2 is $2(\cos 0 + i\sin 0)$, i.e. the polar coordinate is $\begin{pmatrix} 2 \\ 0 \end{pmatrix}$, whence we have

$$r^5 = 2,\ 5\theta = 0 + 2\pi n,\ n \in \mathbb{Z}.$$

Hence $r = \sqrt[5]{2}$ and $\theta = \frac{2n}{5}\pi \quad (n \in \mathbb{Z})$.

When we substitute the values of θ back into the polar form, we may restrict the range of n to the least positive residue modulo 5 since sine and cosine are periodic functions of period 2π:

$$\left\{5\left(\frac{n}{5} - \left[\frac{n}{5}\right]\right)\middle| n \in \mathbb{Z}\right\} = \{0, 1, 2, 3, 4\},$$

where $[\alpha]$ is the integer part of α in Corollary 4.3. Hence, corresponding to five values of $\theta = 0, \frac{2\pi}{5}, \frac{4\pi}{5}, \frac{6\pi}{5}, \frac{8\pi}{5}$, there are five values of z. For $\theta = 0$, $z = \sqrt[5]{2}$. It is instructive to express other values of z using roots. Especially, the number $2\cos\frac{\pi}{5} = \frac{1+\sqrt{5}}{2}$ is known as the golden ratio, cf. Exercise 78.

Remark 1.5. The procedure in Example 1.7 clearly gives all the solutions to the equation $z^n = 1$, i.e. all nth roots of 1: $z = e^{\frac{2\pi ik}{n}}$, $k = 0, 1, \cdots, n-1$. Note that those with $\gcd(n, k) = 1$ are **primitive nth roots** of 1.

Also Example 1.7 gives a method for treating an equation of the form $z^n = b \neq 0$. In the first place the radius is $|z| = |b|^{1/n}$ and there remains the determination of the argument.

Definition 1.6. The inner product in a complex vector space H is defined as follows. For $a, b \in H$, their inner product (a, b) is defined by the following conditions. For $a, b, c \in H$ and $\lambda \in \mathbb{C}$,
(i) $(a, a) \geq 0$ and $(a, a) = 0 \Longleftrightarrow a = 0$.
(ii) $(b, a) = \overline{(a, b)}$ the bar meaning the complex conjugate.
(iii) (linearity in the first factor) $(a+b, c) = (a, b) + (a, c)$, $(\lambda a, b) = \lambda(a, b)$.

It follows that $(a, b+c) = (a, b) + (a, c)$ and $(a, \lambda b) = \bar{\lambda}(a, b)$. Compare the inner product and norm introduced in Definition 3.2 below.

Exercise 26. Show that the inner product in (1.9) satisfies the above conditions. Define the inner product on $\mathbb{C}^n$ and check the defining conditions.

1.7 Maps and their compositions

Definition 1.7. A **map** (or a mapping) from a set X to a set Y denoted $f : X \to Y$ is a rule which assigns to every point x in its domain of definition X a *unique* value $f(x) \in Y$. Writing the value as y we have a familiar notation $y = f(x)$. We often write $f = f(x)$ to mean that f is the label of the mapping and $f(x)$ its value at $x \in X$. We often with the abuse of language write $y = y(x)$. One-valuedness is equivalent to the statement that $x = y$ implies $f(x) = f(y)$ and in this case the map is said to be **well-defined**. This is to be checked especially when there are some distinct ways of expressing an element of the domain. Typically when one defines a map on equivalence classes, since an equivalence class has many representatives, one must show that the definition does not depend on the choice of a representative.

For a mapping $f : X \to Y$ we denote the **image** of by $f(X')$ for $X' \subset X$, the **inverse image** by $f^{-1}(Y')$ for $Y' \subset Y$ and particularly the **image** of f by $f(X)$ or $\mathrm{Im}\, f$:

$$f(X') = \{f(x)|x \in X'\}, \quad f^{-1}(Y') = \{x \in X|f(x) \in Y'\}. \tag{1.74}$$

Lemma 1.1. *Suppose $f : X \to Y$ and A's $\in 2^X$ and B's $\in 2^Y$. Then*
(i) $A_1 \subset A_2 \Longrightarrow f(A_1) \subset f(A_2)$; $B_1 \subset B_2 \Longrightarrow f^{-1}(B_1) \subset f^{-1}(B_2)$.
(ii) $f(A_1 \cup A_2) = f(A_1) \cup f(A_2)$; $f^{-1}(B_1 \cup B_2) = f^{-1}(B_1) \cup f^{-1}(B_2)$.
(ii)′ $f(\cup_{\lambda\in\Lambda} A_\lambda) = \cup_{\lambda\in\Lambda} f(A_\lambda)$; $f^{-1}(\cup_{\lambda\in\Lambda} B_\lambda) = \cup_{\lambda\in\Lambda} f^{-1}(B_\lambda)$.
(iii) $f(A_1 \cap A_2) \subset f(A_1) \cap f(A_2)$; $f^{-1}(B_1 \cap B_2) = f^{-1}(B_1) \cap f^{-1}(B_2)$.
(iii)′ $f(\cap_{\lambda\in\Lambda} A_\lambda) \subset \cap_{\lambda\in\Lambda} f(A_\lambda)$; $f^{-1}(\cap_{\lambda\in\Lambda} B_\lambda) = \cap_{\lambda\in\Lambda} f^{-1}(B_\lambda)$.
(iv) $f^{-1}(f(A)) \supset A$; $f(f^{-1}(B)) = f(X) \cap B$.

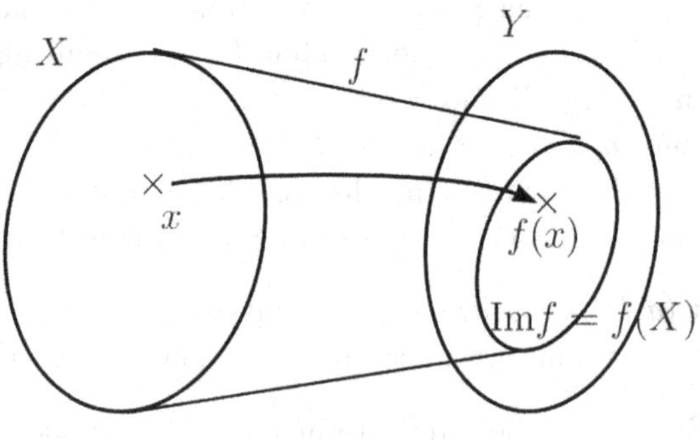

Fig. 1.1: Mapping

Definition 1.8. Given two mappings $g : Y \to Z$ and $f : X \to Y$ (or $f(X) \subset Y$), we may consider their **composition**

$$g \circ f : X \to Z; (g \circ f)(x) = g(f(x)). \tag{1.75}$$

For three mappings f, g, h which can form compositions, the associative law holds:

$$(h \circ g) \circ f = h \circ (g \circ f).$$

Definition 1.9. If a map $f : X \to Y$ sends distinct points to distinct values, i.e. if $f(x) = f(y), x, y \in X$ implies $x = y$, then f is called an injection or an injective map. If every point $y \in Y$ is a value of f at some x, i.e. for $f(X) = Y$, then f is called a surjection or a surjective map. A map which is both injective and surjective is called a **bijection** or a bijective map. When considering algebraic structures, we use the terminologies: monomorphism, epimorphism and isomorphism.

Any bijective map has its **inverse map** f^{-1} given by

$$f^{-1} : Y \to X; \quad f^{-1}(f(x)) = x. \tag{1.76}$$

We have

$$f^{-1} \circ f = I_X, \quad f \circ f^{-1} = I_Y, \tag{1.77}$$

where I_X and I_Y are the **identity maps** on X and Y, respectively.

For a continuous function of a real variable, Theorem 4.3 assures the existence of the inverse function whereby monotonicity is the key. For a differentiable function a concise criterion is the non-vanishing of the derivative.

If f, g are functions of real variables, we often call the composition the composite function of g and f. Since commutative law does not hold in general, the order is meaningful and it is to be understood that one operates the function nearer to the variable. In some disciplines, one may use exponential notation in which case, it is rather apparent: $x^{f \circ g} = x^{f^g}$.

Example 1.8. Given two linear maps $\boldsymbol{z} = \boldsymbol{g}(\boldsymbol{y})$ and $\boldsymbol{y} = \boldsymbol{f}(\boldsymbol{x})$ in the form

$$\begin{pmatrix} z_1 \\ \vdots \\ z_m \end{pmatrix} = \boldsymbol{z} = B\boldsymbol{y} = \begin{pmatrix} \sum_{j=1}^n b_{1j} y_j \\ \vdots \\ \sum_{j=1}^n b_{mj} y_j \end{pmatrix}, \begin{pmatrix} y_1 \\ \vdots \\ y_n \end{pmatrix} = \boldsymbol{y} = A\boldsymbol{x} = \begin{pmatrix} \sum_{k=1}^l a_{1k} x_k \\ \vdots \\ \sum_{k=1}^l a_{nk} x_k \end{pmatrix} \tag{1.78}$$

$\boldsymbol{x} = {}^t\left(x_1, \cdots, x_l\right) \in \mathbb{R}^l$, where A and B are coefficient matrices given by

$$A = (a_{jk})_{\substack{1\le j\le n\\ 1\le k\le l}} = \begin{pmatrix} a_{11} & \cdots & a_{1l} \\ & \cdots & \\ a_{n1} & \cdots & a_{nl} \end{pmatrix}, \quad B = (b_{ij})_{\substack{1\le i\le m\\ 1\le j\le n}}, \tag{1.79}$$

we would like to define the composition $\boldsymbol{g} \circ \boldsymbol{f}$ in terms of coefficient matrices.

(1.78) is equivalent to the systems of equations

$$\begin{cases} z_i = \sum_{j=1}^n b_{ij} y_j & (1 \le i \le m) \\ y_j = \sum_{k=1}^l a_{jk} x_k & (1 \le j \le n) \end{cases} \tag{1.80}$$

under the rule that in the product of an (n, l)-matrix A and a vector $\boldsymbol{x} = {}^t\left(x_1, \cdots, x_l\right)$, the matrix is to be divided into row vectors $\begin{pmatrix} \boldsymbol{a}'_1 \\ \vdots \\ \boldsymbol{a}'_n \end{pmatrix}$ and multiplication of $\boldsymbol{a}'_j$ and $\boldsymbol{x}$ is done $\boldsymbol{a}'_j \boldsymbol{x} = \sum_{k=1}^l a_{jk} x_k$ as in the case of an inner product (1.9). This is one of **block multiplications**

$$A\boldsymbol{x} = \begin{pmatrix} \boldsymbol{a}'_1 \\ \vdots \\ \boldsymbol{a}'_n \end{pmatrix} \boldsymbol{x} = \begin{pmatrix} \boldsymbol{a}'_1 \boldsymbol{x} \\ \vdots \\ \boldsymbol{a}'_n \boldsymbol{x} \end{pmatrix} = \begin{pmatrix} \sum_{k=1}^l a_{1k} x_k \\ \vdots \\ \sum_{k=1}^l a_{nk} x_k \end{pmatrix}. \tag{1.81}$$

Substituting the expression for y_j in the first equality in (1.80), we obtain

$$z_i = \sum_{j=1}^n b_{ij} \sum_{k=1}^l a_{jk} x_k = \sum_{k=1}^l \left(\sum_{j=1}^n b_{ij} a_{jk} \right) x_k \quad (1 \le i \le m). \tag{1.82}$$

The composition $g \circ f$ is

$$\boldsymbol{z} = (g \circ f)(\boldsymbol{x}) = g(f(\boldsymbol{x})) = B(A\boldsymbol{x}) = (BA)\boldsymbol{x},$$

where in the last equality, the "product" BA is to be defined as in (1.82) for $g \circ f$ to be a linear map, i.e. the (i, k)-entry of BA is to be $\sum_{j=1}^n b_{ij} a_{jk}$. This amounts to another of block multiplications of BA in which A is divided into column vectors $A = (\boldsymbol{a}_1, \cdots, \boldsymbol{a}_l)$ and multiplication is conducted as $BA = (B\boldsymbol{a}_1, \cdots, B\boldsymbol{a}_l)$ following the above rule:

$$BA = B(\boldsymbol{a}_1, \cdots, \boldsymbol{a}_l) = (B\boldsymbol{a}_1, \cdots, B\boldsymbol{a}_l) = \begin{pmatrix} \sum_{j=1}^n b_{1j} a_{j1} & \cdots & \sum_{j=1}^n b_{1j} a_{jl} \\ & & \vdots \\ \sum_{j=1}^n b_{mj} a_{j1} & \cdots & \sum_{j=1}^n b_{mj} a_{jl} \end{pmatrix}. \tag{1.83}$$

Definition 1.10. For $m, n \in \mathbb{N}$ define the **Kronecker delta** (symbol) δ_{mn} by

$$\delta_{mn} = \begin{cases} 1, & m = n \\ 0, & m \neq n \end{cases}. \tag{1.84}$$

This is a prototype of the Dirac distribution in Example 15.2. We define the **identity matrix** of degree n by

$$E = E_n = (\delta_{ij}) = \begin{pmatrix} 1 & & O \\ & \ddots & \\ O & & 1 \end{pmatrix} \tag{1.85}$$

with entries in the principal diagonal all 1, other entries being 0. With the product of matrices defined by (1.83), the identity matrix serves as the identity element in the ring of matrices of degree n. We also find the column vector expression for E_n to be $(\boldsymbol{e}_1, \cdots, \boldsymbol{e}_n)$, where the j**th basic unit column vector** is

$$\boldsymbol{e}_j = {}^t(\delta_{1j}, \cdots, \delta_{nj}) = \begin{pmatrix} 0 \\ \vdots \\ 0 \\ 1 \\ 0 \\ \vdots \\ 0 \end{pmatrix} \tag{1.86}$$

with 1 in the jth row. Cf. (C.3) for row vectors.

Exercise 27. Form the compositions of the following functions defined for $x \neq 0, 1$

$$\begin{aligned} \phi_1(x) = x,\ \phi_2(x) = \frac{1}{1-x},\ \phi_3(x) = \frac{x-1}{x}, \\ \phi_4(x) = 1 - x,\ \phi_5(x) = \frac{x}{x-1},\ \phi_6(x) = \frac{1}{x} \end{aligned} \tag{1.87}$$

and arrange them in a multiplication table.

Exercise 28. Given a function

$$r(1) = 2, r(2) = 3, r(3) = 1 \tag{1.88}$$

we write

$$r = \begin{pmatrix} 1 & 2 & 3 \\ r(1) & r(2) & r(3) \end{pmatrix} = \begin{pmatrix} 1\,2\,3 \\ 2\,3\,1 \end{pmatrix}. \tag{1.89}$$

If

$$s = \begin{pmatrix} 1 & 2 & 3 \\ 1 & 3 & 2 \end{pmatrix} \tag{1.90}$$

then their composition $r \circ s$, which denote by rs is

$$rs = \begin{pmatrix} 1 & 2 & 3 \\ (rs)(1) & (rs)(2) & (rs)(3) \end{pmatrix} = \begin{pmatrix} 1 & 2 & 3 \\ 2 & 1 & 3 \end{pmatrix}, \tag{1.91}$$

where $e = \begin{pmatrix} 1 & 2 & 3 \\ 1 & 2 & 3 \end{pmatrix}$ is the identity. Find all the compositions of permutations in $S_3 = \{e, r, r^2, s, sr, rs\}$ and complete the multiplication table.

Exercise 29. Let $E = E_3 = \begin{pmatrix} 1 & 0 & 0 \\ 0 & 1 & 0 \\ 0 & 0 & 1 \end{pmatrix}$,

$$C_3 = \begin{pmatrix} 0 & 1 & 0 \\ 0 & 0 & 1 \\ 1 & 0 & 0 \end{pmatrix}, C_3^2 = \begin{pmatrix} 0 & 0 & 1 \\ 1 & 0 & 0 \\ 0 & 1 & 0 \end{pmatrix} \tag{1.92}$$

and

$$\begin{gathered} \sigma_v(1) = \begin{pmatrix} 1 & 0 & 0 \\ 0 & 0 & 1 \\ 0 & 1 & 0 \end{pmatrix}, \sigma_v(2) = \begin{pmatrix} 0 & 0 & 1 \\ 0 & 1 & 0 \\ 1 & 0 & 0 \end{pmatrix}, \\ \sigma_v(3) = \begin{pmatrix} 0 & 1 & 0 \\ 1 & 0 & 0 \\ 0 & 0 & 1 \end{pmatrix}. \end{gathered} \tag{1.93}$$

Form the multiplication table of G:

$$G = \{E, C_3, C_3^2, \sigma_v(1), \sigma_v(2), \sigma_v(3)\}. \tag{1.94}$$

Exercise 30. Put

$$A_1 = \begin{pmatrix} 1 & 0 \\ 0 & 1 \end{pmatrix}, A_2 = \begin{pmatrix} 0 & -1 \\ 1 & -1 \end{pmatrix}, A_3 = \begin{pmatrix} -1 & 1 \\ -1 & 0 \end{pmatrix}, \tag{1.95}$$

$$A_4 = \begin{pmatrix} -1 & 1 \\ 0 & 1 \end{pmatrix}, A_5 = \begin{pmatrix} 1 & 0 \\ 1 & -1 \end{pmatrix}, A_6 = \begin{pmatrix} 0 & -1 \\ -1 & 0 \end{pmatrix}.$$

Form the multiplication table.

Exercise 31. We write the iterated composition, called **iterates**, of the same function f by f^n where an iterate is defined as the composition $f^2 = f \circ f$. Let

$$f(x) = \begin{cases} 2x & (0 \le x \le \frac{1}{2}) \\ 2(1-x) & (\frac{1}{2} \le x \le 1). \end{cases} \tag{1.96}$$

Plot the graphs of f^n and $\frac{1}{2^{n-1}} f^n$.

Iterates are nested structures that are similar to R_4, the nested set property by Cantor and give rise to a discrete chaos.

Fig. 1.2: Abel

Chapter 2

Continuity of real numbers

2.1 Limit of a sequence

Any set whose cardinality is the same as that of the set $\mathbb{N}$ of natural numbers is called a **countable set**. The set $\mathbb{N}$ itself is an infinite set whose fact follows from the proposition that it is not in one-to-one correspondence with any of its finite subset. It is used as an index set for a sequence $\{a_n\}$ ($n \in \mathbb{N}$). Indeed, it is a function $a : \mathbb{N} \to \mathbb{C}$ as in Definition 1.3. By abuse of language we use the symbol $\{a_n | n \in \mathbb{N}\}$ to mean the set of terms in the sequence as well as the sequence itself. When there are same terms, they do not coincide. E.g. if $a_n = (-1)^{n-1}$, then the set of a_n's is $1, -1$ and the sequence is $1, -1, 1, -1, \cdots$. Since the index set is an infinite set we may speak of the limit

$$\lim_{n\to\infty} a_n = a \tag{2.1}$$

to mean that a_n approaches the number a as n surpasses all limits. To understand this, it would be best to consider its negation. I.e. we consider the case in which however large n we take, there is always a_n which is not near enough to a. This is like a computer game, an $\varepsilon - N$ game. The computer assigns a very small width $\varepsilon > 0$ and the player is to assign a large natural number N such that $n > N$ always implies that a_n lies within the thin band $|a_n - a| < \varepsilon$. If the player can always assign the number $N = N(\varepsilon)$ then he wins the game, which is the case of convergence. More precisely, given an $\varepsilon > 0$ there exists an $N = N(\varepsilon) \in \mathbb{N}$ such that $n > N$ implies

$$|a_n - a| < \varepsilon. \tag{2.2}$$

Here the inequality $n > N$ may be replaced by $n \geq N$. Another notation often used is n_0.

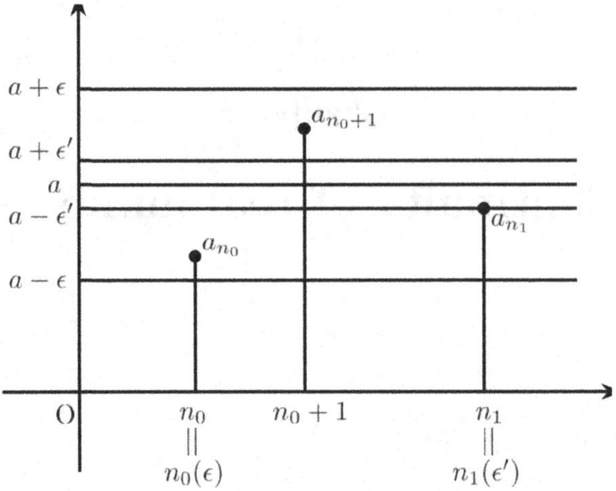

Fig. 2.1: Limit of a sequence

In what follows we shall often use the **Bachmann-Landau notation**: $b_n = o(1)$ to mean that $|b_n| < \varepsilon$ for $n > N$. Then $|a_n - a| < \varepsilon$ may be written as $a_n = a + o(1)$ for $n > N$. The high utility of this notation will become apparent subsequently. This helps in understanding approximations since we treat an inequality in the form of an equality. Similarly, we write $a_n = O(1)$ to mean that $|a_n| \le B$ for a constant $B > 0$. Also the ε may be replaced by $c\varepsilon$ for any constant $c > 0$ since one may think $\frac{1}{c}\varepsilon$ is given and choose an $N \in \mathbb{N}$ for which $n > N$ implies $|a_n - a| < c\frac{1}{c}\varepsilon = \varepsilon$. In the Bachmann-Landau notation, we have

$$co(1) = o(1), \quad cO(1) = O(1) \tag{2.3}$$

for any $c > 0$. We extend the symbol to the case of functions and we say that f is of smaller order than $g > 0$ and write

$$f(x) = o(g(x)) \tag{2.4}$$

in the case where $\lim_{y\to x} \frac{f(y)}{g(y)} = 0$. It is assumed that $g(y) \ne 0$ near x.

Exercise 32. Prove that if a sequence converges, then the limit is unique.

Solution. If $\lim_{n\to\infty} a_n = a$ and $\lim_{n\to\infty} a_n = b$, then $o(1) = a_m - a_n = a - b + o(1)$ for $m, n > N$, so that $a = b$. Or suppose $a < b$ and choose $\varepsilon = \frac{b-a}{2}$. Then there exists an $N \in \mathbb{N}$ such that $|a_n - a| < \frac{b-a}{2}$, $|a_n - b| < \frac{b-a}{2}$. Hence $|a - b| \le |a_n - a| + |b - a_n| < b - a$, a contradiction.

Exercise 33. Prove that a convergent sequence is bounded.

Solution. For a fixed $\varepsilon > 0$ choose $N = N(\varepsilon) \in \mathbb{N}$ so that (2.2) holds. Then putting $\alpha = \min\{a_1, \cdots, a_{N-1}, a_N - \varepsilon\}$ and $\beta = \max\{a_1, \cdots, a_{N-1}, a_N + \varepsilon\}$, we have $\alpha \le a_n \le \beta$.

Lemma 2.1. *The inequality is preserved in the limit, i.e. if*

$$a_n \le b_n \tag{2.5}$$

for $n > n_1$ *and* $\lim_{n\to\infty} a_n = a$ *and* $\lim_{n\to\infty} b_n = b$, *then* $a \le b$. *In particular we have the* **squeezing principle**: *if*

$$a_n \le c_n \le b_n \tag{2.6}$$

and $\lim_{n\to\infty} a_n = \lim_{n\to\infty} b_n = a$, *then* $\lim_{n\to\infty} c_n = a$.

Proof. For if $a > b$, then choosing $\varepsilon = a - b > 0$ and $N = N(\varepsilon) \in \mathbb{N}$ such that for $n > N$, $|a_n - a| < \frac{1}{2}\varepsilon$ and $|b_n - b| < \frac{1}{2}\varepsilon$, so that we have

$$a < a_n + \frac{1}{2}\varepsilon \le b_n + \frac{1}{2}\varepsilon < b + \frac{1}{2}\varepsilon + \frac{1}{2}\varepsilon = b + a - b = a,$$

a contradiction. The latter assertion follows from $|c_n - a| \le |c_n - a_n| + |a_n - a| \le |b_n - a_n| + |a_n - a| \le |b_n - a| + 2|a_n - a| \to 0$ as $n \to \infty$. □

Definition 2.1. If a sequence $\{a_n\}$ satisfies the **Cauchy condition**, it is called a **Cauchy sequence**: Given an $\varepsilon > 0$ there exists an $N \in \mathbb{N}$ such that

$$|a_m - a_n| < \varepsilon \tag{2.7}$$

for all $m, n > N$. This notion may be introduced in a metric space. Cf. §16.6.

Exercise 34. Show that a Cauchy sequence is bounded.

Solution. In Equation (2.7), choose $\varepsilon = 1$ and $m = n_0$. Then for $n \ge n_0$, $|a_n| \le |a_{n_0}| + 1$. Putting $M = \max\{|a_1|, \cdots, |a_{n_0-1}|, |a_{n_0}| + 1\}$, we obtain $|a_n| \le M$ for all $n \in \mathbb{N}$.

Under R_9, this Exercise is a consequence of Exercise 33.

Exercise 35. Prove that any convergent sequence is a Cauchy sequence.

Solution. For if $\lim_{n\to\infty} a_n = a$, then given an $\varepsilon > 0$, there exists an $N \in \mathbb{N}$ such that $|a_n - a| < \varepsilon$ for $n \ge N$. Hence for $m, n \ge N$, we have $|a_n - a_m| \le |a_n - a| + |a - a_m| < 2\varepsilon$.

The converse of the above exercise is R_9, one of the axioms of continuity of real number system. This is useful because it can assure the convergence without knowing the value of the limit in the same way as another axiom of continuity, R_3, bounded monotone sequence theorem.

Theorem 2.1. (Continuity of algebraic operations) *Suppose* $a_n \to a$ *and* $b_n \to b$. *Then we have*
(i) $a_n \pm b_n \to a \pm b$.
(ii) $a_n \cdot b_n \to a \cdot b$.
(iii) *If* $b \neq 0$, *then* $\frac{a_n}{b_n} \to \frac{a}{b}$, *where the quotient is considered only for large* n *for which* $b_n \neq 0$.

Proof. (i) Since $a_n = a + o(1)$ and $b_n = b + o(1)$, it follows that $a_n \pm b_n = a \pm b + o(1)$. (ii) $a_n b_n = (a + o(1))(b + o(1)) = ab + o(1)$. (iii) For some N, $0 \neq b_n = b + o(1)$. Hence $\frac{1}{b_n} = \frac{1}{b} + o(1)$, so that $\frac{a_n}{b_n} = (a + o(1))(\frac{1}{b} + o(1)) = \frac{a}{b} + o(1)$. □

In treating sequences and series the **telescoping series technique** is often useful: for $m < n$

$$\sum_{k=m}^{n-1} (a_{k+1} - a_k) = a_n - a_m. \tag{2.8}$$

Indeed, in light of the difference operator in (13.15), (2.8) amounts to

$$\Delta^{-1}\Delta = I, \tag{2.9}$$

a prototype of the inversion of the differentiation and integration. This will be derived as a special case of the partial summation (or integration by parts with Stieltjes integral) in Remark 10.2.

Many functions are expressed in the form of a (function) series. A series $\sum_{n=1}^{\infty} a_n$ is the sequence of its partial sums $\{S_n\}$, $S_n = a_1 + \cdots + a_n$ and if the limit $\lim_{n\to\infty} S_n$ exists, say S, we define the value $\sum_{n=1}^{\infty} a_n$ to be S:

$$\sum_{n=1}^{\infty} a_n = \lim_{n\to\infty} S_n = S \tag{2.10}$$

and we say that $\sum_{n=1}^{\infty} a_n$ is **convergent** (or converges) to S. The Cauchy condition for the series $\sum_{n=1}^{\infty} a_n$ reads in view of (2.8): Given an $\varepsilon > 0$ there exists an $N \in \mathbb{N}$ such that

$$\left| \sum_{k=m+1}^{n} a_k \right| < \varepsilon \tag{2.11}$$

for all $n > m > N$ or $\lim_{m,n\to\infty} \sum_{k=m+1}^{n} a_k = 0$.

Given a series $\sum_{n=1}^{\infty} a_n$, if the series $\sum_{n=1}^{\infty} |a_n|$ consisting of absolute values is convergent, then it is said to be **absolutely convergent** or we say the series converges absolutely.

Proposition 2.1. *An absolutely convergent series is convergent.*

This is a special case of Weierstrass M-test, Theorem 3.5 and directly follows from the Cauchy condition (2.11):

$$\left|\sum_{k=m+1}^{n} a_k\right| \le \sum_{k=m+1}^{n} |a_k| < \varepsilon.$$

As a necessary condition for convergence of a series $\sum_{n=1}^{\infty} a_n$ we have

$$\lim_{n\to\infty} a_n = 0. \tag{2.12}$$

It is very often the case that the index set is $\mathbb{N}\cup\{0\}$. This is so especially in the case of power series.

The following theorem is a special case of Theorem 12.1 in §12.2.

Theorem 2.2. *For sequences $\{a_n\}$, $\{b_n\}$ let B_n denote the nth partial sum of b_n and suppose the following conditions are satisfied:*
(i) $\lim_{n\to\infty} a_n B_n$ *exists,* (ii) $\sum_{n=1}^{\infty} |a_n - a_{n+1}| < \infty$, (iii) $B_n = \sum_{k=1}^{n} b_k$ *is bounded. Then* $\sum_{n=1}^{\infty} a_n b_n$ *is convergent.*

Proof. Note that $\sum_{k=1}^{n} |a_k - a_{k+1}||B_k| \le M \sum_{k=1}^{n} |a_k - a_{k+1}| < \infty$, where M is a bound of B_n. Hence $\sum_{k=1}^{n} |a_k - a_{k+1}||B_k|$ is an upward bounded monotone increasing sequence, so that it is convergent by R_3 below, or $\sum_{n=1}^{\infty}(a_n - a_{n+1})B_n$ is absolutely convergent. Hence by partial summation, Corollary 10.1, the assertion follows. □

Corollary 2.1. *Suppose $\{a_n\}$ is a monotone decreasing real sequence converging to 0 and that the partial sum of $\{b_n\}$ is bounded. Then $\sum_{n=1}^{\infty} a_n b_n$ is convergent.*

Corollary 2.2. *Suppose $\{a_n\}$ is a monotone converging real sequence and that $\sum_{n=1}^{\infty} b_n$ is convergent. Then $\sum_{n=1}^{\infty} a_n b_n$ is convergent.*

Corollary 2.3. *Suppose $\{a_n\}$ is a monotone decreasing real sequence converging to 0, $a_n > 0$. Then the alternating series $\sum_{n=1}^{\infty} (-1)^n a_n$ is convergent to S. The difference between the nth partial sum S_n and S is*

$$(-1)^n (S_n - S) < a_{n+1}. \tag{2.13}$$

Proof.

$$(-1)^n(S_n - S) = \sum_{k=1}^{\infty}(-1)^{k+1}a_{n+k} = \sum_{k=1}^{\infty}(a_{n+2k-1} - a_{n+2k}) > 0$$

$$(-1)^n(S - S_n) = a_{n+1} - \sum_{k=1}^{\infty}(a_{n+2k} - a_{n+2k+1}) < a_{n+1}.$$

□

2.2 Continuity of real numbers

There are many ways of stating the continuity of real number system reflecting that of the real line.

A (Archimedes principle). $\mathbb{N} \subset \mathbb{R}$ is not bounded above. This means that there exists an arbitrarily large natural number in the sense that given any real number R there exists a natural number which exceeds R and not that there is always a natural number exceeding any given one. This is not a part of continuity of $\mathbb{R}$ but rather its Archimedean nature that it is a maximal Archimedean ordered field. Therefore those axioms below which need A to be equivalent to other axioms are feasible of generalization to general topological spaces. Compactitude and completeness are probably the most important.

R$_1$ (Dedekind section). If two subsets U_1, U_2 satisfy the condition

$$\mathbb{R} = U_1 \cup U_2, \quad U_1 \cap U_2 = \emptyset \tag{2.14}$$

and that for any $a_1 \in U_1$ and $a_2 \in U_2$, we have $a_1 < a_2$, then there exists either $\max U_1$ or $\min U_2$.

R$_2$ (Restricted completeness due to Weierstrass). An upward (downward) bounded set has the supremum (infimum), where the supremum is the least upper bound (lub) and the infimum is the greatest lower bound (glb).

R$_3$ (Monotone sequence theorem). An upward [downward] bounded monotone increasing [decreasing] sequence is convergent (to its sup [inf]).

R$_4$ (Nested set property by Cantor). Any family of nested closed intervals has at least one common point, i.e. if $\mathcal{F} = \{I_n = [a_n, b_n] | n \in \mathbb{N}\}$ satisfies the nested property $I_n \supset I_{n+1}$, then $\cap_{n\in\mathbb{N}} I_n \neq \emptyset$.

R$_5$ (Sequence of shrinking intervals by Bachman. Corollary to R$_4$). If in R$_4$,

$$\lim_{n\to\infty}(b_n - a_n) = 0,$$

then $\cap_{n\in\mathbb{N}} I_n$ is a singleton, i.e. consisting of one element.

R_6 (Heine-Borel covering theorem). Every open covering of a compact set contains a finite subcovering.

R_7 (Bolzano-Weierstrass). Every bounded infinite set contains an accumulation point.

R_8 (Corollary to R_7). Every bounded sequence contains a convergent subsequence.

R_9 (Completeness). Every Cauchy sequence is convergent.

R_{10} (Connectedness). $\mathbb{R}$ (and any of its intervals finite or infinite) is a connected set.

Any one of the axioms R_{10}, R_1, R_2, R_3, R_6, R_8 if assumed, entails the order relation to be Archimedean, i.e. A. R_4 has a generalization (Theorem 16.5) as a property of sets having finite intersection property. R_8 has R_7 as a generalization. R_6 is considered below as a definition of compactness. R_9 is independent of A since there exists a complete space which does not have A property.

Definition 2.2. A point x in a metric space X is called an **accumulation point** (or a cluster point or a limit point) of a set A if every ε-neighborhood $V_\varepsilon(x) = \{a \in X \mid d(a, x) < \varepsilon\}$ contains a point of A other than x. The set of all accumulation points of A is called the derived set and denoted A^d.

The above condition may be replaced by the statement.
Every ε-neighborhood $V_\varepsilon(x) = \{a \in X \mid d(a, x) < \varepsilon\}$ contains infinitely many points of A. For the latter condition implies the former. Conversely if some ε-neighborhood contains only finitely many points $a_1, \cdots, a_n$, then with $\varepsilon_1 = \min_{1 \le j \le n}\{|x - a_j|\}$, the neighborhood $V_{\varepsilon_1}(x)$ does not contain any point of A, a contradiction.

Lemma 2.2. *Archimedes principle implies that the sequence $\{g^{-n}\}$ for $1 < g \in \mathbb{N}$ is a null sequence as well as $\{\frac{1}{n}\}$.*

Proof. Given $\varepsilon > 0$, we may find $N = N(\varepsilon) \in \mathbb{N}$ such that $n > \frac{1}{\varepsilon}$, or $0 < \frac{1}{n} < \varepsilon$ and $\{\frac{1}{n}\}$ is a null sequence. Since

$$g^n \ge 2^n, \tag{2.15}$$

enough to consider $\{2^n\}$ for which we have $2^n = 1 + n + \cdots > n$. Hence $0 < 2^{-n} < \frac{1}{n}$. □

Lemma 2.3. *If a monotone increasing [decreasing] sequence $\{a_n\}$ [$\{b_n\}$] is convergent to a [b], we have $a_n \le a$ [$b_n \ge b$] for all n.*

In the sequel we shall prove equivalence of these axioms. It is sometimes useful to apply the **Weierstrass method of successive halving**, called WSH.

(WSH). Given a condition C (which describes a specific real number) and there exists at least one such a real number α satisfying C. Then there is an interval $I_0 = [a_o, b_0]$ in which this number lies and which should satisfy a condition C' related to C. The mid point $\frac{a_0+b_0}{2}$ divides the interval I_0 into two $I_1 \cup I_1'$ and at least one of them satisfies C'. If both satisfy, then choose the left one, so that we have $I_1 = [a_1, b_1]$, say. Again the mid point $\frac{a_1+b_1}{2}$ divides the interval I_1 into two and we choose the halved interval $I_2 = [a_2, b_2]$. In this way we obtain a nested sequence of intervals satisfying C'.

If we assume A, then by Corollary 2.2, we obtain a sequence of shrinking intervals satisfying C' and R_5 implies the existence of the real number satisfying C.

Theorem 2.3. R_2 *and* R_3 *are equivalent.*

Proof. $\mathrm{R}_2 \implies \mathrm{R}_3$. Suppose $\{a_n\}$ is upward bounded and consider the set $A = \{a_n\}$. Since it is upward bounded, there exists $a = \sup A$. For every $\varepsilon > 0$, $a - \varepsilon$ is not an upper bound, so that there exists an n_0 such that $a - \varepsilon < a_{n_0}$. Since $\{a_n\}$ is monotone increasing, we have for $n > n_0$ $a_n \le a < a_{n_0} + \varepsilon \le a_n + \varepsilon$, or $|a_n - a| < \varepsilon$.

$\mathrm{R}_3 \implies \mathrm{R}_2$. Let $A \subset \mathbb{R}$ be upward bounded and choose an upper bound b_0 and an element a_0 of A. If $b_0 = a_0$, then $b_0 = \sup A$. Hence we suppose $a_0 < b_0$. We apply WSH to the interval $I_0 = [a_0, b_0]$. The arithmetic mean $\frac{a_0+b_0}{2}$ is either an upper bound or not. If it is an upper bound let $b_1 = \frac{a_0+b_0}{2}$ and $a_1 = a_0$. If not, then let $a_1 = \frac{a_0+b_0}{2}$ and $b_1 = b_0$. Hence we have a halved interval $I_1 = [a_1, b_1]$ with condition C that the left end point is smaller than an element of A and the right one is an upper bound of A. We continue this process to obtain a sequence $\{a_n\}$ and $\{b_n\}$ satisfying

$$a_0 \le a_1 \le \cdots \le a_n \le \cdots \le b_n \le \cdots \le b_1 \le b_0. \tag{2.16}$$

Hence by R_3, both are convergent: $\lim_{n\to\infty} a_n = a$ and $\lim_{n\to\infty} b_n = b$. Then $a \le b$. We contend that $a = b$. For suppose $a < b$. Since both $\{a_n\}$ and $\{b_n\}$ are convergent, given $\varepsilon > 0$ we may find $N = N(\varepsilon)$ such that $n > N$ implies $|a - a_n| < \varepsilon$, $|b - b_n| < \varepsilon$. By Lemma 2.3, for $n > N$, $0 \le a_{n+1} - a_n \le a - a_n < \varepsilon$, $0 \le b_n - b_{n+1} \le b_n - b < \varepsilon$. By our way of constructing sequences, $b_{n+1} - a_{n+1}$ is either $a_{n+1} - a_n$ or $b_n - b_{n+1}$. Hence in either case, $0 < b - a \le b_{n+1} - a_{n+1} < \varepsilon$, which is impossible.

Now we prove that $a = \sup A$. For if a is not an upper bound of A, there exists an $\alpha \in A$ such that $a < \alpha$. Let $\varepsilon = \alpha - a$. Then choosing $N = N(\varepsilon)$ such that for $n > N$, $|b_n - b| < \varepsilon$, we have $b_n < b + \varepsilon = b + \alpha - a = \alpha$, which contradicts the fact that $\{b_n\}$ is a sequence of upper bounds. To prove a is the least upper bound, suppose there is an upper bound $\beta < a$. Then arguing similarly as before, we put $\varepsilon = a - \beta$ and choose $N = N(\varepsilon)$ such that for $n > N$, $a - a_n < \varepsilon$, we have $a_n > a - \varepsilon = a + \beta - a = \beta$, which contradicts the fact that β is an upper bound of A. □

Exercise 36. Prove that $\mathrm{R}_5 + \mathrm{A} \Longrightarrow \mathrm{R}_2$.

Solution. By the way of constructing the sequence (2.16) we have

$$|b_n - a_n| = \frac{b_0 - a_0}{2^n} \tag{2.17}$$

and so by Corollary 2.2 the intervals $[a_n, b_n]$ are shrinking and they determine the unique real number $\lim_{n\to\infty} a_n = \lim_{n\to\infty} b_n = a$. The rest is as in the proof of Theorem 2.3.

Theorem 2.4. R_3 *implies* R_4*, which in turn implies* R_5.

Proof. As in the proof of Theorem 2.3, (2.16) holds and so $a_m \le b_n$ for all m and $\lim_{n\to\infty} a_n = c$ exists. Hence $c \in I_n$. □

Theorem 2.5. R_3 *implies* R_7.

Proof depends on the following

Lemma 2.4. *From every sequence we may choose a monotone subsequence.*

Proof. If a term in a sequence is not less than all the subsequent terms, it is called a dragon head. If there are infinitely many dragons, then we can choose a monotone decreasing subsequence. If there are only finitely many dragons, let a_{ν_1} be the consecutive term immediately after the last dragon head. Then since it is not a dragon head, there is a $\nu_2 > \nu_1$ so that $a_{\nu_1} \le a_{\nu_2}$. Since a_{ν_2} is not a dragon head, there is a $\nu_3 > \nu_2$ so that $a_{\nu_2} \le a_{\nu_3}$. In this way we obtain a monotone increasing subsequence. □

The proof of Theorem 2.5 follows immediately from the lemma since a chosen monotone subsequence is bounded and so it must converge. We may also formulate a version of R_7 in terms of lim inf and lim sup of the sequence.

Theorem 2.6. R_5 *implies* R_6.

Proof. The proof is a typical application of WSH. Suppose that an open covering $\mathcal{U}$ of the interval $I = [a, b]$ does not have a finite subcovering. By the mid point $m = \frac{a+b}{2}$, I is divided into two. If both can be covered by finitely many U's in $\mathcal{U}$, then so is I, which is impossible. Hence one of them cannot have a finite covering. If $[a, m]$ is so, then we put $a_1 = a, b_1 = m$. If not, then $a_1 = m, b_1 = b$. Continuing this process, we obtain an infinite sequence of subintervals $\{[a_n, b_n] \mid n \in \mathbb{N}\}$ for which

$$[a_1, b_1] \supset [a_2, b_2] \supset \cdots, \quad b_n - a_n = \frac{b_{n-1} - a_{n-1}}{2}. \tag{2.18}$$

Hence $\lim_{n\to\infty}(b_n - a_n) = 0$. Hence R_5 implies $\cap_{n\in\mathbb{N}}[a_n, b_n] = \{x\}$, say. Since $x \in [a, b] \subset \cup\,\mathcal{U}$, it follows that there exists a $\mathcal{U} \ni U = (c, d) \ni x$. Since $c < x$, we have $c \notin \cap_{n\in\mathbb{N}}[a_n, b_n]$, whence there exists an $n_1 \in \mathbb{N}$ such that $c < a_{n_1}$. Similarly there exists an $n_2 \in \mathbb{N}$ such that $b_{n_2} < d$. Choosing $N = \max\{n_1, n_2\}$, we have for $n \geq N$, $c < a_{n_1} \leq a_n$ and an $n_1 \in \mathbb{N}$ such that $b_n \leq b_{n_2} < d$. Hence $[a_n, b_n] \subset (c, d)$, which means that $[a_n, b_n]$ is covered by one open set, a contradiction. . □

Theorem 2.7. R_9 *and* A *imply* R_2.

Proof. By the way of constructing the sequence (2.16) we have

$$|a_{n+1} - a_n| \leq \frac{b_0 - a_0}{2^{n+1}}, \quad |b_{n+1} - b_n| \leq \frac{b_0 - a_0}{2^{n+1}}, \quad |b_n - a_n| = \frac{b_0 - a_0}{2^n}. \tag{2.19}$$

For $n > m$ by telescoping series technique (2.8), we have

$$|a_n - a_m| = \left|\sum_{k=m}^{n-1}(a_{k+1} - a_k)\right| \leq \sum_{k=m}^{n-1}|a_{k+1} - a_k| \leq \sum_{k=m}^{n-1}\frac{b_0 - a_0}{2^{k+1}}. \tag{2.20}$$

Now the sum $\sum_{k=m}^{n-1} 2^{-k-1} \leq \sum_{k=m}^{\infty} 2^{-k-1} \leq 2^{-m} \to 0$ as $m \to \infty$ by Corollary 2.2. Hence $\{a_n\}$ is a Cauchy sequence and so is $\{b_n\}$. Hence by R_9, they are convergent, i.e.

$$a_n = a + o(1), \quad b_n = b + o(1) \tag{2.21}$$

for $n \geq n_0$. We must have $a = b$. For

$$|a - b| = b - a \leq b_n - a_n = \frac{b_0 - a_0}{2^n} \to 0 \tag{2.22}$$

by (2.19) and Lemma 2.2. Now we may repeat the same argument as in the proof of Theorem 2.3. □

Theorem 2.8. R_{10} *and* A *imply* R_1.

Proof. We remark that under the situation in R_1 it does not happen that both $a_1 = \max U_1$ and $a_2 = \min U_2$. For then the mid point $a_0 = \frac{a_1+a_2}{2}$ should belong to either U_1 or U_2 and both cases lead to contradiction.

Suppose the situation in R_1. If for any $a_2 \in U_2$, there is an $a_1 = a_1(a_2) \in U_1$ such that the mid point $a_1 < a_0 = \frac{a_1+a_2}{2} < a_2$ belongs to U_2. If it belongs to U_2, then letting $0 < \varepsilon < a_2 - a_0$, we have $V_{\varepsilon(a_2)} \subset U_2$, so that U_2 is an open set. Similarly, if for any $a_1 \in U_1$, there is an $a_2 = a_2(a_1) \in U_2$ such that the mid point $a_0 = \frac{a_1+a_2}{2}$ belongs to U_1, then U_1 is an open set. By R_{10}, both cannot happen and so we suppose the negation of one of them. Suppose there is an $a_2 \in U_2$ such that for any $a_1 \in U_1$, the mid point $b_1 = \frac{a_1+a_2}{2}$ belongs to U_1. Then with this b_1 the mid point $b_2 = \frac{b_1+a_2}{2}$ belongs to U_1 again. Continuing this process, we obtain a sequence $\{b_n\} \subset U_1$ converging to a_2 under A. If there is an $x \in U_2$ such that $x < a_2$, then there is a b_n lying between x and a_2, which contradicts the condition that $b_n < x$. Hence $a_2 = \min U_2$. □

Theorem 2.9. R_2 *implies the following assertion which in turn entails* R_{10}. *For* X, $\emptyset \neq X \subset \mathbb{R}$, *to be connected it is necessary and sufficient that for any* $x, y \in X$, $x < y$, *the closed interval* $[x, y] \subset X$.

Proof. Necessity. Suppose there exist x, y such that $x < y$, $[x, y] \not\subset X$. Choose an $a \in [x, y] - X$. Then

$$U_1 = (-\infty, a) \cap X, \quad U_2 = (a, \infty) \cap X$$

are open subsets of X satisfying the condition (16.4), whence X is not connected.

Sufficiency. Suppose X is not connected. There are open sets U_1, U_2 and $x, y \in X$ $(x < y)$ such that

$$X = U_1 \cup U_2, \quad U_1 \cap U_2 = \emptyset, \quad x \in U_1, \quad y \in U_2.$$

If $[x, y] \subset X$, put $V_i = U_i \cap [x, y]$ $(i = 1, 2)$. Then $V_1 \cup V_2 = X, V_1 \cap V_2 = \emptyset$. Let $a = \sup V_1$, we can assume $a \neq x, y$. If $a \in V_1 \subset U_1$ there exists an $\varepsilon > 0$, $V_\varepsilon(a) \subset U_1 \cup [x, y] = V_1$, a contradiction. Similarly the case $a \in V_2$ is a contradiction. Hence $[x, y] \not\subset X$. □

Theorem 2.10. R_7 *implies* R_8.

Proof. Suppose $A := \{a_n | n \in \mathbb{N}\}$ is an infinite set. Then by R_7 it has an accumulation point x and for every $\varepsilon > 0$, $V_\varepsilon(x) \cap A$ is an infinite set. Choose $\varepsilon = \frac{1}{n}$, $n \in \mathbb{N}$ and set

$$N_n = \{k \in \mathbb{N} \mid a_k \in V_{1/n} \cap A\} \tag{2.23}$$

which is an infinite set. Let $n(1) = \min N_1$ and inductively set

$$n(\nu) = \min\{N_\nu - \{n(1), \cdots, n(\nu - 1)\}\}.$$

Then $\{n(\nu)\} \subset \mathbb{N}$ is a strictly increasing sequence satisfying

$$a_{n(\nu)} \in V_{1/\nu} \cap A,$$

or $|a_{n(\nu)} - x| < \frac{1}{\nu}$, i.e. $\{a_{n(\nu)}\}$ is convergent.
Suppose $A = \{a_n | n \in \mathbb{N}\}$ is a finite set, say $A = \{\alpha_1, \cdots, \alpha_n\}$. If for every j, $1 \leq j \leq n$, the set $\{n \in \mathbb{N} | a_n = \alpha_j\}$ is finite, then their union is also finite contrary to the infinitude of $\mathbb{N}$. Hence for some j, the set $B_j := \{n \in \mathbb{N} | a_n = \alpha_j\}$ is an infinite set. Arranging this as $\{a_{n(\nu)}\}$, we infer that $a_{n(\nu)} \to \alpha_j$. □

Theorem 2.11. R$_8$ *implies* R$_9$.

Proof. In (2.7) choose $m = N + 1$. Then $|a_n| < |a_{N+1}| + \varepsilon$ for $n > N$. Hence

$$|a_n| \leq \max\{|a_1|, \cdots, |a_N|, |a_{N+1}| + \varepsilon\}$$

for all $n \in \mathbb{N}$, so that $\{a_n\}$ is bounded. R$_8$ implies the existence of a subsequence $\{a_{n(\nu)}\}$ converging to a, say. For every $\varepsilon > 0$ choose $N_1 \in \mathbb{N}$ so that

$$|a_{n(\nu)} - a| < \varepsilon$$

for $\nu \geq N_1$. Choose another $N_2 \in \mathbb{N}$ such that (2.7) holds true. Take $N_3 = \max\{N_1, N_2\}$. Then since $n(\nu) \geq \nu$, it follows that for $n(\nu), n > N_3$ we have

$$|a_n - a| < |a_n - a_{n(\nu)}| + |a_{n(\nu)} - a| < 2\varepsilon.$$

Hence $\{a_n\}$ is convergent. □

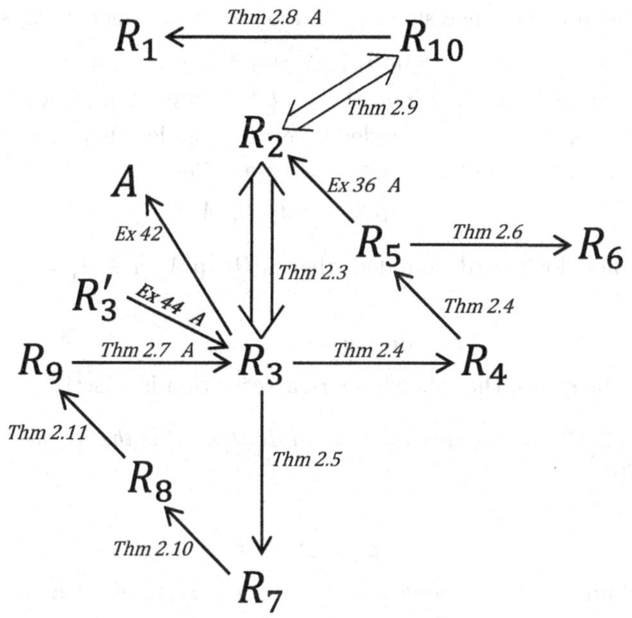

Fig. 2.2: Logical scheme

2.3 Limes principals

We shall expound the two defining conditions for limes principals. Let $\{a_n\} = \{a_n\}_{n=1}^{\infty}$ be a real sequence, which we abbreviate as $\{a_n\} \subset \mathbb{R}$, by interpreting $\{a_n\}$ to mean the set $\{a_n | n \in \mathbb{N}\}$ of all terms of the sequence. The set $\{a_n\}$ may be finite.

We adopt the monotone sequence theorem (MST) R_3 and R_2 as our axioms for continuity of real numbers.

Since as can be seen $\liminf_{n\to\infty} a_n = -\limsup_{n\to\infty}(-a_n)$, we shall confine ourselves to the case of $\limsup a_n$.

First there are two cases to consider:

(i) $\{a_n\}$ is not upward bounded, i.e. for any large $M > 0$, there exists an $N = N(M) \in \mathbb{N}$ such that $a_N > M$. In this case we define

$$\limsup_{n\to\infty} a_n = \infty. \tag{2.24}$$

(ii) $\{a_n\}$ is upward bounded. Then each subset

$$A_n = \{a_n, a_{n+1}, \cdots\}$$

of $\{a_n | n \in \mathbb{N}\}$ has a finite supremum: $\sup A_n$, in view of R_2.

Now the new sequence $\sup A_n$ is a monotone non-increasing sequence:

$$\sup A_{n+1} \le \sup A_n.$$

There are two cases possible: the set $\{A_n | n \in \mathbb{N}\}$ is downward bounded or not. If it is downward bounded, then R_3 implies that the limit exists and is equal to the inf of that set. Then we define

$$\limsup_{n\to\infty} a_n = \inf\{\sup A_n\}. \tag{2.25}$$

If it is not downward bounded, then $\inf\{\sup A_n | n \in \mathbb{N}\} = -\infty$ and we define

$$\limsup a_n = -\infty. \tag{2.26}$$

For applications, the following characterization is effective.

Theorem 2.12. *If $A = \limsup a_n$ is finite, then it is the number satisfying two conditions*
(i) *For any $\varepsilon > 0$*

$$a_n < A + \varepsilon \tag{2.27}$$

for almost all n, or any number $> A$ is an upper bound for all a_n for some $n_0 = n_0(\varepsilon)$ onwards, i.e. (2.27) *holds for $n \ge n_0$.*
(ii) *For infinitely many n, we have*

$$A - \varepsilon < a_n. \tag{2.28}$$

No number $< A$ can be a lower bound for $\{\sup A_n\}$, i.e. (2.28) *holds for infinitely many n.*

Proof. For any $\varepsilon > 0$, there exists an n_0 such that
(i)′ $\quad \sup A_{n_0} < A + \varepsilon$,
and
(ii)′ $\quad A \le \sup A_n$, $\forall n \in \mathbb{N}$. We shall prove that (i) and (ii) are equivalent to (i)′ and (ii)′ thereby our theorem.

Since $\sup A_n \le \sup A_{n_0}$ for $n \ge n_0$, it follows that $a_n \le \sup A_n \le \sup A_{n_0} < A + \varepsilon$, whence (i) follows.

By the definition of sup, for any $\varepsilon > 0$, $\sup A_n - \varepsilon$ cannot be an upper bound for $\{A_n\}$: $\exists n_0 = n_0(\varepsilon) \in \mathbb{N}$ such that

$$A - \varepsilon \le \sup A_n - \varepsilon < a_{n_0}, \quad n \ge n_0,$$

whence (ii).

Conversely, assume (i) and (ii). Then (i) implies that $\sup A_{n_0} < A + \varepsilon$, i.e. (i)′. That $A > \sup A_{n_0}$, *for some* $n_0 \in \mathbb{N}$ means that A is not a lower bound for the set $\{A_n | n \in \mathbb{N}\}$, a contradiction. Hence (ii)′ holds. This completes the proof. □

Exercise 37. Let $\{a_n\} \subset \mathbb{R}$ be a bounded sequence and let $a = \limsup a_n$ (or $a = \liminf a_n$). Then prove that there is a subsequence $\{a_{n(\nu)}\}$ converging to a.

Solution. Suppose we have chosen $a_{n(1)}, \cdots, a_{n(\nu)}, \cdots$ so that

$$a - \frac{1}{\nu} \le a_{n(\nu)} \le a + \frac{1}{\nu}.$$

Then there are infinitely many n's for which $a - \frac{1}{\nu+1} \le a_n \le a + \frac{1}{\nu+1}$. Choose one of them and label it by $a_{n(\nu+1)}$ with $n(\nu+1) \ge n(\nu)$.

Exercise 38. Prove that a sequence $\{a_n\} \subset \mathbb{R}$ is bounded if and only if from any subsequence, we can choose a convergent subsequence.

Solution. If $\{a_n\}$ is bounded, then we apply Exercise 37. If not bounded, then we may choose a subsequence $\{a_{n(1)}, \cdots, a_{n(\nu)}\}, \cdots$ so that

$$a_{n(\nu)} \ge \nu, \quad \nu \in \mathbb{N}.$$

Hence any subsequence of this sequence is not bounded whence not convergent by Exercise 33.

Exercise 39. Let $A + B$ denote the set of all sums of two elements from two sets $A, B \subset \mathbb{R}$. Prove the following

$$\sup(A + B) = \sup A + \sup B \tag{2.29}$$

and for $c \in \mathbb{R}$

$$\sup(cA) = \begin{cases} c \sup A, & c > 0 \\ 0, & c = 0 \\ c \inf A, & c < 0. \end{cases}$$

Corresponding results hold for inf.

Fig. 2.3: Archimedes

Chapter 3

Limits and continuity

In this and the next chapters we state basic properties of limit concept and continuity of functions, using rather freely some knowledge and notation from Chapter 16 in which all terminologies about topological spaces are stated.

Definition 3.1. The following holds true for metric spaces but for concreteness, we state the case $\boldsymbol{f} : X(\subset \mathbb{R}^n) \to Y(\subset \mathbb{R}^m)$. If $\boldsymbol{a}$ is an accumulation point of X and $\boldsymbol{b} \in \mathbb{R}^m$, then

$$\lim_{\boldsymbol{x}\to\boldsymbol{a}} \boldsymbol{f}(\boldsymbol{x}) = \boldsymbol{b} \tag{3.1}$$

means that for every neighborhood $V = V(\boldsymbol{b}) \in \mathcal{U}(\boldsymbol{b}) \subset 2^{\mathbb{R}^m}$ there exists a neighborhood $U = U(\boldsymbol{a}) \in \mathcal{U}(\boldsymbol{a}) \subset 2^{\mathbb{R}^n}$ such that $\boldsymbol{x} \in U'(\boldsymbol{a}) \cap X$ implies $\boldsymbol{f}(\boldsymbol{x}) \in V(\boldsymbol{b})$, where $U'(\boldsymbol{a}) = U(\boldsymbol{a}) \setminus \{\boldsymbol{a}\}$ is the deleted neighborhood. This may also be expressed as

$$\boldsymbol{f}(U'(\boldsymbol{a}) \cap X) \subset V(\boldsymbol{b}) \tag{3.2}$$

or as

$$U'(\boldsymbol{a}) \cap X \subset \boldsymbol{f}^{-1}(V(\boldsymbol{b})) \tag{3.3}$$

by Lemma 1.1.

Example 3.1. We state convergence in the form of ε-δ argument. (3.1) for $\boldsymbol{f} : X(\subset \mathbb{R}^n) \to Y(\subset \mathbb{R}^m)$ means that given $\varepsilon > 0$, there exists a $\delta = \delta(\boldsymbol{a}, \varepsilon) > 0$ such that $0 < |\boldsymbol{x} - \boldsymbol{a}| < \delta$ and $\boldsymbol{x} \in X$ implies $|\boldsymbol{f}(\boldsymbol{x}) - \boldsymbol{b}| < \varepsilon$. In the case of a sequence $\{\boldsymbol{f}_n(\boldsymbol{x})\}$, $\lim_{n\to\infty} \boldsymbol{f}_n(\boldsymbol{x}) = \boldsymbol{f}(\boldsymbol{x})$ $(\boldsymbol{x} \in X)$ means that given $\varepsilon > 0$, there exists an $N = N(\boldsymbol{x}, \varepsilon) \in \mathbb{N}$ such that $n > N$ implies $|\boldsymbol{f}_n(\boldsymbol{x}) - \boldsymbol{f}(\boldsymbol{x})| < \varepsilon$. If $N = N(\varepsilon) \in \mathbb{N}$ can be chosen independently of $\boldsymbol{x} \in X$, then $\boldsymbol{f}_n$ is said to be **uniformly convergent to $\boldsymbol{f}$** on X. This may be stated more concisely as follows. Let

$$\|\boldsymbol{f}\|_X = \sup\{|\boldsymbol{f}(\boldsymbol{x})| \big| \boldsymbol{x} \in X)\}. \tag{3.4}$$

Then $\|\boldsymbol{f}-\boldsymbol{g}\|<\varepsilon$ means that $\boldsymbol{g}$ approximates $\boldsymbol{f}$ within $\varepsilon>0$ uniformly in X. Now we may state: a sequence $\{\boldsymbol{f}_n(\boldsymbol{x})\}$ is uniformly convergent to $\boldsymbol{f}$ if and only if

$$\lim_{n\to\infty}\|\boldsymbol{f}_n-\boldsymbol{f}\|_X=0. \tag{3.5}$$

$\|\boldsymbol{f}\|_X$ satisfies the properties of norm.

Definition 3.2. The **norm** in a complex vector space is defined by the following conditions.
(i) $\|a\|\geq 0$ and $\|a\|=0 \Longleftrightarrow a=0$.
(ii) $\|\lambda a\|=|\lambda|\|a\|$.
(iii) (triangular inequality) $\|a+b\|\leq\|a\|+\|b\|$.

Exercise 40. State the relation between the norm and inner product in Definition 1.6.

Definition 3.3. Suppose $\boldsymbol{f}(\boldsymbol{a})$ is defined and choose $\boldsymbol{b}=\boldsymbol{f}(\boldsymbol{a})$. Then Definition 3.1 amounts to the following. Given any neighborhood $V=V(\boldsymbol{b})=V_\varepsilon(\boldsymbol{b})$, the ε-neighborhood, say, there exists a $U=U(\boldsymbol{a})=U_\delta(\boldsymbol{a})$, the δ-neighborhood, say, such that $\boldsymbol{x}\in U_\delta(\boldsymbol{a})\cap X$ implies $\boldsymbol{f}(\boldsymbol{x})\in V_\varepsilon(\boldsymbol{b})$. Or correspondingly to (3.3)

$$U(\boldsymbol{a})\cap X\subset \boldsymbol{f}^{-1}(V(\boldsymbol{b})). \tag{3.6}$$

We say that $\boldsymbol{f}$ is **continuous** at $\boldsymbol{a}$. If $\boldsymbol{a}$ is not an accumulation point, we say that it is continuous only if $\boldsymbol{f}(\boldsymbol{a})$ is defined. A mapping $f:X\to Y$ which is continuous at every point of X is said to be continuous on X.

Example 3.2. We state continuity in the form of ε-δ argument. A mapping $f:X\to Y$ between metric spaces is said to be continuous at $a\in X$ if given $\varepsilon>0$, there exists a $\delta=\delta(a)>0$ such that $d(x,a)<\delta$ implies $d(f(x),f(a))<\varepsilon$. Specified to $\boldsymbol{f}:X(\subset\mathbb{R}^n)\to Y(\subset\mathbb{R}^m)$, this amounts to $|\boldsymbol{f}(\boldsymbol{x})-\boldsymbol{f}(\boldsymbol{a})|<\varepsilon$ if $|\boldsymbol{x}-\boldsymbol{a}|<\delta$ and $\boldsymbol{x}\in X$. If $\delta>0$ can be chosen independently of $a\in X$, then f is said to be **uniformly continuous** on X.

Example 3.3. Choosing $X=\mathbb{N}$ and $\boldsymbol{a}=\infty$ specifies Definition 3.1 to the following. Given any neighborhood $V=V(\boldsymbol{b})=V_\varepsilon(\boldsymbol{b})$, the ε-neighborhood, say, there exists an $N=N(\varepsilon)$-neighborhood $U=U(N)=\{n>N\}$ of ∞ such that $n\in U$ implies $\boldsymbol{f}(n)\in V_\varepsilon(\boldsymbol{b})$. Or given $\varepsilon>0$, there exists an $N=N(\varepsilon)\in\mathbb{N}$ such that $|\boldsymbol{f}(n)-\boldsymbol{b}|<\varepsilon$ for $n>N$, i.e. the ordinary definition of convergence of the sequence $\{\boldsymbol{f}(n)\}$.

Proposition 3.1. *If $\boldsymbol{a}$ is an accumulation point of $X \subset \mathbb{R}^m$, then there exists a sequence $\{\boldsymbol{x}_n\}$, $\boldsymbol{x}_n$ being all distinct and $\boldsymbol{x}_n \neq \boldsymbol{a}$ such that $\lim_{n\to\infty} \boldsymbol{x}_n = \boldsymbol{a}$. Conversely, if there exists a sequence $\{\boldsymbol{x}_n\}$, $\boldsymbol{x}_n$ being all distinct and $\boldsymbol{x}_n \neq \boldsymbol{a}$ such that $\lim_{n\to\infty} \boldsymbol{x}_n = \boldsymbol{a}$, then $\boldsymbol{a} \in X^d$.*

In what follows we often use V to denote a neighborhood.

Theorem 3.1. *Let $X \in \mathcal{O}(\mathbb{R}^m)$, let $\boldsymbol{f} : X \to \mathbb{R}^k$ and let $\boldsymbol{a}$ be an accumulation point of X. Let $\{\boldsymbol{x}_n\} \subset X$ be such that $\boldsymbol{x}_n \neq \boldsymbol{a}$ and $\lim_{n\to\infty} \boldsymbol{x}_n = \boldsymbol{a}$. Then*
(i) *If $\lim_{\boldsymbol{x}\to\boldsymbol{a}} \boldsymbol{f}(\boldsymbol{x}) = \boldsymbol{b}$, then $\lim_{n\to\infty} \boldsymbol{f}(\boldsymbol{x}_n) = \boldsymbol{b}$.*
(ii) *Conversely, suppose all such sequences converge to the same limit: $\lim_{n\to\infty} \boldsymbol{f}(\boldsymbol{x}_n) = \boldsymbol{b}$ and $\lim_{\boldsymbol{x}\to\boldsymbol{a}} \boldsymbol{f}(\boldsymbol{x}) = \boldsymbol{b}$. I.e. continuous limit and discrete limit are the same.*

Proof. (i) For any $V(\boldsymbol{b}) = V_\varepsilon(\boldsymbol{b})$, there exists a $V(\boldsymbol{a}) = V_\delta(\boldsymbol{a})$ such that $\boldsymbol{x} \in V_\delta'(\boldsymbol{a}) \cap X$ implies $\boldsymbol{f}(\boldsymbol{x}) \in V(\boldsymbol{b})$. For this $V_\delta(\boldsymbol{b})$, there exists a $N \in \mathbb{N}$ such that $n > N$ implies $\boldsymbol{x}_n \in V'(\boldsymbol{a})$, so that $\boldsymbol{f}(\boldsymbol{x}_n) \in V(\boldsymbol{b})$, or $\lim_{n\to\infty} \boldsymbol{f}(\boldsymbol{x}_n) = \boldsymbol{b}$.
(ii) We show that $\lim_{\boldsymbol{x}\to\boldsymbol{a}} \boldsymbol{f}(\boldsymbol{x})$ exists and is equal to $\boldsymbol{b}$. Suppose contrary. Then there exists a $V(\boldsymbol{b}) = V_\varepsilon(\boldsymbol{b})$ such that for any $V'(\boldsymbol{a}) = V_\delta'(\boldsymbol{a})$ there exists a $\boldsymbol{x} \in V'(\boldsymbol{a}) \cap X$ such that $\boldsymbol{f}(\boldsymbol{x}) \notin V(\boldsymbol{b})$. For each $n \in \mathbb{N}$ choose $\delta = \frac{1}{n}$ we conclude that there exists a $\boldsymbol{x}_n \in V_{1/n}'(\boldsymbol{a}) \cap X$ such that $\boldsymbol{f}(\boldsymbol{x}_n) \notin V(\boldsymbol{b})$, i.e. $\lim_{n\to\infty} \boldsymbol{f}(\boldsymbol{x}_n) \neq \boldsymbol{b}$. But this sequence $\{\boldsymbol{x}_n\} \subset X$ is such that $\boldsymbol{x}_n \neq \boldsymbol{a}$ and $\lim_{n\to\infty} \boldsymbol{x}_n = \boldsymbol{a}$, a contradiction. □

Exercise 41. Prove that in Theorem 3.1 we may only suppose that any such sequence converges. I.e. they converge to the same limit: $\lim_{n\to\infty} \boldsymbol{f}(\boldsymbol{x}_n) = \boldsymbol{b}$, say.

Solution. Let $\{\boldsymbol{x}_n\}, \{\boldsymbol{y}_n\} \subset X$ be such that $\boldsymbol{x}_n, \boldsymbol{y}_n \neq \boldsymbol{a}$, $\lim_{n\to\infty} \boldsymbol{x}_n = \lim_{n\to\infty} \boldsymbol{y}_n = \boldsymbol{a}$ and $\lim_{n\to\infty} \boldsymbol{f}(\boldsymbol{x}_n) = \boldsymbol{x}, \lim_{n\to\infty} \boldsymbol{f}(\boldsymbol{y}_n) = \boldsymbol{y}$. Then $\boldsymbol{x} = \boldsymbol{y}$. For if not, define a new sequence $\boldsymbol{z}_n$ by

$$\boldsymbol{z}_n = \begin{cases} \boldsymbol{x}_n & \text{if} \quad n \equiv 0 \mod 2 \\ \boldsymbol{y}_n & \text{if} \quad n \equiv 1 \mod 2. \end{cases}$$

Then since $\{\boldsymbol{z}_n\} \subset X$ is one of the sequences in question, the limit exists: $\lim_{n\to\infty} \boldsymbol{f}(\boldsymbol{z}_n) = \boldsymbol{z}$. Hence for any $V(\boldsymbol{z}) = V_\varepsilon(\boldsymbol{b})$, there exists a $V(\boldsymbol{a}) = V_\delta(\boldsymbol{a})$ such that $\boldsymbol{z}_n \in V_\delta'(\boldsymbol{a}) \cap X$ implies $\boldsymbol{f}(\boldsymbol{z}_n) \in V(\boldsymbol{z})$. For this $V_\delta(\boldsymbol{b})$, there exists a $N \in \mathbb{N}$ such that $n > N$ implies $\boldsymbol{x}_n, \boldsymbol{y}_n \in V'(\boldsymbol{a})$, so that

$\boldsymbol{f}(\boldsymbol{x}_n), \boldsymbol{f}(\boldsymbol{y}_n) \in V(\boldsymbol{b})$. Hence

$$\boldsymbol{x} = \lim_{n\to\infty} \boldsymbol{f}(\boldsymbol{x}_n) = \lim_{n\to\infty} \boldsymbol{f}(\boldsymbol{y}_n) = \boldsymbol{y}$$

contradictory to the assumption.

Theorem 3.2. *In metric (topological) spaces X, Y, the following are equivalent.*
(i) $f : X \to Y$ *is continuous.*
(ii) $\{f^{-1}(V) \mid V \in \mathcal{O}(Y)\} \subset \mathcal{O}(X)$, *i.e. the inverse image of any open set in Y is an open set in X.*

Proof. (i)$\Longrightarrow$(ii). We show that any point $x \in f^{-1}(V)$ is an inner point for all $V \in \mathcal{O}(Y)$. Since $f(x) \in V$, there exists an ε-neighborhood $V_\varepsilon(f(x)) \subset V$. We may choose a δ-neighborhood $U_\delta(x)$ satisfying (3.6): $U_\delta(x) \cap X \subset f^{-1}(V_\varepsilon(f(x)) \subset f^{-1}(V)$. Hence $f^{-1}(V) \in \mathcal{O}(X)$.
(ii)$\Longrightarrow$(i). For any $x \in X$ and $\varepsilon > 0$, the ε-neighborhood $V_\varepsilon(f(x)) \in \mathcal{O}(Y)$, its inverse image is an open set in X: $f^{-1}(V_\varepsilon(f(x)) \in \mathcal{O}(X)$. Hence there exists a δ-neighborhood $U_\delta(x)$ satisfying (3.6), whence f is continuous. □

Remark 3.1. By this theorem *we may express continuity of a mapping on X in terms of open sets,* and *a fortiori*, in terms of other types of sets endowing topology on X. For a mapping f between topological spaces X, Y, Theorem 3.2, (ii) serves as a definition of continuity of f.

Theorem 3.3. (Cauchy condition for uniform convergence) *Suppose the sequence $\{f_n\}$ is defined on a set $X \subset \mathbb{R}^n$ and satisfies*

$$\lim_{m,n\to\infty} \|f_n - f_m\|_X = 0. \tag{3.7}$$

Then there exists a function $f(\boldsymbol{x})$ to which $\{f_n\}$ converges uniformly.

Proof. For each $\boldsymbol{x} \in X$, we have $\lim_{m,n\to\infty} |f_n(\boldsymbol{x}) - f_m(\boldsymbol{x})| = 0$. Hence $\{f_n(\boldsymbol{x})\}$ is a Cauchy sequence and so by R_9 it converges to a function $f(\boldsymbol{x})$. To prove that the convergence is uniform we note that given $\varepsilon > 0$ there exists an $N = N(\varepsilon) \in \mathbb{N}$ such that for $n, m > N$

$$\begin{aligned} |f_n(\boldsymbol{x}) - f(\boldsymbol{x})| &\le |f_n(\boldsymbol{x}) - f_m(\boldsymbol{x})| + |f_m(\boldsymbol{x}) - f(\boldsymbol{x})| \\ &\le \|f_n - f_m\|_X + |f_m(\boldsymbol{x}) - f(\boldsymbol{x})| \le \varepsilon + |f_m(\boldsymbol{x}) - f(\boldsymbol{x})|. \end{aligned} \tag{3.8}$$

For each $\boldsymbol{x} \in X$. Choose $N < m = m(\varepsilon, \boldsymbol{x}) \in \mathbb{N}$ above so that $|f_m(\boldsymbol{x}) - f(\boldsymbol{x})| < \varepsilon$. Hence in (3.8), the far-right side is $< 2\varepsilon$, so that (3.5) holds and the convergence is uniform. □

Given a function series $\sum_{n=1}^{\infty} f_n(\boldsymbol{x})$, we denote its nth partial sum by

$$S_n = S_n(\boldsymbol{x}) = \sum_{k=1}^{n} f_k(\boldsymbol{x}).$$

Then the Cauchy condition (3.7) for the series is $\lim_{m,n\to\infty} \|S_n - S_m\|_X = 0$ and we have

Theorem 3.4. (Cauchy condition for uniform convergence) *Suppose the sequence $\{f_n\}$ is defined on a set $X \subset \mathbb{R}^n$ and the Cauchy condition for $n > m$*

$$\lim_{m,n\to\infty} \left\| \sum_{k=m+1}^{n} f_k \right\|_X = 0 \tag{3.9}$$

holds. Then the series $\sum_{n=1}^{\infty} f_n(\boldsymbol{x})$ converges uniformly.

Definition 3.4. For a given sequence $\{a_n\}$, the sequence of positive numbers $\{M_n\}$ is called its Majorant if for $n > n_0$, we have $|a_n| \leq M_n$. In the case of a function series $\sum_{n=1}^{\infty} f_n(\boldsymbol{x})$ defined on a subset $X \subset \mathbb{R}^n$, each term of the Majorant series satisfies

$$\|f_n\|_X \leq M_n, \tag{3.10}$$

i.e. $|f_n(x)| \leq M_n$ for $n \in \mathbb{N}$ and for all $x \in X$.

Theorem 3.5. (Weierstrass M-test) *Given a sequence of complex-valued functions f_n defined on a subset $X \subset \mathbb{R}^n$ (or any complete vector space), suppose that it has a convergent Majorant series $\sum_{n=1}^{\infty} M_n < \infty$ satisfying* (3.10). *Then $\sum_{n=1}^{\infty} f_n(x)$ is absolutely and uniformly convergent. We often say by "absolute convergence" to mean uniform convergence.*

Proof. It suffices to check (3.9). Since for a given $\varepsilon > 0$, there exists an $N = N(\varepsilon) \in \mathbb{N}$ such that for $n > m > N$ for all $x \in X$, $|f_{m+1}(x)| + \cdots + |f_n(x)| \leq M_{m+1} + \cdots + M_n < \varepsilon$. Hence $\sum_{n=1}^{\infty} f_n(x)$ is absolutely and uniformly convergent. □

Theorem 3.6. (D'Alembert test) (i) *Suppose for a series $\sum_{n=1}^{\infty} a_n$ with non-zero terms, there exists a constant $0 \leq \lambda < 1$ such that $\left|\frac{a_{n+1}}{a_n}\right| < \lambda$ for $n \geq n_0$. Then $\sum_{n=1}^{\infty} a_n$ is absolutely convergent.*
(ii) *If there exists a constant $1 \leq \lambda$ and $n_0 \in \mathbb{N}$ such that $a_n \neq 0$, $\left|\frac{a_{n+1}}{a_n}\right| \geq \lambda$ for $n \geq n_0$, then the series is divergent.*

Proof. (i) Since $|a_{n+1}| \leq \lambda |a_n| \leq \cdots \leq \lambda^{n-n_0+1} |a_{n_0}| = \lambda^{-n_0} |a_{n_0}| \lambda^{n+1}$ for $n \geq n_0$, it follows that

$$\sum_{n=1}^{\infty} |a_n| \leq \sum_{n=1}^{n_0} |a_n| + \lambda^{-n_0} |a_{n_0}| \sum_{n=n_0+1}^{\infty} \lambda^n. \tag{3.11}$$

Since the Majorant, the geometric series, is convergent, so is the series and the convergence is absolute and uniform. (ii) Since $|a_{n+1}| \geq \lambda |a_n| \geq |a_n| \geq \cdots \geq |a_{n_0}| > 0$, it follows that $\lim_{n\to\infty} a_n \neq 0$. Hence by (2.12), the series is not convergent. □

Chapter 4

Properties of continuous mappings

Theorem 4.1. *If a topological space X is connected, then for any continuous mapping $f : X \to Y$ into a topological space Y, the image $f(X)$ (in relative topology) is also connected.*

Proof. Suppose $f(X)$ is not connected. Then there exist open sets U_1, U_2 of Y such that

$$f(X) \subset U_1 \cup U_2, \quad U_1 \cap U_2 \cap f(X) = \emptyset, \quad U_i \cap f(X) \neq \emptyset \; (i = 1, 2). \tag{4.1}$$

Then

$$X \subset f^{-1}(f(X)) \subset f^{-1}(U_1 \cup U_2) = f^{-1}(U_1) \cup f^{-1}(U_2) \subset X,$$
$$f^{-1}(U_1) \cap f^{-1}(U_2) = f^{-1}(U_1 \cap U_2) = \emptyset, \quad f^{-1}(U_i) \neq \emptyset \; (i = 1, 2),$$

so that X is not connected. □

Corollary 4.1. (Intermediate value theorem) *Suppose $f : [a, b] \to \mathbb{R}$ is a continuous function with $f(a) < f(b)$. Then for any $y \in [f(a), f(b)]$ there exists an $x \in [a, b]$ such that $f(x) = y$.*

Proof. By Theorem 2.9, $[a, b]$ is connected. Hence by Theorem 4.1, $Y = f([a, b])$ is connected. Hence by Theorem 2.9, $[f(a), f(b)] \subset Y$ and the result follows. □

Another proof of Corollary 4.1 follows from

Lemma 4.1. (Bolzano) *If $f : [a, b] \to \mathbb{R}$ is continuous and $f(a)f(b) < 0$, then there is a $c \in (a, b)$ such that $f(c) = 0$.*

Proof. We may suppose $f(a) > 0, f(b) < 0$. Let $A = \{x \in [a, b] | f(x) \geq 0\}$. Then since $a \in A$ and $A \subset [a, b]$, A is a non-empty bounded set. By R_2

there exists $c = \sup A$ which belongs to (a, b) since near a, f is positive and near b, it is negative. Suppose $f(c) \neq 0$. Then $|f(c)| > 0$. For $0 < \varepsilon < |f(c)|$ there exists a $\delta > 0$ such that $x \in V_\delta(c) \cap [a, b]$ implies $|f(x) - f(c)| < \varepsilon$. Hence we have $f(c)f(x) > 0$. It follows that

$$\begin{cases} V_\delta(c) \cap [a, b] \subset A & \text{if} \quad f(c) > 0 \\ V_\delta(c) \cap A = \emptyset & \text{if} \quad f(c) < 0 \end{cases}. \tag{4.2}$$

It contradicts to $c = \sup A$. Hence $f(c) = 0$. □

Another proof of Corollary 4.1 follows from the following observation. For any $y \in (f(a), f(b))$ put $g(x) = f(x) - y$ which satisfies the conditions in Lemma 4.1 and so the corollary follows.

Theorem 4.2. *Let $f : X \to Y$ be a continuous map from a topological space X to Y. Then the image $f(A)$ of a compact set $A(\subset X)$ is compact.*

Proof. Suppose $\mathcal{U} \subset \mathcal{O}(Y)$ is an open covering of $f(A)$. Then $\{f^{-1}(U) | U \in \mathcal{U}\}$ is an open covering of A. Hence there exist finitely many U_j's $1 \leq j \leq n$ such that $f^{-1}(U_1) \cup \cdots \cup f^{-1}(U_n) \supset A$. By Lemma 1.1, $(U_1 \cap f(X)) \cdots (U_n \cap f(X)) \supset f(A)$, whence $\{U_j \mid 1 \leq j \leq n\}$ is a finite covering of $f(A)$. □

Corollary 4.2. (Weierstrass' theorem) *Suppose X is a compact space. Then the image $f(X)$ of a continuous function $f : X \to \mathbb{R}$ has its max and min, i.e. there exist x_1, x_2 such that $f(x_1) = \min f(X)$ and $f(x_2) = \max f(X)$.*

Proof. By Theorem 4.2, $f(X)$ is compact. By Corollary 16.2, it is bounded and closed. Hence by Exercise 135, it has max and min. □

Definition 4.1. A function $f : X(\subset \mathbb{R}) \to \mathbb{R}$ is said to be monotone increasing [decreasing] if $x_1 < x_2$ implies $f(x_1) < f(x_2)$ $[f(x_1) > f(x_2)]$. A bijection (one-to-one, onto mapping) f between topological spaces X, Y is called a homeomorphism if both f and f^{-1} are continuous and denoted: $f : X \approx Y$.

Lemma 4.2. *Let $A \subset \mathbb{R}$, $B \subset \mathbb{R}$ and $f : A \to B$. For f to be continuous it is necessary and sufficient that for any (a, b), $f^{-1}((a, b) \cap B) \in \mathcal{O}(A)$. In particular, for $f : \mathbb{R} \to \mathbb{R}$ to be continuous it is necessary and sufficient that for any (a, b), $f^{-1}((a, b)) \in \mathcal{O}(\mathbb{R})$.*

Proof. By Exercise 133, any open set $U \in \mathcal{O}(\mathbb{R})$ is a union of open intervals I_λ. Since $f^{-1}(\cup_{\lambda\in\Lambda} I_\lambda) = \cup_{\lambda\in\Lambda} f^{-1}(I_\lambda)$ by Exercise 1.1, it suffices to check one interval in the case of $\mathbb{R}$. For $f : A \to B$ we recall that an open set in B is given in the form $U \cap B$ with $U \in \mathcal{O}(\mathbb{R})$. Since $f^{-1}((\cup I_\lambda) \cap B) = \cup f^{-1}(I_\lambda \cap B)$, it suffices to check the case $f^{-1}((a,b) \cap B)$. □

Theorem 4.3. (Existence and continuity of an inverse function) *The image of a continuous increasing mapping $f : [a,b] \to \mathbb{R}$ is a closed interval $[c,d]$ and f is a homeomorphism $f : [a,b] \to [c,d]$. And the inverse function exists uniquely and is continuous.*

Proof. Let $f(a) = c$ and $f(b) = d$. Then for any $y \in [c,d]$, there is an $x \in [a,b]$ such that $y = f(x)$ by Corollary 4.1. This x is unique by monotonicity of f. Hence $f([a,b]) = [c,d]$. To prove that the inverse function $f^{-1} : [c,d] \to [a,b]$ is continuous, it suffices to prove that for any open interval $(x,y) \subset \mathbb{R}$ the inverse image of $(x,y) \cap [a,b]$ is an open set in $[c,d]$ by Lemma 4.2. But these are of the form $(x,y), [a,y), (x,b]$ and their inverse images can be found to be $(f(x), f(y)), [c, f(y)), (f(x), b]$ as above, which are open sets in $[c,d]$. □

Definition 4.2. In a metric space X, the diameter of a subset $A \subset X$ is defined by

$$\delta(A) = \sup\{d(x,y) | x, y \in A\}. \tag{4.3}$$

Proposition 4.1. *Let X be a compact metric space and let $\mathcal{U}$ be its open covering. One may choose a suitable constant $\rho > 0$ such that if the diameter of a set $A \subset X$ is $< \rho$, then A is contained in one set $\in \mathcal{U}$. Such a constant is called the Lebesgue number of $\mathcal{U}$.*

Proof. For each $x \in X$ let $U_x \in \mathcal{U}$ and choose an $\varepsilon(x)$-neighborhood such that $V_{\varepsilon(x)}(x) \subset U_x$. Then $\{V_{\varepsilon(x)/2}(x) | x \in X\}$ is an open covering of X. Hence there are finitely many $x_1, \cdots, x_n \in X$ such that $V_{\varepsilon(x_1)/2}(x_1) \cup \cdots \cup V_{\varepsilon(x_n)/2}(x_n) = X$. Then let $\rho = \min\{\varepsilon(x_1)/2, \cdots, \varepsilon(x_n)/2\}$ and suppose $\delta(A) < \rho$.

Fix an $x \in A$. Then $A \subset V_\rho(x)$. Choose x_j such that $x \in V_{\varepsilon(x_j)/2}(x_j)$. Then for any $y \in A$, we have $y \in V_\rho(x)$ and $d(y, x_j) \le d(y,x) + d(x, x_j) < \rho + \varepsilon(x_j)/2 \le \varepsilon(x_j)$. Hence $y \in V_{\varepsilon(x_j)}(x_j) \subset U_{x_j}$. □

Theorem 4.4. *If X is a compact metric space, then a continuous mapping between metric spaces $f : X \to Y$ is uniformly continuous.*

Proof. Given $\varepsilon > 0$, consider the open covering

$$\mathcal{U} = \{f^{-1}(V_{\varepsilon/2}(f(x)))|x \in X\} \tag{4.4}$$

and let $\rho = \rho(\mathcal{U})$ be the Lebesgue number which is determined according to ε. Then if $x, y \in X$ satisfies $d(x, y) < \rho$, then x, y belongs to one set $f^{-1}(V_{\varepsilon/2}(f(x_j)))$. Hence $d(f(x), f(y)) \le d(f(x), f(x_j)) + d(f(x_j), f(y)) < \varepsilon$. Hence f is uniformly continuous. □

Exercise 42. Prove $\mathrm{R}_3 \Longrightarrow \mathrm{A}$.

Solution. If $\mathbb{N}$ is upward bounded, then since the sequence $\{n\}$ is monotone increasing, it converges: $\lim_{n\to\infty} n = \alpha > 0$. For an ε, $0 < \varepsilon < \frac{1}{2}$, there must exist an $n_0 \in \mathbb{N}$ such that $|n - \alpha| < \varepsilon$ for $n \ge n_0$, whence $1 \le |n+1-\alpha| + |n-\alpha| < 2\varepsilon < 1$, which is impossible.

Corollary 4.3. *Archimedes principle* A *implies the existence of the greatest integer not exceeding the real number x called the* **integral part** *of x and denoted $[x]$ (or also called the* **greatest integer function,** *the* **Gaussian symbol** *of x; in computer science it is often called the floor function denoted $\lfloor x \rfloor$). In particular,*

$$x - 1 < [x] \le x. \tag{4.5}$$

Proof. If $x = 0$, we may take $[x] = 0$. If $x > 0$, then Archimedes principle assures the existence of an $n \in \mathbb{N}$ such that $n + 1 > x$. Then $X = \{m \in \mathbb{N} \mid x - 1 < m\} \subset \mathbb{N}$ is a non-empty subset of $\mathbb{N}$. By Theorem 1.1, there exists the minimal element, say $x - 1 < q \in X$. If $q > x$, then $q - 1 > x - 1$ and $q-1 < q$ is an integer smaller than q, contradicting the minimality of q. Hence $q \le x$. If $x < 0$, then $-x > 0$ and we may apply the above argument to this and conclude that $q = [-x]$ exists. Then we have $q \le -x < q + 1$, so that $-q - 1 < x \le -q$. Hence $[x] = -q$ or $[x] = -q - 1$. □

Another solution. Suppose $x > 1$ and let

$$A = \{m \in \mathbb{N}|m \le x\} \tag{4.6}$$

and put $[x] = \max A$. Then $[x] \le x$. Since $[x] < [x] + 1 \in \mathbb{N}$, $[x] + 1$ cannot be in A, so that $x < [x] + 1$.

Remark 4.1. To find $[x]$ the following remark is often useful. If $m \in \mathbb{Z}$ and $0 \le \alpha < 1$, then

$$[m + \alpha] = m. \tag{4.7}$$

If one introduces the ceiling function $\lceil x \rceil$ to be the unique integer q satisfying $q - 1 < x \le q$. Then the last passage of the above proof gives $\lfloor x \rfloor = -\lceil -x \rceil$.

In the first proof we may also argue that since $X = \{m \in \mathbb{N} \mid x - 1 < m \le n\} \subset \mathbb{N}$ is a finite subset, there exists the minimal element q.

Theorem 4.5. *In the ring of rational integers $\mathbb{Z}$ the Euclidean division holds true: For any $m \in \mathbb{Z}, n \in \mathbb{N}$ there exist $q, r \in \mathbb{Z}$ such that*

$$m = nq + r, \tag{4.8}$$

where $0 \le r < n$ is called the residue and q the quotient.

Proof. Choose $q = \left[\frac{m}{n}\right] \in \mathbb{Z}$ and put $r = m - nq \in \mathbb{Z}$. Then since

$$q \le \frac{m}{n} < q + 1,$$

it follows that $0 \le r < n$. □

Exercise 43. Suppose $1 < n \in \mathbb{N}$ is not a perfect square. Prove that $\sqrt{n} \notin \mathbb{Q}$.

Solution. Suppose the contrary and that

$$\sqrt{n} = \frac{\ell}{m}, \quad \ell, m \in \mathbb{N} \tag{4.9}$$

in their lowest terms. By the Euclidean division, let $q = \left[\frac{l}{m}\right]$ and put $m_1 = \ell - mq$, $0 \le m_1 < m$. By (4.9), we have

$$m^2 n = \ell^2. \tag{4.10}$$

If $m_1 = 0$, $\ell = mq$ and $n = q^2$, which is a contradiction. Hence $m_1 \ne 0$. Subtracting ℓqm from both sides we deduce that

$$m(mn - \ell q) = \ell(\ell - qm) = \ell m_1. \tag{4.11}$$

Letting $\ell_1 = mn - lq$, we have $m\ell_1 = m_1\ell$. Since $m_1 < m$ we must have $\ell_1 < \ell$ which contradicts the minimality of the expression.

If one does not want to appeal to the minimality, then we remove the "in their lowest terms" condition and continue the process of making the infinite sequences of integers $m > m_1 > \cdots$ and $\ell > \ell_1 > \cdots$, which is impossible.

But the most standard way is to apply the UFD property of $\mathbb{Z}$ to (4.10) to arrive at a contradiction. Recall that the UFD property is a consequence of the Euclidean division.

Definition 4.3. Let $1 < g \in \mathbb{N}$ and let $\{c_n\}$ be a given sequence satisfying

$$0 \le c_n < g, \quad n \in \mathbb{N}. \tag{4.12}$$

Let $\alpha_n = c_0 + c_1 g^{-1} + \cdots + c_n g^{-n} = c_0.c_1 \cdots c_n$, where $c_0 \in \mathbb{N} \cup \{0\}$. Since $\{\alpha_n\}$ is an upward bounded monotone increasing sequence, it converges by R_3 and we define this limit to be the sum of the g-adic expansion

$$c_0.c_1 \cdots c_n \cdots . \tag{4.13}$$

R_3' (g-adic expansion). All g-adic expansions (4.13) are convergent and constitute the nonnegative real number system up to uniqueness.

Exercise 44. Prove that R_3' and A imply R_3.

Solution. Suppose $\{a_n\}$ is an upward bounded monotone increasing sequence expressed in g-adic expansions:

$$\begin{aligned} a_1 &= c_{10}.c_{11} \cdots c_{1n} \cdots \\ a_2 &= c_{20}.c_{21} \cdots c_{2n} \cdots \\ &\cdots . \end{aligned} \tag{4.14}$$

We consider sequences consisting of entries in each column. Since $\{a_n\}$ is bounded, there exists an N_0 such that $c_{N_0,0} = \max C_0 = \sup\{c_{10}, c_{20}, \cdots\}$ for the first column sequence C_0. Since $\{c_{i0}\}$ is monotone increasing, we must have $c_{i0} = c_{N_0,0}$ for $i \geq N_0$. Then let C_1 be the part of the second column sequence starting from $c_{N_0,1}$. Since these are g-digits, it is a finite sequence and $N_1 \geq N_0$ we have $c_{N_1,1} = \max C_1$ and $c_{i1} = c_{N_1,1}$ for $i \geq N_1$.

In this way we obtain a sequence $\{b_n = c_{N_0,0}.c_{N_1,1} \cdots c_{N_n,n}\}$ converging to $b = c_{N_0,0}.c_{N_1,1} \cdots c_{N_n,n} \cdots$. Since $|a_m - b| < g^{-n}$ for $N_n \leq m \in \mathbb{N}$. Lemma 2.2 shows that $a_m - b$ is a null sequence, i.e. $\lim_{m\to\infty} a_m = b$.

Remark 4.2. We need to assume A which entails Lemma 2.2 which in turn implies, as is very well-known, that here is ambiguity of g-adic expansion of the form

$$n.0 = (n-1).(g-1)(g-1)(g-1) \cdots . \tag{4.15}$$

Indeed, the right-hand side is $n-1+(g-1)\sum_{k=0}^{\infty} g^{-k} = n-1+(g-1)\frac{1}{g-1} = n$. This ambiguity may be removed by adopting the expression containing infinitely many non-zero terms. We therefore treat the g-adic expansion as a handy expression.

Exercise 45. Prove that $(0,1] \times (0,1]$ and $(0,1]$ have the same cardinality.

Solution. Here the decimal expansion can be used effectively. Let the decimal expansion of $a, b \in (0,1]$ be

$$a = 0.a_1 \cdots a_n \cdots , \quad b = 0.b_1 \cdots b_n \cdots ,$$

which is (4.13) with $g = 10$. Then putting e.g. $c_{2n-1} = a_n, c_{2n} = b_n$, the correspondence $(a,b) \leftrightarrow c = 0.c_1 \cdots c_n \cdots$ is one-to-one.

Chapter 5

Differentiation

The limit concept in the Euclidean space is stated in Definition 3.1. Let $\boldsymbol{f} : X \to \mathbb{R}^m$ be a vector-valued function defined on an open set $X \subset \mathbb{R}^n$: $\boldsymbol{y} = \boldsymbol{f}(\boldsymbol{x})$. This is a concise vector expression for

$$\begin{pmatrix} y_1 \\ \vdots \\ y_m \end{pmatrix} = \boldsymbol{y} = \boldsymbol{f}(\boldsymbol{x}) = \begin{pmatrix} y_1(\boldsymbol{x}) \\ \vdots \\ y_m(\boldsymbol{x}) \end{pmatrix} = \begin{pmatrix} y_1(x_1, \cdots, x_n) \\ \vdots \\ y_m(x_1, \cdots, x_n) \end{pmatrix}, \tag{5.1}$$

$\boldsymbol{x} = {}^t\left(x_1, \cdots, x_n\right) \in X$ which is equivalent to the system of equations

$$\begin{cases} y_1 = y_1(\boldsymbol{x}) = y_1(x_1, \cdots, x_n) \\ \vdots \\ y_m = y_m(\boldsymbol{x}) = y_m(x_1, \cdots, x_n). \end{cases} \tag{5.2}$$

Definition 5.1. The vector-valued function $\boldsymbol{y} = \boldsymbol{f}(\boldsymbol{x})$ is said to be **totally differentiable** (or Frechét differentiable) at $\boldsymbol{x} \in X$ if the increment of $\boldsymbol{y}$ is approximated by a linear map $(\mathrm{d}\boldsymbol{f})$ in the increment $\boldsymbol{h} \to \boldsymbol{o}$, i.e. there exists an $m \times n$ matrix A such that

$$\boldsymbol{f}(\boldsymbol{x} + \boldsymbol{h}) = \boldsymbol{f}(\boldsymbol{x}) + A\boldsymbol{h} + o(|\boldsymbol{h}|) \tag{5.3}$$

as $\boldsymbol{h} = {}^t(h_1, \cdots, h_n) \to \boldsymbol{o}$, i.e. $|\boldsymbol{h}| \to 0$. The vector increment $\boldsymbol{h}$ is often denoted by $\Delta\boldsymbol{x}$. The matrix A is called the **gradient** (or the Jacobi matrix) of $\boldsymbol{f}$, denoted $\nabla\boldsymbol{f}$ or sometimes $\operatorname{grad}\boldsymbol{f}$. The linear map $\mathrm{d}\boldsymbol{f}$ is called the **differential** of $\boldsymbol{f}$ at $\boldsymbol{x}$ and its value at $\boldsymbol{h}$ is

$$\mathrm{d}\boldsymbol{f}\boldsymbol{h} = \nabla\boldsymbol{f}\boldsymbol{h} = A\boldsymbol{h}, \tag{5.4}$$

which is a generalization of (5.21).

We remark that Definition 5.1 and Theorem 5.1 entail all the other definitions of derivatives as special cases most of which have a division counterpart. However (5.3) *cannot in general have a division counterpart.*

Theorem 5.1. *If $\boldsymbol{y} = \boldsymbol{f}(\boldsymbol{x})$ is totally differentiable at $\boldsymbol{x}$, then it is continuous there. It is also partially differentiable in each variable and the Jacobi matrix is expressed as*

$$A=\nabla \boldsymbol{f}=\nabla\begin{pmatrix} y_1\\ \vdots \\ y_m\end{pmatrix}=\begin{pmatrix} \nabla y_1\\ \vdots \\ \nabla y_m\end{pmatrix}=\begin{pmatrix} \frac{\partial y_1}{\partial x_1} & \cdots & \frac{\partial y_1}{\partial x_n}\\ & \cdots & \\ \frac{\partial y_m}{\partial x_1} & \cdots & \frac{\partial y_m}{\partial x_n}\end{pmatrix}=\begin{pmatrix} y_{1_{x_1}} & \cdots & y_{1_{x_n}}\\ & \cdots & \\ y_{m_{x_1}} & \cdots & y_{m_{x_n}}\end{pmatrix}. \tag{5.5}$$

Proof. Proof amounts to the case of a scalar function in n variables. I.e. if $y_i = 0$ for all $i \neq k$ i.e. $\boldsymbol{y} = {}^t(0, \cdots, y_k, \cdots, 0)$, then writing $y = y_k$, (5.3) entails

$$y(\boldsymbol{x}+\boldsymbol{h}) = y(\boldsymbol{x}) + (\nabla y)\boldsymbol{h} + o(|\boldsymbol{h}|). \tag{5.6}$$

In order to determine the components of ∇y, we specify further that $h_i = 0$ for all $i \neq j$ i.e. $\boldsymbol{h} = {}^t(0, \cdots, h_j, \cdots, 0)$, then (5.6) entails

$$y(x_1, \cdots, x_j+h_j, \cdots, x_n) = y_k(x_1, \cdots, x_j, \cdots, x_n)+\frac{\partial y}{\partial x_j}h_j+o(h_j), \tag{5.7}$$

where ∇y in this case is written as $\frac{\partial y}{\partial x_j}$ called the **partial derivative** of y with respect to x_j, i.e. the instantaneous ratio of change of y in x_j with other variables remaining constant. It is also denoted y_{x_j}.

It has a division counterpart

$$\frac{\partial y}{\partial x_j} = y_{x_j} = \lim_{h\to 0}\frac{y(x_1, \cdots, x_j+h_j, \cdots, x_n) - y(x_1, \cdots, x_j, \cdots, x_n)}{h_j}. \tag{5.8}$$

Hence the jth component of ∇y in (5.6) is $\frac{\partial y}{\partial x_j}$, so that

$$\nabla y = \left(\frac{\partial y}{\partial x_1}, \cdots, \frac{\partial y}{\partial x_n}\right) = \left(y_{x_1}, \cdots, y_{x_n}\right). \tag{5.9}$$

Since this is the kth component of $\nabla \boldsymbol{y}$, (5.4) follows. □

Definition 5.2. In case $m = n$ the determinant of the Jacobi matrix (5.5) of $\boldsymbol{y} = \boldsymbol{f}(\boldsymbol{x})$ is called the **Jacobian** denoted

$$J_{\boldsymbol{f}} = J_{\boldsymbol{y}} = \frac{\partial(y_1, \cdots, y_n)}{\partial(x_1, \cdots, x_n)} = \begin{vmatrix} \frac{\partial y_1}{\partial x_1} & \cdots & \frac{\partial y_1}{\partial x_n}\\ & \cdots & \\ \frac{\partial y_n}{\partial x_1} & \cdots & \frac{\partial y_n}{\partial x_n}\end{vmatrix} = \begin{vmatrix} y_{1_{x_1}} & \cdots & y_{1_{x_n}}\\ & \cdots & \\ y_{n_{x_1}} & \cdots & y_{n_{x_n}}\end{vmatrix}. \tag{5.10}$$

As we shall see below, this plays the role of the derivative $f'(x)$.

We shall state the important special cases of Theorem 5.1 in detail.

First, if $\boldsymbol{y} = \boldsymbol{f}(\boldsymbol{x})$ is a scalar function $y = f(x)$, then ∇y is denoted by $f'(x)$ and

$$f(x+h) = f(x) + f'(x)h + o(h), \tag{5.11}$$

as $h \to 0$ (where by convention we often omit the absolute value sign), which is equivalent to the usual definition (**division counterpart**) of the **differential coefficient** of a function of one real or complex variable at the point x_0

$$f'(x_0) = \lim_{h \to 0} \frac{f(x_0 + h) - f(x_0)}{h}, \tag{5.12}$$

which in the case of real variable is the slope of a **tangent** (line) (of the graph of the function $y = f(x)$) at the point $(x_0, f(x_0))$. Hence at the point x_0, the tangential line is given by

$$y = f'(x_0)(x - x_0) + f(x_0). \tag{5.13}$$

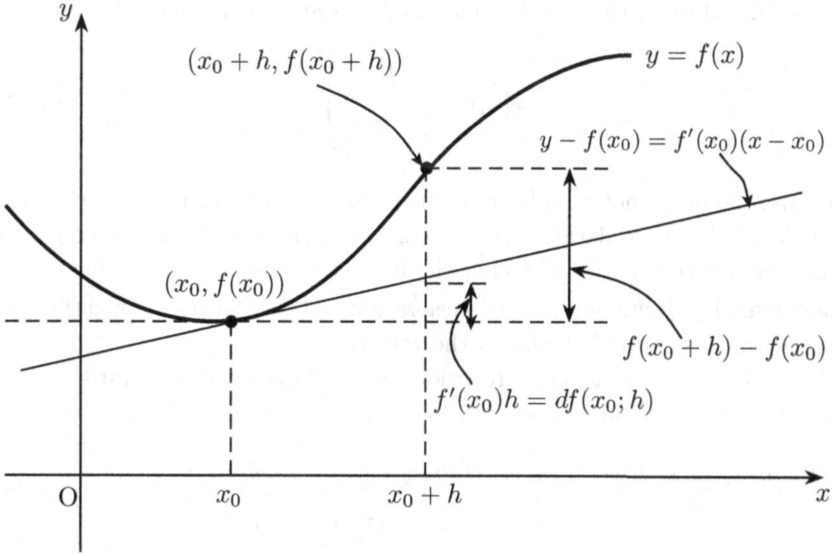

Fig. 5.1: Differential

If f has the differential coefficient at each point $x \in X$, then there arises a new function with the differential coefficient as its value at each point, called the **derivative** (or derived function) of f and denoted f', $\frac{\mathrm{d}f}{\mathrm{d}x}$, y', $\frac{\mathrm{d}y}{\mathrm{d}x}$ etc. In this case, the differential $\mathrm{d}f$ is a $(1,1)$-matrix $(f'(x))$ and its value at $h = \mathrm{d}x$ amounts to the product $f'(x)\mathrm{d}x$.

Secondly, if the variable is a scalar, then we denote it by t (t for time) and the vector-valued function $\boldsymbol{y} = \boldsymbol{f}(t)$ is an n-dimensional curve or a position vector, often denoted

$$\boldsymbol{x}(t) = {}^t(x_1(t), \cdots, x_n(t)) \tag{5.14}$$

indicating the point on the curve or the position of the particle at time t. Equation (5.3) reads

$$\boldsymbol{x}(t+h) = \boldsymbol{x}(t) + \frac{\mathrm{d}}{\mathrm{d}t}\boldsymbol{x}(t)h + o(h) \tag{5.15}$$

which has a division counterpart

$$\frac{\mathrm{d}}{\mathrm{d}t}\boldsymbol{x}(t_0) = \dot{\boldsymbol{x}}(t_0) = \lim_{h\to 0} \frac{\boldsymbol{x}(t_0+h) - \boldsymbol{x}(t_0)}{h}. \tag{5.16}$$

In the case of a curve, this is the **tangent vector** at $\boldsymbol{x}(t_0)$ and is the vector in the direction of the tangent line of the curve. In component form

$$\dot{\boldsymbol{x}}(t_0) = \begin{pmatrix} \dot{x}_1(t_0) \\ \vdots \\ \dot{x}_n(t_0) \end{pmatrix}. \tag{5.17}$$

We use the prime notation for the differentiation with respect to arc length. See §14.3 for more details. In the case of a particle this corresponds to the velocity vector $\boldsymbol{v}(t) = \frac{\mathrm{d}}{\mathrm{d}t}\boldsymbol{x}(t)$ the instantaneous ratio of change of the position. Its derivative is the acceleration $\boldsymbol{a}(t) = \frac{\mathrm{d}}{\mathrm{d}t}\boldsymbol{v}(t) = \frac{\mathrm{d}^2}{\mathrm{d}t^2}\boldsymbol{x}(t)$, the instantaneous ratio of change of the velocity.
We state the case of a scalar function in n variables which is stated in the proof as an independent corollary for its importance.

Corollary 5.1. *If $y = f(\boldsymbol{x})$ is (totally) differentiable at $\boldsymbol{x}$, i.e.*

$$y(\boldsymbol{x}+\boldsymbol{h}) = y(\boldsymbol{x}) + (\nabla y)\boldsymbol{h} + o(|\boldsymbol{h}|), \tag{5.18}$$

then it is continuous there. It is also partially differentiable in each variable and we have

$$\nabla y = \left(\frac{\partial y}{\partial x_1}, \cdots, \frac{\partial y}{\partial x_n}\right) = \left(y_{x_1}, \cdots, y_{x_n}\right). \tag{5.19}$$

We write

$$\boldsymbol{h} = \mathrm{d}\boldsymbol{x} = {}^t(\mathrm{d}x_1, \cdots, \mathrm{d}x_n) \tag{5.20}$$

with $\mathrm{d}x_j$ denoting the increment in x_j (which will be a basis of differential forms in §14.1). Then

$$\mathrm{d}y = \mathrm{d}y(\mathrm{d}\boldsymbol{x}) = \nabla y \mathrm{d}\boldsymbol{x} = \sum_{k=1}^{n} \frac{\partial y}{\partial x_k} \mathrm{d}x_k, \tag{5.21}$$

which is a (differential) 1-form in n-dimension. This will appear as the result of derivation of a 0-form in (14.29).

Remark 5.1. In the cases where **higher order derivatives** exist, we introduce them inductively.

$$f^{(n)} = \left(f^{(n-1)}\right)', \quad f^{(0)} = f, \quad f^{(1)} = f', \quad f^{(2)} = f'', \cdots. \tag{5.22}$$

Recall Leibniz's rule in Exercise 7. Similarly for partial derivatives. Cf. Definition 7.2 for the class of continuously differentiable functions. Note that being differential operators, they are applied nearer to the variable. E.g. $\frac{\partial^2 y}{\partial x_2 \partial x_1} = \frac{\partial}{\partial x_2}\left(\frac{\partial y}{\partial x_1}\right)$. However, by Schwarz's or Young's theorem below, they are interchangeable. In another notation, this amounts to $y_{x_1 x_2}$.

Another proof of Corollary 5.2 will be given in Exercise 102 in §11.8.

Exercise 46. Anticipating §11.1, prove the following

$$\int x^w \log^k x \,\mathrm{d}x = \frac{\partial^k}{\partial w^k}\left(\int x^w \,\mathrm{d}x\right) = \frac{\partial^k}{\partial w^k}\left(\frac{x^{w+1}}{w+1}\right) \tag{5.23}$$

the far-right side member giving the left side indefinite integral. Differentiate in w and substitute the value of w. Prove the following.

$$\int \log x \,\mathrm{d}x = x\log x - x + C, \quad \int x\log x \,\mathrm{d}x = \frac{1}{2}x^2\log x - \frac{1}{4}x^2 + C,$$
$$\int x^2 \log x \,\mathrm{d}x = \frac{1}{3}x^3 \log x - \frac{1}{9}x^3 + C, \tag{5.24}$$
$$\int \log^2 x \,\mathrm{d}x = x(\log^2 x - 2\log x + 2) + C.$$

These are to be added to the formulas in §11.5.

As an example of higher order derivatives, anticipating Corollary 7.4 we state

Example 5.1. Among those indefinite integrals which can be expressed by elementary functions, arctan-function in (11.73) is the most ubiquitous as well as the arcsin, arcsinh-functions in (11.76). In this example we give an independent treatment of the arctan-function. We find successive differential coefficients of $\arctan x$ at $x = 0$ and recover (11.56). $\tan x = \frac{\sin x}{\cos x}$ is defined in §11.3 and is strictly increasing in the interval $\left(-\frac{\pi}{2}, \frac{\pi}{2}\right)$. Hence the inverse function $\arctan x$ exists and is differentiable in this interval, called the principal branch. First by the differentiation rule for quotients, we have

$$(\tan x)' = \frac{1}{\cos^2 x}. \tag{5.25}$$

Since

$$y = \tan x \Longleftrightarrow x = \arctan y, \tag{5.26}$$

it follows that $\tan(\arctan y) = y$. Hence by the chain rule,

$$1 = (\tan(\arctan y))'(\arctan y)' = \frac{1}{\cos^2 x}(\arctan y)' = (1+\tan^2 x)(\arctan y)',$$

so that

$$(\arctan x)' = \frac{1}{1+x^2} \tag{5.27}$$

on changing the variables. To find successive derivatives, it is easier to use the partial fraction expansion

$$\frac{1}{1+x^2} = \frac{1}{2i}\left(\frac{1}{x-i} - \frac{1}{x+i}\right) = \frac{1}{2i}\left((x-i)^{-1} - (x+i)^{-1}\right). \tag{5.28}$$

Hence

$$\begin{aligned}(\arctan x)^{(n)} &= (-1)^{n-1}(n-1)!\frac{1}{2i}\left((x-i)^{-n} - (x+i)^{-n}\right) \\ &= (-1)^{n-1}(n-1)!\frac{1}{(x^2+1)^n}\operatorname{Im}(x+i)^n.\end{aligned} \tag{5.29}$$

It follows that

$$\operatorname{Im}(x+i)^n = \sum_{\substack{j=1 \\ j \text{ odd}}}^{n} \binom{n}{j} x^{n-j}(-1)^{\frac{j-1}{2}}. \tag{5.30}$$

Hence

$$(\arctan 0)^{(n)} = \begin{cases} 0 & n \quad \text{even} \\ (-1)^{\frac{n-1}{2}}(n-1)! & n \quad \text{odd} \end{cases}. \tag{5.31}$$

Hence by Theorem 7.5 we may write

$$\arctan x = \sum_{k=1}^{n} \frac{(-1)^{k-1}}{2k-1} x^{2k-1} + R_{2n}(x), \tag{5.32}$$

where

$$R_{2n}(x) = -\frac{1}{2n}\frac{x^{2n}}{(\theta^2 x^2 + 1)^{2n}}\operatorname{Im}(\theta x + i)^{2n}.$$

Hence

$$|R_{2n}(x)| \leq \frac{1}{2n}\left(\frac{|x|}{\sqrt{(\theta x)^2+1}}\right)^{2n} \leq \frac{1}{2n}|x|^{2n}.$$

Hence for $|x| \leq 1$, $R_{2n}(x) \to 0$ as $n \to \infty$, whereby (5.32) leads to (11.56):

$$\arctan x = \sum_{k=1}^{\infty} \frac{(-1)^{k-1}}{2k-1} x^{2k-1}, \quad |x| \leq 1. \tag{5.33}$$

For historical comments on the numerical value $\arctan 1 = \frac{\pi}{4}$ cf. Remark 6.1.

Exercise 47. In the disc $|z| < \frac{\pi}{2}$ we have the Maclaurin expansion

$$\tan z = \sum_{n=0}^{\infty} \frac{A_{2n+1}}{(2n+1)!} z^{2n+1}, \quad |z| < \frac{\pi}{2}. \tag{5.34}$$

The coefficients $\{A_{2n+1}\}$ are called **tangent numbers**.
Verify

$$A_1 = 1, A_3 = 2, A_5 = 16, A_7 = 272, A_9 = 7936, A_{11} = 353792, \cdots. \tag{5.35}$$

Hence (5.34) reads

$$\tan x = x + \frac{1}{3}x^3 + \frac{2}{15}z^5 + \frac{17}{315}z^7 + \cdots. \tag{5.36}$$

Solution. This is expounded in [Chakraborty *et al.* (2016), Exercise 1.6, pp. 17–18] together with the radius of convergence being π. Thus we give only indications. The most naive method would be to use the expansions for sin and cos to divide the former by the latter, or writing $\cos \tan = \sin$ and comparing the coefficients.

The second method uses the equality

$$f' = 1 + f^2, \tag{5.37}$$

which is equivalent to the Pythagoras theorem $1 + \tan^2\theta = \frac{1}{\cos^2\theta}$. Hence $f(0) = 0, f'(0) = 1$. In computing higher derivatives, we substitute (5.37) whenever we encounter f'.

The third method relies on Euler numbers. Cf. (13.89) for the relation between them.

Example 5.2. (i) In the case of the vector-valued function (1.72) arising from the polar coordinates (1.72), we have

$$\nabla\phi = \begin{pmatrix} \frac{\partial x}{\partial r} & \frac{\partial x}{\partial \theta} \\ \frac{\partial y}{\partial r} & \frac{\partial y}{\partial \theta} \end{pmatrix} = \begin{pmatrix} \cos\theta & -r\sin\theta \\ \sin\theta & r\cos\theta \end{pmatrix}. \tag{5.38}$$

Hence the Jacobian is

$$J_{\phi} = |\nabla \phi| = \begin{vmatrix} \frac{\partial x}{\partial r} & \frac{\partial x}{\partial \theta} \\ \frac{\partial y}{\partial r} & \frac{\partial y}{\partial \theta} \end{vmatrix} = \begin{vmatrix} \cos\theta & -r\sin\theta \\ \sin\theta & r\cos\theta \end{vmatrix} = r. \tag{5.39}$$

(ii) The spherical coordinates (or the polar coordinates in the space) is given by

$$\begin{cases} x = r\cos\theta\sin\varphi \\ y = r\sin\theta\sin\varphi \\ z = r\cos\varphi \end{cases} \tag{5.40}$$

where ${}^t(x, y, z) = \boldsymbol{x}$ is a point P on the sphere S of radius r and center at the origin O, φ is the angle subtended by OP and ON with N the north pole and θ is the angle subtended by the positive x-axis and OP', where P' is the projection of P onto the xy-plane. Hence

$$0 \le \theta \le 2\pi, \quad 0 \le \varphi \le \pi. \tag{5.41}$$

As a vector-valued function,

$$\begin{aligned} &{}^t(x, y, z) = \boldsymbol{x} = \boldsymbol{\phi}(\boldsymbol{t}), \boldsymbol{t} = {}^t(r, \theta, \varphi) = (r\cos\theta\sin\varphi, r\sin\theta\sin\varphi, r\cos\varphi), \\ &\boldsymbol{\phi} : \mathbb{R}_{\ge 0} \times [0, 2\pi] \times [0, \pi] \to S. \end{aligned} \tag{5.42}$$

This is the special case of Example 14.1 with radius r fixed. Correspondingly to (5.38), we have

$$\nabla \boldsymbol{\phi} = \begin{pmatrix} \frac{\partial x}{\partial r} & \frac{\partial x}{\partial \theta} & \frac{\partial x}{\partial \varphi} \\ \frac{\partial y}{\partial r} & \frac{\partial y}{\partial \theta} & \frac{\partial y}{\partial \varphi} \\ \frac{\partial z}{\partial r} & \frac{\partial z}{\partial \theta} & \frac{\partial z}{\partial \varphi} \end{pmatrix} = \begin{pmatrix} \cos\theta\sin\varphi & -r\sin\theta\sin\varphi & r\cos\theta\cos\varphi \\ \sin\theta\sin\varphi & r\cos\theta\sin\varphi & r\sin\theta\cos\varphi \\ \cos\varphi & 0 & -r\sin\varphi \end{pmatrix}. \tag{5.43}$$

Correspondingly to (5.39), we have

$$J_{\phi} = |\nabla \boldsymbol{\phi}| = \begin{vmatrix} \frac{\partial x}{\partial r} & \frac{\partial x}{\partial \theta} & \frac{\partial x}{\partial \varphi} \\ \frac{\partial y}{\partial r} & \frac{\partial y}{\partial \theta} & \frac{\partial y}{\partial \varphi} \\ \frac{\partial z}{\partial r} & \frac{\partial z}{\partial \theta} & \frac{\partial z}{\partial \varphi} \end{vmatrix} = -r^2 \sin\varphi. \tag{5.44}$$

These will be often used in the computation of integrals by the change of variable.

As a converse to Theorem 5.1, we state a sufficient condition for total differentiability.

Theorem 5.2. *Let $D \subset \mathbb{R}^n$ be an open set. Suppose $\boldsymbol{f} = {}^t(f_1, \cdots, f_m) : D \to \mathbb{R}^m$ of class $C^1(D)$, i.e. all $D_k f_i = \frac{\partial f_j}{\partial x_k}$ exist and are continuous $(1 \le i \le m, 1 \le k \le n)$. Then $\boldsymbol{f}$ is totally differentiable at each $\boldsymbol{x} \in D$.*

Proof. It suffices to show that each component $f_i = f$, say is totally differentiable at $\boldsymbol{x} \in D$. By assumption, each $D_k f$ is continuous at $\boldsymbol{x}$. Hence, given $\varepsilon > 0$, there exists a $\delta > 0$ such that

$$|D_k f(\boldsymbol{y}) - D_k f(\boldsymbol{x})| < \frac{\varepsilon}{n} \tag{5.45}$$

as long as $\boldsymbol{y} \in V_{\delta/2}(\boldsymbol{x})$. We write

$$\boldsymbol{y} - \boldsymbol{x} = \lambda \boldsymbol{u}, \quad |\boldsymbol{u}| = 1, \quad 0 \neq \lambda \in \mathbb{R}, \quad \lambda \to 0$$

and $(1 \leq k \leq n)$

$$\boldsymbol{u}_k = \boldsymbol{u}_{k-1} + a_k \boldsymbol{e}_k = \sum_{j=1}^{k} a_j \boldsymbol{e}_j, \quad \boldsymbol{u}_n = \boldsymbol{u}, \quad \boldsymbol{u}_0 = \boldsymbol{o},$$

so that

$$0 < |\lambda| = |\boldsymbol{y} - \boldsymbol{x}| < \frac{\delta}{2}, \quad |a_k| \leq |\boldsymbol{u}_k| \leq |\boldsymbol{u}| = 1. \tag{5.46}$$

Then by telescoping,

$$\sum_{k=1}^{n} (f(\boldsymbol{x} + \lambda \boldsymbol{u}_k) - f(\boldsymbol{x} + \lambda \boldsymbol{u}_{k-1})) = f(\boldsymbol{x} + \lambda \boldsymbol{u}) - f(\boldsymbol{x}) = f(\boldsymbol{y}) - f(\boldsymbol{x}). \tag{5.47}$$

Noting $|\lambda \boldsymbol{u}_k| \leq |\lambda| < \frac{\delta}{2}$, we put for each k for which $a_k \neq 0$, $\boldsymbol{x} + \lambda \boldsymbol{u}_k \in V_{\delta/2}(\boldsymbol{x})$

$$g(t) = f(\boldsymbol{x} + \lambda \boldsymbol{u}_{k-1} + t\boldsymbol{e}_k), \quad t \in I_k = [-|\lambda a_k|, |\lambda a_k|]. \tag{5.48}$$

Then g is continuous on I_k and $g'(t) = D_k f(\boldsymbol{x} + \lambda \boldsymbol{u}_{k-1} + t\boldsymbol{e}_k)$. The kth summand in (5.47) is $g(\lambda a_k) - g(0)$. Hence appealing to the mean value theorem, Theorem 7.3 below, we deduce that $g(\lambda a_k) - g(0) = \lambda a_k g'(\alpha_k)$ with $\alpha_k \in I_k$. Hence (5.47) amounts to

$$f(\boldsymbol{y}) - f(\boldsymbol{x}) = \sum_{k=1}^{n} \lambda a_k D_k f(\boldsymbol{x} + \lambda \boldsymbol{u}_{k-1} + \alpha_k \boldsymbol{e}_k). \tag{5.49}$$

Recalling

$$(\nabla f)(\boldsymbol{y} - \boldsymbol{x}) = (D_1 f, \cdots, D_n f) \lambda \sum_{k=1}^{n} a_k \boldsymbol{e}_k = \lambda \sum_{k=1}^{n} a_k D_k f(\boldsymbol{x} + \lambda \boldsymbol{u}_k),$$

we find that (5.49) may be written as

$$|f(\boldsymbol{y}) - f(\boldsymbol{x}) - (\nabla f)(\boldsymbol{y} - \boldsymbol{x})| \leq \sum_{k=1}^{n} \lambda |a_k| R_k, \tag{5.50}$$

where $R_k = |D_k f(\boldsymbol{x} + \lambda \boldsymbol{u}_{k-1} + \alpha_k \boldsymbol{e}_k) - D_k f(\boldsymbol{x} + \lambda \boldsymbol{u}_k)|$. By (5.46), $|\lambda \boldsymbol{u}_{k-1} + \alpha_k \boldsymbol{e}_k| \leq |\lambda| + |\alpha_k| \leq |\lambda| + |\lambda||a_k| \leq 2\lambda < \delta$, so that we may apply (5.45) to estimate R_k. Hence it follows that the right-hand side of (5.50) is $\leq |\lambda| \sum_{k=1}^{n} |a_k| \frac{\varepsilon}{n} \leq \varepsilon |\lambda| = \varepsilon |\boldsymbol{y} - \boldsymbol{x}|$ and the proof is complete. □

Theorem 5.3. (Schwarz) *If a function $f(\boldsymbol{x})$ is of class C^1 and $f_{x_1x_2} = \frac{\partial^2 f}{\partial x_2 \partial x_1}$ exists and is continuous on an open set $D \subset \mathbb{R}^n$. Then $f_{x_2x_1}$ exists and*

$$f_{x_2x_1} = f_{x_1x_2} \tag{5.51}$$

throughout D.

Corollary 5.2. (Young) *If a function $f(x)$ is of class C^2 on an open set D. Then* (5.51) *holds.*

Proof. Proof is in the lines of the proof of Theorem 5.2. Assume that $D_{21}f = f_{x_1x_2}$ exists and is continuous. Correspondingly to (5.47) with $n = 2$ and $h_1, h_2 \to 0$, we consider at any point (x_1, x_2)

$$\begin{aligned}\Delta = \Delta(h_1, h_2) &:= g_{h_2}(x_1 + h_1) - g_{h_2}(x_1) \\ &= f(x_1 + h_1, x_2 + h_2) - f(x_1 + h_1, x_2) - (f(x_1, x_2 + h_2) - f(x_1, x_2))\end{aligned} \tag{5.52}$$

where correspondingly to (5.48)

$$g_{h_2}(t) = f(t, x_2 + h_2) - f(t, x_2). \tag{5.53}$$

Since $g'_{h_2}(t) = D_1 f(t, x_2 + h_2) - D_1 f(t, x_2)$, it follows from the mean value theorem that

$$\begin{aligned}\Delta = g_{h_2}(x_1 + h_1) - g_{h_2}(x_1) &= h_1 g'_{h_2}(x_1 + \theta_1 h_1) \\ &= h_1(D_1 f(x_1 + \theta_1 h_1, x_2 + h_2) - D_1 f(x_1 + \theta_1 h_1, x_2))\end{aligned} \tag{5.54}$$

with $0 < \theta_1 < 1$. By assumption, $D_{21}f$ exists and is continuous, we may apply the mean value theorem again to transform (5.54) into

$$\Delta = h_1 h_2 D_{21} f(x_1 + \theta_1 h_1, x_2 + \theta_2 h_2) \tag{5.55}$$

with $0 < \theta_2 < 1$. By continuity of $D_{21}f$, it follows from this that

$$D_{21} f(x_1, x_2) = \lim_{h_1, h_2 \to 0} D_{21} f(x_1 + \theta_1 h_1, x_2 + \theta_2 h_2) = \lim_{h_1, h_2 \to 0} \frac{\Delta}{h_1 h_2}. \tag{5.56}$$

Now we change the role of the first and the second variables. With (5.53) replaced by

$$g_{h_1}(u) = f(x_1 + h_1, u) - f(x_1, u), \tag{5.57}$$

(5.52) reads

$$\Delta = g_{h_1}(x_2 + h_2) - g_{h_1}(x_2). \tag{5.58}$$

Hence (5.54) reads

$$\Delta = h_2 g'_{h_1}(x_2 + \theta_4 h_2) = h_2(D_2 f(x_1 + h_1, x_2 + \theta_4 h_2) - D_2 f(x_1, x_2 + \theta_4 h_2)) \tag{5.59}$$

with $0 < \theta_4 < 1$. If $D_{12}f$ exists, then (5.59) becomes the counterpart of (5.55):

$$\Delta = h_1 h_2 D_{12} f(x_1 + \theta_3 h_1, x_2 + \theta_4 h_2) \tag{5.60}$$

with $0 < \theta_3 < 1$. If $f \in C^2$, then the limit $\lim_{h_1,h_2 \to 0} \frac{\Delta}{h_1 h_2}$ in (5.56) exists and is equal to $D_{12}f(x_1, x_2)$, whence Corollary 5.2 follows.

For the proof of theorem, we need to show the existence (and coincidence with D_{21}) of $D_{12}f(x_1, x_2)$. For $h_1 \to 0$, let

$$h_1 F(h_1) := D_2 f(x_1 + h_1, x_2) - D_2 f(x_1, x_2). \tag{5.61}$$

By definition, as $h_2 \to 0$

$$h_1 h_2 F(h_1) = \Delta(h_1.h_2) + o(h_2),$$

so that invoking (5.55), we have

$$h_1 h_2 F(h_1) = h_1 h_2 D_{21} f(x_1 + \theta_1 h_1, x_2 + \theta_2 h_2) + o(h_2). \tag{5.62}$$

Hence, with fixed $h_1 \to 0$,

$$F(h_1) = \lim_{h_2 \to 0} D_{21} f(x_1 + \theta_1 h_1, x_2 + \theta_2 h_2). \tag{5.63}$$

Now the situation is similar to (5.45) with D_k replaced by D_{21}: Given $\varepsilon > 0$, there exists a $\delta > 0$ such that

$$|D_{21} f(x_1 + \theta_1 h_1, x_2 + \theta_2 h_2) - D_{21} f(x_1, x_2)| < \frac{\varepsilon}{2} \tag{5.64}$$

as long as $(x_1 + \theta_1 h_1, x_2 + \theta_2 h_2) \in V_\delta(x_1.x_2)$, which is the case if $0 < |h_1|, |h_2| < \frac{\delta}{2}$. Hence for $0 < |h_1| < \frac{\delta}{2}$, let $h_2 \to 0$ in (5.64), then by (5.63),

$$|F(h_1) - D_{21} f(x_1, x_2)| \le \frac{\varepsilon}{2}. \tag{5.65}$$

Recalling (5.61), (5.65) shows that $D_{12}f(x_1, x_2)$ exists and is equal to $D_{21}f(x_1, x_2)$. □

Fig. 5.2: Borel

Chapter 6

Properties of differentiable functions

In this section we shall state important properties of differentiable functions.

Theorem 6.1. (Chain rule) *Suppose $D \subset \mathbb{R}^n$ and $D_1 \subset \mathbb{R}^m$ be open sets. Suppose that*

$$\boldsymbol{g} : D_1 \to \mathbb{R}^l; \quad \boldsymbol{f} : D \to \mathbb{R}^m; \quad f(D) \subset D_1, \tag{6.1}$$

that $\boldsymbol{f}$ is totally differentiable at $\boldsymbol{x}_0 \in D$ and that $\boldsymbol{g}$ is totally differentiable at $\boldsymbol{f}(\boldsymbol{x}_0) \in D'$. Then $\boldsymbol{g} \circ \boldsymbol{f} : D \to \mathbb{R}^l$ is totally differentiable at $\boldsymbol{x}_0$ and the **chain rule** *holds true:*

$$\nabla(\boldsymbol{g} \circ \boldsymbol{f})(\boldsymbol{x}_0) = \nabla \boldsymbol{g}(\boldsymbol{y}_0) \nabla \boldsymbol{f}(\boldsymbol{x}_0), \tag{6.2}$$

where we write $\boldsymbol{y}_0 = \boldsymbol{f}(\boldsymbol{x}_0)$. Writing $\boldsymbol{f} = {}^t(f_1, \cdots, f_m)$, $\boldsymbol{g} = {}^t(g_1, \cdots, g_l)$, the chain rule (6.2) *reads*

$$\frac{\partial g_i}{\partial x_k} = \sum_{j=1}^{m} \frac{\partial g_i}{\partial y_j} \frac{\partial y_j}{\partial x_k} \tag{6.3}$$

for $1 \le i \le l, 1 \le k \le n$.

Proof. We have

$$\boldsymbol{f}(\boldsymbol{x} + \boldsymbol{h}) = \boldsymbol{f}(\boldsymbol{x}) + \nabla \boldsymbol{f}(\boldsymbol{x}) \boldsymbol{h} + o(|\boldsymbol{h}|) \tag{6.4}$$

as $\boldsymbol{h} = {}^t(h_1, \cdots, h_n) \to \boldsymbol{o}$ and

$$\boldsymbol{g}(\boldsymbol{y} + \boldsymbol{k}) = \boldsymbol{g}(\boldsymbol{y}) + \nabla \boldsymbol{g}(\boldsymbol{y}) \boldsymbol{k} + o(|\boldsymbol{k}|) = \boldsymbol{g}(\boldsymbol{y}) + (\nabla \boldsymbol{g}(\boldsymbol{y}) + o(1)) \boldsymbol{k} \tag{6.5}$$

as $\boldsymbol{k} = {}^t(k_1, \cdots, k_m) \to \boldsymbol{o}$. Writing

$$\boldsymbol{k} = \boldsymbol{f}(\boldsymbol{x} + \boldsymbol{h}) - \boldsymbol{f}(\boldsymbol{x}) = \nabla \boldsymbol{f}(\boldsymbol{x}) \boldsymbol{h} + o(|\boldsymbol{h}|) \tag{6.6}$$

and substituting this in the second equality in (6.5), we deduce that

$$\begin{aligned} \boldsymbol{g}(\boldsymbol{f}(\boldsymbol{x} + \boldsymbol{h})) &= \boldsymbol{g}(\boldsymbol{f}(\boldsymbol{x}) + \boldsymbol{k}) \\ &= \boldsymbol{g}(\boldsymbol{f}(\boldsymbol{x})) + (\nabla \boldsymbol{g}(\boldsymbol{f}(\boldsymbol{x})) + o(1))(\nabla \boldsymbol{f}(\boldsymbol{x}) \boldsymbol{h} + o(|\boldsymbol{h}|)) \\ &= \boldsymbol{g}(\boldsymbol{f}(\boldsymbol{x})) + \nabla \boldsymbol{g}(\boldsymbol{f}(\boldsymbol{x})) \nabla \boldsymbol{f}(\boldsymbol{x}) \boldsymbol{h} + o(|\boldsymbol{h}|) \end{aligned} \tag{6.7}$$

where in the last step we used the estimate in Exercise 49. □

Corollary 6.1. *In the setting of Theorem 6.1 let $l = m = n$ and $\boldsymbol{h} = \boldsymbol{g} \circ \boldsymbol{f}$ and suppose that $\boldsymbol{f}$ and $\boldsymbol{g}$ are totally differentiable at $\boldsymbol{x}$ and $\boldsymbol{f}(\boldsymbol{x})$, respectively. Then the* **multiplication formula for the Jacobians** *holds true:*

$$J_{\boldsymbol{g}\circ\boldsymbol{f}}(\boldsymbol{x}) = J_{\boldsymbol{h}}(\boldsymbol{x}) = J_{\boldsymbol{g}}(\boldsymbol{f}(\boldsymbol{x}))J_{\boldsymbol{f}}(\boldsymbol{x}). \tag{6.8}$$

Example 6.1. Suppose $f = f(\boldsymbol{x})$ is differentiable at $\boldsymbol{x}_0 \in \mathbb{R}^n$. Let $\boldsymbol{o} \neq \boldsymbol{a} \in \mathbb{R}^n$. Then

$$D_{\boldsymbol{a}}f(\boldsymbol{x}_0) := \frac{\mathrm{d}}{\mathrm{d}t}f(\boldsymbol{x}_0 + t\boldsymbol{a}) \tag{6.9}$$

is called the **directional derivative** of f in the direction of $\boldsymbol{a} = {}^t(a_1, \cdots, a_n)$. Since this is a composite function $f(\boldsymbol{x}(t))$, $\boldsymbol{a}(t) = \boldsymbol{x}_0 + t\boldsymbol{a}$, Theorem 6.1 entails

$$D_{\boldsymbol{a}}f(\boldsymbol{x}_0) = (\nabla f)\boldsymbol{a} = \sum_{k=1}^{n} a_k \frac{\partial f}{\partial x_k}. \tag{6.10}$$

Or this may be thought of as a special case of Corollary 5.1 with $\boldsymbol{h} = \boldsymbol{a}$. Then this may be also written as $\mathrm{d}\,f(\boldsymbol{a}) = \mathrm{d}\,f(\boldsymbol{x}_0; \boldsymbol{a})$.

The special case $\boldsymbol{u} = \frac{1}{|\boldsymbol{a}|}\boldsymbol{a}$ suffices to show its meaning in $\mathbb{R}^3$. For $\boldsymbol{u} = {}^t(\cos\alpha_1, \cos\alpha_2, \cos\alpha_3)$, the directional cosines (cf. Exercise 146) and (6.9) means the instantaneous change of ratio in the direction of $\boldsymbol{u}$. If $(\boldsymbol{u} = (\boldsymbol{e}_k))$, it reduces to the ordinary partial differential coefficient

$$D_{\boldsymbol{e}_k}f(\boldsymbol{x}_0) = \frac{\partial f}{\partial x_k}(\boldsymbol{x}_0). \tag{6.11}$$

The one variable case of the chain rule is more well-known as the differentiation of composite functions and will appear as a formula for change of variable in Theorem 11.2.

Corollary 6.2. (Differentiation of composite functions) *Suppose $f : (a, b) \to \mathbb{R}$ is differentiable at $x_0 \in (a, b)$ and $g : (\alpha, \beta) \to \mathbb{R}$ is differentiable at $f(x_0) \in (\alpha, \beta)$, where $(\alpha, \beta) \supset f(a, b)$. Then the composite function $g \circ f : (a, b) \to \mathbb{R}$ is also differentiable at x_0 and the chain rule holds true:*

$$(g \circ f)'(x)\big|_{x=x_0} = g'(f(x_0))f'(x_0). \tag{6.12}$$

Exercise 48. Compare Example 1.8 with (6.2). Write down the cases $l = 1$, $l = n = 1$ and $l = m = 1$.

Exercise 49. Let $\boldsymbol{f} : \mathbb{R}^n \to \mathbb{R}^m$ be a linear map $\boldsymbol{y} = \boldsymbol{y}(\boldsymbol{x}) = A\boldsymbol{x}$, where $A = (a_{ij})_{1 \le i \le m, 1 \le j \le n}$:

$$y_k = \sum_{j=1}^{n} a_{kj} x_j, \qquad 1 \le k \le m. \tag{6.13}$$

$\boldsymbol{x} = {}^t(x_1, \cdots, x_n) \in X$. Prove that $\boldsymbol{y} = O(|\boldsymbol{x}|)$.

Solution.

$$|y_k|^2 = \left| \sum_{j=1}^{n} a_{kj} x_j \right|^2 \le \left(\sum_{j=1}^{n} |a_{kj}||x_j| \right)^2 \le \sum_{j=1}^{n} |a_{kj}|^2 \sum_{i=1}^{n} |x_i|^2 \tag{6.14}$$

by the Cauchy-Schwarz inequality. Adding these over $k = 1, \cdots, m$, we have

$$|\boldsymbol{y}|^2 = \sum_{k=1}^{m} |y_k|^2 \le \sum_{k=1}^{m} \sum_{j=1}^{n} |a_{kj}|^2 |\boldsymbol{x}|^2, \tag{6.15}$$

whence $|\boldsymbol{y}| \le B|\boldsymbol{x}|$ with $B = \sum_{k,j} |a_{kj}|^2 = \|A\|^2$.

Exercise 50. (i) Let

$$\boldsymbol{z} = \boldsymbol{g}(\boldsymbol{x}) = \begin{pmatrix} g_1(\boldsymbol{x}) \\ g_2(\boldsymbol{x}) \end{pmatrix}, \begin{pmatrix} x \\ y \end{pmatrix} = \boldsymbol{x} = \boldsymbol{f}(t) = \begin{pmatrix} \varphi(t) \\ \psi(t) \end{pmatrix}.$$

Find $\frac{\mathrm{d}g_j}{\mathrm{d}t}$, $j = 1, 2$.

(i′) Let

$$z = g(\boldsymbol{x}), \begin{pmatrix} x \\ y \end{pmatrix} = \boldsymbol{x} = \boldsymbol{f}(t) = \begin{pmatrix} \varphi(t) \\ \psi(t) \end{pmatrix}.$$

Find the derivative $\frac{\mathrm{d}z}{\mathrm{d}t}$. (This is the special vase of (i).)
Find the derivatives $\frac{\mathrm{d}z}{\mathrm{d}t}$ in the following special cases (save for singular points).

(1) $z = g(\boldsymbol{x}) = x^2 - 2xy + y^2, \boldsymbol{x} = \boldsymbol{f}(t) = \begin{pmatrix} (t+1)^2 \\ (t-1)^2 \end{pmatrix}$.

(2) $z = g(\boldsymbol{x}) = \arctan xy, \boldsymbol{x} = \boldsymbol{f}(t) = \begin{pmatrix} \sin t \\ \sec t \end{pmatrix}$.

(3) $z = g(\boldsymbol{x}) = x \sin y, \boldsymbol{x} = \boldsymbol{f}(t) = \begin{pmatrix} \frac{1}{t} \\ \arctan t \end{pmatrix}$.

(4) $z = g(\boldsymbol{x}) = \log \sqrt{\frac{x-y}{x+y}}, \boldsymbol{x} = \boldsymbol{f}(t) = \begin{pmatrix} \sec t \\ \tan t \end{pmatrix}$.

(ii) Let $z = g(\boldsymbol{x})$, $\begin{pmatrix} x \\ y \end{pmatrix} = \boldsymbol{x} = \boldsymbol{f}(t) = \begin{pmatrix} t \\ \varphi(t) \end{pmatrix}$. Find the derivative $\frac{\mathrm{d}z}{\mathrm{d}t}$.

Find the derivatives $\frac{\mathrm{d}z}{\mathrm{d}t}$ in the following special cases (save for singular points).

(1) $z = g(\boldsymbol{x}) = \frac{x-y}{1-y}, \boldsymbol{x} = \boldsymbol{f}(t) = \begin{pmatrix} t \\ t^{-2} \end{pmatrix}$.

(2) $z = g(\boldsymbol{x}) = \arctan \frac{x}{y}, \boldsymbol{x} = \boldsymbol{f}(t) = \begin{pmatrix} t \\ \sqrt{1+t^2} \end{pmatrix}$.

(ii′) Let

$$z = g(\boldsymbol{x}), \begin{pmatrix} x \\ y \\ z \end{pmatrix} = \boldsymbol{x} = \boldsymbol{f}(t) = \begin{pmatrix} t \\ \varphi(t) \\ \psi(t) \end{pmatrix}.$$

Find the derivative $\frac{\mathrm{d}z}{\mathrm{d}t}$.

Find the derivatives $\frac{\mathrm{d}z}{\mathrm{d}t}$ in the following special cases (save for singular points).

(3)

$$z = g(\boldsymbol{x}) = \log(x^2 + y^2 + z^2), \boldsymbol{x} = \boldsymbol{f}(t) = \begin{pmatrix} t \\ t\sin t \\ t\cos t \end{pmatrix}.$$

(4)

$$z = g(\boldsymbol{x}) = \frac{x+y}{x+z}, \boldsymbol{x} = \boldsymbol{f}(t) = \begin{pmatrix} t \\ \log t \\ \log \frac{1}{t} \end{pmatrix}.$$

(iii) Let

$$w = g(x, y, z), x = x(r, s), y = y(r, s), z = z(r, s).$$

Find $\frac{\partial w}{\partial r}$ and $\frac{\partial w}{\partial s}$.

Theorem 6.2. (Inverse function theorem for one variable) *Suppose $f : I = (a, b) \to \mathbb{R}$ is differentiable, strictly monotone and $f'(x) \neq 0$ on I. Then the* **inverse function** *$f^{-1} : f(I) \to I$ is differentiable and (on putting $y = f(x)$)*

$$\frac{\mathrm{d}}{\mathrm{d}y} f^{-1}(y) = \frac{1}{\frac{\mathrm{d}}{\mathrm{d}x} f(x)}; \quad \frac{\mathrm{d}x}{\mathrm{d}y} = \frac{1}{\frac{\mathrm{d}y}{\mathrm{d}x}}. \tag{6.16}$$

Proof. By Corollary 7.3, f is strictly monotone, so that by Theorem 4.3, the inverse function exists which is continuous and monotone. Noting that $f(I)$ is an open interval, we take any $y \in f(I)$ and choose $h \neq 0$ small enough that $y + h \in f(I)$. Then putting $k = f^{-1}(y+h) - f^{-1}(y)$ we see that $k \neq 0$ and $= o(1)$ as $h \to 0$. Since $f^{-1}(y+h) = k + f^{-1}(y)$ is the same as $y + h = f(k + f^{-1}(y)) = f(x+k)$, it follows that $h = f(x+k) - f(x) = f'(x)k(1 + o(1))$. Hence

$$f^{-1}(y+h) - f^{-1}(y) = k = \frac{1}{f'(x)} h \frac{1}{1+o(1)} = \frac{1}{f'(x)} h(1 + o(1))$$

and (6.16) follows. □

For a several variables version, cf. Theorem 8.3.

6.1 Inverse trigonometric functions

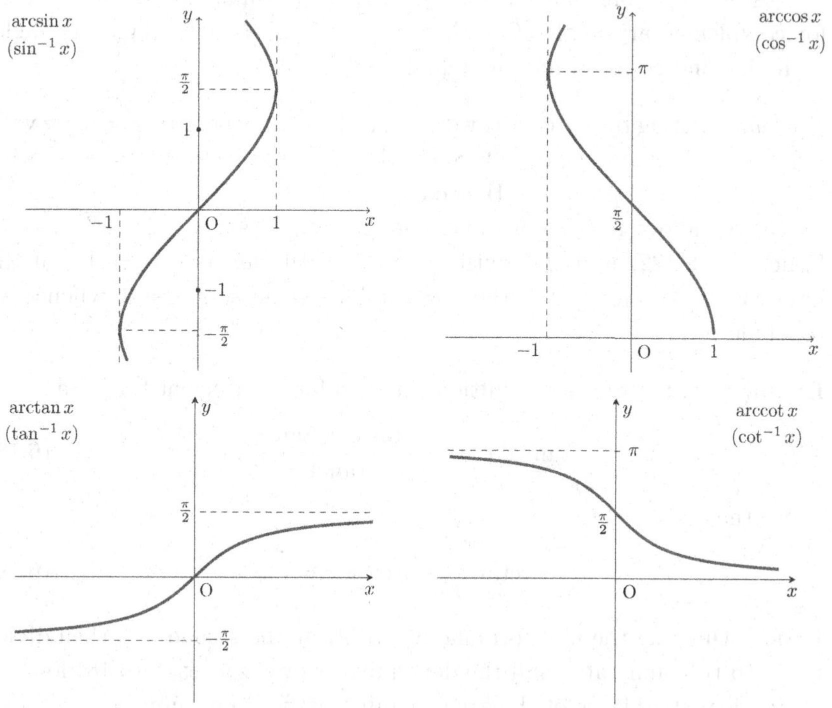

Fig. 6.1: Inverse trigonometric functions

In this subsection we collect some useful formulas among inverse trigonometric functions which exist as inverse functions of the trigonometric functions. The introduction of trigonometric functions themselves as functions of arc length is rather tedious and we understand them as defined as power series or as the inverse functions of inverse trigonometric functions. The latter will be introduced in §11.3 by Klein's method.

Exercise 51. Simplify the expression $\cos(\arcsin x)$.

Solution. Recall that $\alpha = \arcsin x$ is equivalent to$x = \sin\alpha$ with $-\frac{\pi}{2} \le \alpha \le \frac{\pi}{2}$. What we want to simplify is $\cos\alpha$ which is ≥ 0 in $-\frac{\pi}{2} \le \alpha \le \frac{\pi}{2}$. Hence $\cos\alpha = \sqrt{1-\sin^2\alpha}$, and so $\cos(\arcsin x) = \sqrt{1-x^2}$.

Exercise 52. Prove the identity

$$\arcsin x + \arccos x = \frac{\pi}{2} \tag{6.17}$$

for $x \in [-1, 1]$. Note that for $0 < \alpha, \beta < \frac{\pi}{2}$, (6.17) amounts to the fact that for complementary angles (i.e. their sum is $\frac{\pi}{2}$) in a right-angled triangle, their sine and cosine values are equal.

Solution. Setting $\alpha = \arcsin x$, we see that it is equivalent to $x = \sin\alpha$ with $-\frac{\pi}{2} \le \alpha \le \frac{\pi}{2}$, where $\cos\alpha \ge 0$. Similarly, $\beta = \arccos x \iff x = \cos\beta$ with $0 \le \beta \le \pi$, where $\sin\beta \ge 0$. Hence $\cos\alpha = \sin\beta$.

By the above remark, we may contend that $\cos\alpha = \sqrt{1-x^2} = \sin\beta$. Hence by (8.122) and the Pythagoras theorem, $\sin(\alpha+\beta) = 1$. In the interval $-\frac{\pi}{2} \le \alpha+\beta \le \frac{3}{2}\pi$, there is a unique value $\frac{\pi}{2}$ of $\alpha+\beta$, whence we conclude (6.17).

Example 6.2. Prove the addition theorem for the tangent function

$$\tan(\alpha+\beta) = \frac{\tan\alpha + \tan\beta}{1 - \tan\alpha\tan\beta}. \tag{6.18}$$

Prove that

$$\arctan\frac{1}{x} + \arctan x = \frac{\pi}{2}. \tag{6.19}$$

Proof. Dividing the first formula in (8.122) by the second and then dividing both the numerator and the denominator by $\cos\alpha\cos\beta$, (6.18) follows.

In view of oddness of the tangent function, we may suppose $x > 0$ and in view of the reciprocals, we may further suppose that $0 < x \le 1$. The case $x = 1$ being clear, we suppose $0 < x < 1$. Putting $\alpha = \arctan x \iff$

$x = \tan\alpha$, we have $0 < \alpha < \frac{\pi}{4}$ and $\beta = \arctan\frac{1}{x} \iff \frac{1}{x} = \tan\beta$, with $\frac{\pi}{4} < \beta < \frac{\pi}{2}$.

It then follows that $\tan(\alpha+\beta) = \infty$ with $\frac{\pi}{4} < \alpha+\beta < \frac{3\pi}{4}$, whence we conclude (6.19). □

Exercise 53. Prove the identity

$$\arctan\frac{1}{2} + \arctan\frac{1}{3} = \frac{\pi}{4}. \tag{6.20}$$

Solution. We set

$$\alpha = \arctan\frac{1}{2} \iff \frac{1}{2} = \tan\alpha, \quad 0 < \alpha < \frac{\pi}{4}$$
$$\beta = \arctan\frac{1}{3} \iff \frac{1}{3} = \tan\beta, \quad 0 < \beta < \frac{\pi}{4}.$$

By (6.18), we find that $\tan(\alpha+\beta) = \frac{\frac{1}{2}+\frac{1}{3}}{1-\frac{1}{2}\cdot\frac{1}{3}} = 1$.

Since

$$0 < \alpha+\beta < \frac{\pi}{2},$$

we may solve $\tan(\alpha+\beta) = 1$ to conclude (6.20).

Exercise 54. Prove that

$$\arctan x + \arctan y = \arctan\frac{x+y}{1-xy} \tag{6.21}$$

is equivalent to $xy \neq 1$ with the range of the argument to be checked. (6.19) may be thought of as the extremal case $xy = 1$. E.g. $\arctan 2 + \arctan 3 = \frac{3}{4}\pi$.

Exercise 55. (i) Prove that $\arcsin\frac{12}{13} = \arccos\frac{5}{13}$.
(ii) (Vega) Prove that $5\arctan\frac{1}{7} + 2\arctan\frac{3}{79} = \frac{\pi}{4}$.
(iii) (Machin) Prove that $4\arctan\frac{1}{5} - \arctan\frac{1}{239} = \frac{\pi}{4}$.
(iv) (Somer) Prove that $6\arctan\frac{1}{8} + 2\arctan\frac{1}{57} + \arctan\frac{1}{239} = \frac{\pi}{4}$.
(v) (Rutherford) Prove that $4\arctan\frac{1}{5} - \arctan\frac{1}{70} + \arctan\frac{1}{99} = \frac{\pi}{4}$.

Exercise 56. Prove Euler's formula $\arctan\frac{1}{p} = \arctan\frac{1}{p+q} + \arctan\frac{q}{p^2+pq+1}$ for $p, q \in \mathbf{N}$.

Remark 6.1. Since the series in (5.33), (11.56) is slowly convergent, formulas in Exercise 55 and Exercise 56 were found in order to make a more efficient computation of the value

$$\arctan 1 = \frac{\pi}{4}. \tag{6.22}$$

Fig. 6.2: Dedekind

Chapter 7

Extremal values, mean value theorems and Taylor expansions in one variable

Definition 7.1. Let $f : I = (a, b) \to \mathbb{R}$ and $x_0 \in I$. $f(x_0)$ is called a local maximum of f and x_0 a local maximum point if there exists a $\delta > 0$ such that

$$|x - x_0| < \delta, x \in I \Longrightarrow f(x) \leq f(x_0).$$

Similarly, local minimum can be defined. These are called **extremal values** and the points are called extremal points.

Lemma 7.1. *Suppose $f : I = (a, b) \to \mathbb{R}$ is differentiable at $x_0 \in I$ and takes on an extremal value. Then*

$$f'(x_0) = 0. \tag{7.1}$$

Such a point is called a **stationary point**.

Proof. Suppose it is a local maximum point and that for some $\delta > 0$ and $|h| < \delta$ we have $f(x_0 + h) \leq f(x_0)$. Then

$$\begin{aligned} 0 \geq \lim_{h \to 0+} \frac{f(x_0 + h) - f(x_0)}{h} &= f'_+(x_0) \\ = f'_-(x_0) = \lim_{h \to 0-} \frac{f(x_0 + h) - f(x_0)}{h} &\geq 0. \end{aligned} \tag{7.2}$$

□

Exercise 57. Find an example of a stationary point which is not an extremal point.

Theorem 7.1. (Rolle's theorem) *Suppose f is continuous on $[a, b]$ and differentiable on (a, b) and $f(a) = f(b)$, where at the end points continuity means one-sided continuity. Then there exists a $c \in (a.b)$ such that $f'(c) = 0$.*

Proof. Since f is continuous on the bounded closed set $[a,b]$, it has max M and min m by Weierstrass's theorem, Corollary 4.2. For all $x \in [a,b]$, $m \le f(x) \le M$. If it so happens that m, M are values of f at end point, e.g. $m = f(a), M = f(b)$. Then f is a constant and $f'(c) = 0$ for any $c \in (a,b)$. If one of them is a value of f at inner point, say $m = f(c)$, $c \in (a,b)$. Then by Lemma 7.1 $f'(c) = 0$. □

Another proof. Apply (7.2) to the extremal point.

Theorem 7.2. (Mean value theorem in Lagrange's form) *Suppose f and g are continuous on $[a,b]$ and differentiable on (a,b). Then there exists a $c \in (a.b)$ such that*

$$f'(c)(g(b) - g(a)) = g'(c)(f(b) - f(a)). \tag{7.3}$$

Proof. Putting

$$F(x) = (g(b) - g(a))f(x), \quad G(x) = (f(b) - f(a))g(x),$$

we infer that F, G are continuous on $[a,b]$ and differentiable on (a,b), whence so is $h(x) = F(x) - G(x)$. Moreover, we have $h(a) = h(b)$ and Rolle's theorem applies to give $h'(c) = 0$ for some c, or $F'(c) = G'(c)$, i.e. (7.3). □

Corollary 7.1. (L'Hospital's rule) *Suppose f, g are differentiable on $I = (a,b)$, $g'(x) \neq 0$, $\lim_{x\to a+} f(x) = 0, \lim_{x\to a+} g(x) = 0$ and that the limit*

$$\lim_{x\to a+} \frac{f'(x)}{g'(x)} \tag{7.4}$$

exists, then the indefinite limit $\lim_{x\to a+} \frac{f(x)}{g(x)}$ also exists and we have L'Hospital's rule

$$\lim_{x\to a+} \frac{f(x)}{g(x)} = \lim_{x\to a+} \frac{f'(x)}{g'(x)}. \tag{7.5}$$

L'Hospital's rule has many variations and under similar conditions, indefinite limits may be reduced to those of the quotient of derivatives.

Proof. Define $F(a) = G(a) = 0$. Then F, G are continuous on $[a,b)$. For any $x \in I$, F, G are continuous on $[a,x]$ and differentiable on $(a.x)$. Hence by Theorem 7.2, there exists a $c \in (a,x)$ such that

$$(g(x) - g(a))f'(c) = (f(x) - f(a))g'(c),$$

which means that $g(x)f'(c) = f(x)g'(c)$. Hence for any sequence $\{x_n\} \subset I$, $\lim_{n\to\infty} x_n = a$, we have

$$\frac{f(x_n)}{g(x_n)} = \frac{f'(c_n)}{g'(c_n)}. \tag{7.6}$$

Since the limit (7.4) exists and $\lim_{n\to\infty} c_n$, $c_n > a$, Theorem 3.1 implies that the limits of both sides of (7.6) exist. Then Theorem 3.1 implies that the limit $\lim_{x\to a+} \frac{f(x)}{g(x)}$ exists and is equal to the limit of the left-hand side of (7.6), which is then equal to the right-hand side limit and finally to (7.4). Hence (7.6) follows. □

Example 7.1. (i) We have

$$\lim_{x\to 0+} x\log x = \lim_{x\to 0+} \frac{\log x}{\frac{1}{x}} = \lim_{x\to 0+} \frac{\frac{1}{x}}{-\frac{1}{x^2}} = 0. \tag{7.7}$$

Cf. (11.106) below.
(ii) We have

$$\lim_{x\to 0} \frac{\arcsin x}{x} = \lim_{x\to 0} \frac{\frac{1}{\sqrt{1-x^2}}}{1} = 1. \tag{7.8}$$

For (i), L'Hospital's rule is useful. However, in most cases, it is superseded by the Taylor expansion. E.g. once we invoke the Maclaurin expansion (11.57), (ii) is immediate.

Theorem 7.3. (Mean value theorem in Lagrange's form) *Suppose f is continuous on $[a, b]$ and differentiable on (a, b). Then there exists a $c \in (a.b)$ such that*

$$f(b) - f(a) = f'(c)(b-a). \tag{7.9}$$

These two theorems will be generalized to the Taylor expansion in Cauchy's form (Theorem 7.1) and the Taylor expansion in Lagrange's form (Theorem 7.5), respectively. The several-variable case is given in Corollary 8.1.

Exercise 58. Prove the mean value theorem by applying Rolle's theorem to the function

$$F(x) := f(b) - f(x) - \lambda(b-x). \tag{7.10}$$

Solution. $\lambda = \frac{f(b)-f(a)}{b-a}$.

Corollary 7.2. (**Newton-Leibniz principle**) *Suppose f is differentiable on (a, b) and $f'(x) = 0$ for every $x \in (a, b)$. Then $f(x) = c$ (constant).*

Proof. For any $x_1 \neq x_2 \in (a, b)$, there exists a c between x_1, x_2 such that $f(x_2) - f(x_1) = f'(c)(x_2 - x_1) = 0$ by the mean value theorem, Theorem 7.3. □

Exercise 59. Use the Newton-Leibniz principle to solve the DE (differential equation)

$$f'(x) = \lambda f(x), \tag{7.11}$$

where λ is a constant and we suppose the initial value is $f(0) = a$.

Corollary 7.3. (The first derivative test) *Suppose f is continuous on $[a,b]$ and differentiable on (a,b). Then if $f'(x) > 0$ on (a,b), it is monotone strictly increasing and if $f'(x) < 0$ on (a,b), it is monotone strictly decreasing, where strictly means there is no equality for different values of x.*

Exercise 60. After proving the inequalities, plot the graphs.
(i) Prove **Jordan's inequality** for $0 < x < \frac{\pi}{2}$

$$\sin x > \frac{2}{\pi} x. \tag{7.12}$$

(ii) Prove the inequality for $x > 0$

$$\frac{x}{x^2+1} < \arctan x < x.$$

(iii) Prove the inequality for $x > 1$

$$1 - \frac{1}{x} < \log x < x - 1.$$

Consider the case $0 < x < 1$.
(iv) Prove the inequality for $x > 1$, $\alpha > 1$

$$\alpha(x-1) < x^\alpha - 1 < \alpha x^{\alpha-1}(x-1).$$

Exercise 61. For $p > 1$, $\frac{1}{p} + \frac{1}{q} = 1$ prove that for $x \geq 0$

$$\frac{1}{p} x^p + \frac{1}{q} \geq x.$$

Then replace x by $xy^{-\frac{q}{p}}$ to deduce

$$\frac{1}{p} x^p + \frac{1}{q} y^q \geq xy \tag{7.13}$$

and plot the graphs of members of the inequalities.

Exercise 62. For n-dimensional (real) vectors $\boldsymbol{a} = \begin{pmatrix} a_1 \\ \vdots \\ a_n \end{pmatrix}$, $\boldsymbol{b} = \begin{pmatrix} b_1 \\ \vdots \\ b_n \end{pmatrix}$ define their p-norm and q-norm respectively by $X = |\boldsymbol{a}|_p = \sum_{k=1}^n |a_k|^p$, $Y = |\boldsymbol{b}|_q = \sum_{k=1}^n |b_k|^q$. Using (7.13) in Exercise 61, prove

$$\sum_{k=1}^{n} |a_k b_k| \leq \frac{1}{p}|\boldsymbol{a}|_p + \frac{1}{q}|\boldsymbol{b}|_q. \tag{7.14}$$

Prove that (7.14) may be rewritten as

$$\sum_{k=1}^{n} |a_k b_k| \leq \frac{1}{p}X^p\lambda^p + \frac{1}{q}Y^q\lambda^{-q}$$

for $\lambda > 0$. Prove that the function on the left $f(\lambda) := \frac{1}{p}X^p\lambda^p + \frac{1}{q}Y^q\lambda^{-q}$ has its minimum XY at $\lambda = \frac{Y^{\frac{1}{p}}}{X^{\frac{1}{q}}}$, thereby establishing **Hölder's inequality**

$$|\boldsymbol{a} \cdot \boldsymbol{b}| \leq \sum_{k=1}^{n} |a_k b_k| \leq |\boldsymbol{a}|_p |\boldsymbol{b}|_q. \tag{7.15}$$

Note that Hölder's inequality (7.15) reduces to the Cauchy-Schwarz inequality (1.11) for $p = q = 2$.

Example 7.2. Using elementary means only, we find the extremal values of the function

$$f(x, y) = \left(x^2 + y^2\right) e^{x^2 - y^2}$$

by an elementary method of using the Taylor expansion of the exponential function only. We let $h, k \to 0$ independently. Then consider $f(x+h, y+k)$. Noting that

$$f(x+h, y+k) = \left(x^2 + y^2 + 2xh + 2yk + h^2 + k^2\right) e^{x^2-y^2} e^{2xh-2yk+h^2-k^2},$$

we substitute the Taylor expansion of the exponential function, omitting those higher order infinitesimal terms,

$$\begin{aligned}
f(x+h, y+k) &= e^{x^2-y^2}\left(x^2+y^2+2xh+2yk+h^2+k^2\right)\\
&\cdot(1+2xh-2yk+h^2-k^2+2x^2h^2-4xyhk+2y^2k^2+\cdots)\\
&= e^{x^2-y^2}\left(x^2+y^2+2xh+2yk+h^2+k^2\right)\\
&\cdot(1+2xh-2yk+(2x^2+1)h^2-4xyhk+(2y^2-1)k^2+\cdots)\\
&= e^{x^2-y^2}\Bigg(x^2+y^2+2x(x^2+y^2+1)h-2y(x^2+y^2-1)k\\
&+(4x^2+1+(x^2+y^2)(2x^2+1))h^2-4(x^2+y^2)xyhk\\
&+(-4y^2+1+(x^2+y^2)(2y^2-1))k^2+\cdots\Bigg)\\
&= f(\boldsymbol{x})+(\nabla f)\boldsymbol{h}+\frac{1}{2!}A[\boldsymbol{h}]+\cdots,
\end{aligned}$$

say. Hence $\nabla f = (2x(x^2+y^2+1)e^{x^2-y^2}, -2y(x^2+y^2-1)e^{x^2-y^2})$ and the stationary points are $\begin{pmatrix}0\\0\end{pmatrix}$, $\begin{pmatrix}0\\\pm1\end{pmatrix}$. At $\begin{pmatrix}0\\\pm1\end{pmatrix}$, we have $\frac{1}{2!}A[\boldsymbol{h}] = 2(h^2-k^2)e^{-1}$, so that they are saddle points. At $\begin{pmatrix}0\\0\end{pmatrix}$, we have $\frac{1}{2!}A[\boldsymbol{h}] = h^2+k^2 > 0$. Hence $\begin{pmatrix}0\\0\end{pmatrix}$ is the minimal point.

Proposition 7.1. *If f is partially differentiable at an extremal point $\boldsymbol{x}_0$, it is a stationary point, i.e.*

$$\nabla f(\boldsymbol{x}_0) = \boldsymbol{o}. \tag{7.16}$$

Exercise 63. Generalize Definition 7.1 to $\mathbb{R}^n$ and prove the proposition by Lemma 7.1.

7.1 Taylor expansions

Definition 7.2. A function is said to be **of class** C^r in a neighborhood of a point x_0 or r **times continuously differentiable** if $f^{(k)}$, $0 \le k \le r$ exist and are continuous. Similarly for partial derivatives. If all orders of (partial derivatives) of f exist and are continuous, then it is called an **infinitely many times differentiable** function and denoted $f \in C^\infty$.

Theorem 7.4. (nth Taylor expansion in Cauchy's form) *Suppose* $f, g \in C^{n-1}(V(x_0))$. *Then for any* $x \in V'(x_0)$ *there exists an* $x_1 \in V'(x_0)$ *such that*

$$\left(f(x) - \sum_{k=0}^{n-1} \frac{f^{(k)}(x_0)}{k!}(x - x_0)^k\right) g^{(n)}(x_1) \tag{7.17}$$

$$= \left(g(x) - \sum_{k=0}^{n-1} \frac{g^{(k)}(x_0)}{k!}(x - x_0)^k\right) f^{(n)}(x_1).$$

Proof. For simplicity, we suppose $x_0 < x$. We put for $t \in [x_0, x]$

$$F(t) = f(t) + \sum_{k=1}^{n-1} \frac{f^{(k)}(t)}{k!}(x - t)^k \tag{7.18}$$

$$G(t) = g(t) + \sum_{k=1}^{n-1} \frac{g^{(k)}(t)}{k!}(x - t)^k.$$

By Theorem 7.2 there exists an $x_1 \in (x_0, x)$ such that

$$(F(x) - F(x_0))G'(x_1) = (G(x) - G(x_0))F'(x_1). \tag{7.19}$$

By the telescoping series technique (2.8),

$$F'(t) = f'(t) + \sum_{k=1}^{n-1} \left(\frac{f^{(k+1)}(t)}{k!}(x - t)^k - \frac{f^{(k)}(t)}{(k-1)!}(x - t)^{k-1}\right)$$

$$= \frac{f^{(n)}(t)}{(n-1)!}(x - t)^{n-1} \tag{7.20}$$

and the same for $G(t)$. □

Choosing $g(x) = x^n$, we deduce the special case of the above theorem, which is a generalization of Theorem 7.3. For the most general Roche-Schlömilch remainder, cf. Exercise 64.

Theorem 7.5. (nth Taylor expansion in Lagrange's form) *Suppose* $f \in C^{n-1}(V(x_0))$ *and* $f^{(n)}$ *exists. Then for any* $x \in V'(x_0)$ *there exists an* $x_1 \in V'(x_0)$ *such that*

$$f(x) = \sum_{k=0}^{n-1} \frac{f^{(k)}(x_0)}{k!}(x - x_0)^k + R_n(x), \tag{7.21}$$

where the remainder term $R_n = R_n(x)$ *is given by*

$$R_n = R_n(x) = \frac{f^{(n)}(x_1)}{n!}(x - x_0)^n \tag{7.22}$$

called the **Lagrange remainder**.

Corollary 7.4. *If $f(x) \in C^\infty$ and*

$$\exists M > 0 \text{ s.t. } f^{(n)}(x) = O(M^n)$$

holds in an interval I, then $f(x)$ is **real analytic** *in that interval (can be expanded into a power series):*

$$f(x) = \sum_{n=0}^{\infty} \frac{f^{(n)}(x_0)}{n!}(x - x_0)^n,$$

which is called the **Taylor expansion** *around x_0. In particular, the Taylor expansion around $x_0 = 0$ is called the* **Maclaurin expansion**.

Proof. For any two points x_0, x on I, the remainder term is

$$R_n = O\left(\frac{M^n}{n!}|x - x_0|^n\right). \tag{7.23}$$

Hence, noting Exercise 65 and letting $n \to \infty$, we conclude the assertion. □

Exercise 64. Prove Theorem 7.5 by applying Rolle's theorem to the function

$$F(x) = f(b) - f(x) - \sum_{k=1}^{n-1} \frac{f^{(k)}(x)}{k!}(b - x)^k - \lambda(b - x)^p, \tag{7.24}$$

where $0 < p \leq n$ with the Roche-Schlömilch remainder (7.29). This is a degree k version of Exercise 58.

Solution. For $F(a) = F(b)$ to hold we must have

$$0 = F(b) = F(a) = f(b) - f(a) - \sum_{k=1}^{n-1} \frac{f^{(k)}(a)}{k!}(b - a)^k - \lambda(b - a)^p,$$

whence we choose

$$\lambda(b - a)^p = -\sum_{k=1}^{n-1} \frac{f^{(k)}(a)}{k!}(b - a)^k. \tag{7.25}$$

Note that

$$F'(x) = -f'(x) - \sum_{k=1}^{n-1}\left(\frac{f^{(k+1)}(x)}{k!}(b - x)^k - \frac{f^{(k)}(x)}{(k-1)!}(b - x)^{k-1}\right) + \lambda p(b - x)^{p-1}, \tag{7.26}$$

which, by the telescoping series technique, amounts to

$$F'(x) = -\frac{f^{(n)}(x)}{(n-1)!}(b - x)^{n-1} + \lambda p(b - x)^{p-1}. \tag{7.27}$$

By Rolle's theorem there exists a $c \in (a, b)$ such that $F'(c) = 0$ or

$$\lambda p(b-c)^{p-1} = \frac{f^{(n)}(c)}{(n-1)!}(b-c)^{n-1}. \tag{7.28}$$

Comparing this with (7.25), we conclude the Roche-Schlömilch remainder term in (7.32):

$$R_n = \frac{1}{(n-1)!p} f^{(n)}(c)(b-a)^p (b-c)^{n-p}. \tag{7.29}$$

(7.22) is the case $p = n$ $(b = x, a = x_0)$.

Exercise 65. Prove

$$\lim_{n\to\infty} \frac{z^n}{n!} = 0$$

for $\forall z \in \mathbb{C}$.

Solution. There are five solutions. It is enough to consider the case of $r = |z| > 0$. (i) Since this is the $n+1$st term of the absolutely convergent series $\sum_{n=0}^{\infty} \frac{z^n}{n!}$, it must converge to 0. (ii) Apply the Stirling formula to find that $\frac{r^n}{n!} \sim \frac{1}{\sqrt{2\pi n}}\left(\frac{er}{n}\right)^n$. (iii) The ratio is $\frac{a_{n+1}}{a_n} = \frac{r}{n+1}$. Hence for $n > r-1$, the sequence is decreasing and bounded from below by 0 and R_3 applies. (iv) We first establish a weaker version of the Stirling formula

$$n! \geq \sqrt{n}^n. \tag{7.30}$$

For we write

$$(n!)^2 = \prod_{k=1}^{n} (n-k+1)k. \tag{7.31}$$

Each term $(n-k+1)k = (n-k)k + k \geq n - k + k = n$ and so $(n!)^2 \geq n^n$, whence (7.30). Since $\frac{r^n}{n!} \leq \left(\frac{r}{\sqrt{n}}\right)^n$, the assertion follows. (v) Choose $M \in \mathbb{N}$ such that $M > 2r$ and let $n > M$. Then

$$\frac{r^n}{n!} = \frac{r^{n-M}}{n(n-1)\cdots(n-(n-M-1))}\frac{r^M}{M!} = O(2^{M-n}) = O(2^{-n}) = o(1),$$

$n \to \infty$.

In Exercise 64 (cf. also Exercise 86, the remainder term in the nth Taylor expansion is written in the form of **Roche-Schlömilch remainder term**: For $1 \leq \forall p \leq n$ (not necessarily an integer), with x_1 lying between x_0, x

$$R_n = \frac{1}{(n-1)!p} f^{(n)}(x_1)(x-x_0)^p (x-x_1)^{n-p}. \tag{7.32}$$

Exercise 66. Write down two special cases of the Roche-Schlömilch remainder term with $p = 1$ (the **Cauchy remainder term**) and with $p = n$ (the **Lagrange remainder term**).

Exercise 67. Prove that x_1 in the Roche-Schlömilch remainder term may be written as

$$x_1 = x_0 + \theta\,(x - x_0)$$

with $0 < \exists\theta < 1$. Further, putting $h = x - x_0$, then it can be also written as

$$R_n = \frac{1}{(n-1)!p} f^{(n)}\,(x_0 + \theta h)\,(1-\theta)^{n-p}\,h^n.$$

Work out with the Cauchy and Lagrange remainder terms.

- The expansion of e^x:

$$e^x = \sum_{n=0}^{\infty} \frac{x^n}{n!}. \tag{7.33}$$

  For in view of $(e^x)^{(n)} = e^x$, it suffices to note that when x lies in any finite interval, $e^x = O\,(1)$ i.e. with $(M = 1)$.

- The expansions of $\sin x,\ \cos x$:

$$\sin x = \sum_{n=0}^{\infty} \frac{(-1)^n}{(2n+1)!} x^{2n+1}, \quad \cos x = \sum_{n=0}^{\infty} \frac{(-1)^n}{(2n)!} x^{2n}. \tag{7.34}$$

  It suffices to note that for $\forall x \in \mathbb{R}$ $\left|\begin{pmatrix}\sin\\ \cos\end{pmatrix}^{(n)}(x)\right| = \left|\begin{pmatrix}\sin\\ \cos\end{pmatrix}(x + \frac{\pi}{2}n)\right| = O(1)$.

If f is twice differentiable at x_0, we may use the second derivative test as opposed to the first derivative test in Corollary 7.3.

Example 7.3. (Second derivative test) Theorem 7.5 with $n = 2$ reads (assuming $f''(x_0) \neq 0$; if this is 0 choose the first non-zero derivative)

$$f(x) = f(x_0) + f'(x_0)(x - x_0) + \frac{1}{2} f''(x_0)(x - x_0)^2 + o(|x - x_0|^2) \tag{7.35}$$

in a neighborhood $V(x_0)$. If x_0 is an extremal point, then (7.35) reads

$$f(x) = f(x_0) + \frac{1}{2} f''(x_0)(x - x_0)^2 + o(|x - x_0|^2) \tag{7.36}$$

and if $f''(x_0) > 0$, $f(x)$ has a local minimum at $x = x_0$. If it is negative, then it has a local maximum.

Chapter 8

Extremal values, mean value theorems and Taylor expansions in several variables

Definition 8.1. Suppose $f : X \subset \mathbb{R}^n \to \mathbb{R}$ has all partial derivatives in a neighborhood of $V(\boldsymbol{x}) \subset X$ up to order $k \in \mathbb{N}$. For the variable vector $\boldsymbol{h} = {}^t(h_1, \cdots, h_n)$ and $\boldsymbol{x}$ define the operator

$$\boldsymbol{h} \cdot \nabla = \sum_{j=1}^{n} h_j \frac{\partial}{\partial x_j} \tag{8.1}$$

and the **differential of order** k by applying it k times

$$\mathrm{d}^k f(\boldsymbol{x}; \boldsymbol{h}) = (\boldsymbol{h} \cdot \nabla)^k f(\boldsymbol{x}) = \sum_{i_1, \cdots, i_k = 1}^{n} D_{i_1, \cdots, i_k} f(\boldsymbol{x}) h_{i_1} \cdots h_{i_k}, \tag{8.2}$$

where

$$D_{i_1, \cdots, i_k} f = \frac{\partial^k}{\partial x_{i_1} \cdots \partial x_{i_k}} f \tag{8.3}$$

and by convention we define

$$\mathrm{d}^0 f(\boldsymbol{x}; \boldsymbol{h}) = f(\boldsymbol{x}). \tag{8.4}$$

We denote the **line segment** joining $\boldsymbol{a}$ and $\boldsymbol{b}$, not being equal, by

$$L(\boldsymbol{a}, \boldsymbol{b}) = \{\boldsymbol{p}(t) = (1-t)\boldsymbol{a} + t\boldsymbol{b}, t \in [0, 1]\}. \tag{8.5}$$

Theorem 8.1. *Suppose $X \subset \mathbb{R}^n$ is an open set, $f \in C^{m-1}(V(X))$ and all partial derivatives of order m $\frac{\partial^m}{\partial x_{i_1} \cdots \partial x_{i_m}} f$ exist and that $\boldsymbol{a}, \boldsymbol{b}, L(\boldsymbol{a}, \boldsymbol{b}) \subset X$. Then there exists a $\boldsymbol{z} \in L(\boldsymbol{a}, \boldsymbol{b})$ such that*

$$f(\boldsymbol{b}) = \sum_{k=0}^{m-1} \frac{1}{k!} \mathrm{d}^k f(\boldsymbol{a}; \boldsymbol{b} - \boldsymbol{a}) + \frac{1}{m!} \mathrm{d}^m f(\boldsymbol{z}; \boldsymbol{b} - \boldsymbol{a}). \tag{8.6}$$

Proof. Shifting the $k = 0$ term to the left, (8.6) gives an expression for $f(\boldsymbol{b}) - f(\boldsymbol{a})$ in terms of the right-hand side sum starting from $k = 1$. Set $g(t) = f(\boldsymbol{p}(t))$. Then $g(0) = f(\boldsymbol{a})$, $g(1) = f(\boldsymbol{b})$. By Theorem 7.5 there exists a $\theta \in (0, 1)$ such that

$$f(\boldsymbol{b}) - f(\boldsymbol{a}) = g(1) - g(0) = \sum_{k=1}^{m-1} \frac{g^{(k)}(0)}{k!} 1^k + \frac{g^{(m)}(\theta)}{m!}. \tag{8.7}$$

Then we may prove that

$$g^{(k)}(t) = \mathrm{d}^k f(\boldsymbol{p}(t); \boldsymbol{b} - \boldsymbol{a}), \quad 1 \le k \le m. \tag{8.8}$$

Indeed, putting

$$\boldsymbol{p}(t) = {}^t(p_1(t), \cdots, p_n(t)),$$

we have

$$p_k(t) = (1 - t)a_k + tb_k, \quad p_k'(t) = b_k - a_k.$$

We apply the chain rule to the following composite function

$$g(t) = f(\boldsymbol{p}(t)), \quad 0 \le t \le 1.$$

Then

$$g'(t) = \sum_{j=1}^{n} \frac{\partial}{\partial x_j} f(\boldsymbol{p}(t)) p_j'(t) = \sum_{j=1}^{n} \frac{\partial}{\partial x_j} f(\boldsymbol{p}(t))(b_j - a_j) = \mathrm{d}f(\boldsymbol{p}(t); \boldsymbol{b} - \boldsymbol{a}). \tag{8.9}$$

Applying the chain rule to (8.9), we obtain

$$g''(t) = \sum_{i=1}^{n} \sum_{j=1}^{n} \frac{\partial^2}{\partial x_i \partial x_j} f(\boldsymbol{p}(t))(b_j - a_j)(b_i - a_i) = \mathrm{d}^2 f(\boldsymbol{p}(t); \boldsymbol{b} - \boldsymbol{a}). \tag{8.10}$$

Hence inductively we arrive at (8.8).

Since

$$g^{(k)}(0) = \mathrm{d}^k f(\boldsymbol{p}(0); \boldsymbol{b} - \boldsymbol{a}) = \mathrm{d}^k f(\boldsymbol{a}; \boldsymbol{a} - \boldsymbol{b}), \quad 1 \le k \le m,$$
$$g^{(m)}(\theta) = \mathrm{d}^m f(\boldsymbol{p}(\theta); \boldsymbol{b} - \boldsymbol{a}),$$

putting $\boldsymbol{z} = \boldsymbol{p}(\theta) = (1 - \theta)\boldsymbol{a} + \theta \boldsymbol{b}$, we complete the proof. □

The following special case of the above theorem corresponds to Corollary 7.3 and plays an important role as the latter does in one variable case.

Corollary 8.1. (Mean value theorem) *Suppose $X \subset \mathbb{R}^n$ is an open set, $f \in C^m(V(X))$ and that $\boldsymbol{x}, \boldsymbol{y} \in X, L(\boldsymbol{x}, \boldsymbol{y}) \subset X$. Then there exists a $\boldsymbol{z} \in L(\boldsymbol{x}, \boldsymbol{y})$ such that*

$$f(\boldsymbol{y}) = f(\boldsymbol{x}) + \nabla f(\boldsymbol{z}) \cdot (\boldsymbol{y} - \boldsymbol{z}). \tag{8.11}$$

Corollary 8.2. *Suppose $X \subset \mathbb{R}^n$ is an open set, $f \in C^m(V(X))$ and that $\{\boldsymbol{x}, \boldsymbol{x}_0\}, L(\boldsymbol{x}, \boldsymbol{x}_0) \subset X$. If $\boldsymbol{x}$ is near $\boldsymbol{x}_0$, f may be approximated by the degree m polynomial:*

$$f(\boldsymbol{x}) = \sum_{k=0}^{m} \frac{1}{k!} \mathrm{d}^k f(\boldsymbol{x}_0; \boldsymbol{x} - \boldsymbol{x}_0) + o(|\boldsymbol{x} - \boldsymbol{x}_0|^m). \tag{8.12}$$

The special case $m = 2$ reads on putting ${}^t(h_1, \cdots, h_n) = \boldsymbol{h} = \boldsymbol{x} - \boldsymbol{x}_0$

$$f(\boldsymbol{x}) = f(\boldsymbol{x}_0) + \nabla f(\boldsymbol{x}_0) \cdot \boldsymbol{h} + \frac{1}{2!} \sum_{i.j=1}^{n} \frac{\partial^2}{\partial x_i \partial x_j} f(\boldsymbol{x}_0) h_i h_j + o(|\boldsymbol{h}|^2). \tag{8.13}$$

Definition 8.2. The matrix

$$A = (a_{ij})_{1 \le i,j \le n}, \quad a_{ij} = \frac{\partial^2}{\partial x_i \partial x_j} f(\boldsymbol{x}_0) \tag{8.14}$$

is called the **Hess matrix** and its determinant is called the **Hessian**. The differential of order 2:

$$\mathrm{d}^2 f(\boldsymbol{x}_0; \boldsymbol{h}) = \sum_{i,j=1}^{n} a_{ij} h_i h_j = {}^t\boldsymbol{h} A \boldsymbol{h} = A[\boldsymbol{h}] \tag{8.15}$$

is a **quadratic form**. According to the number of variables, they are called **binary** quadratic forms, ternary quadratic forms, etc.

For applications we only need (8.13) in the form

$$f(\boldsymbol{x}) = f(\boldsymbol{x}_0) + \nabla f(\boldsymbol{x}_0) \cdot \boldsymbol{h} + \frac{1}{2!} A[\boldsymbol{h}] + o(|\boldsymbol{h}|^2),$$

which amounts to at a stationary point $\boldsymbol{x}_0$

$$f(\boldsymbol{x}) = f(\boldsymbol{x}_0) + \frac{1}{2!} A[\boldsymbol{h}] + o(|\boldsymbol{h}|^2). \tag{8.16}$$

Correspondingly to Example 7.3, we have the **second derivative test**: If $A[\boldsymbol{h}]$ is a definite quadratic form (we also say A is definite), we may find extremal values. If $A[\boldsymbol{h}]$ is an indefinite quadratic form, then the point is called a saddle point, which has a lot of applications.

Definition 8.3. Given n samples (data) $x_1, \cdots, x_n$ and $y_1, \cdots, y_n$ we write

$$\bar{x} = \frac{1}{n} \sum_{k=1}^{n} x_k, \quad s_x^2 = \frac{1}{n} \sum_{k=1}^{n} (x_k - \bar{x})^2, \quad s_{xy} = \frac{1}{n} \sum_{k=1}^{n} (x_k - \bar{x})(y_k - \bar{y}) \tag{8.17}$$

called the **sample mean**, **sample variance** and **sample covariance**, respectively.

Theorem 8.2. (Method of least squares) *The line nearest to given n distinct points* $(x_1, y_1), \cdots, (x_n, y_n)$ *in the sense of least mean square is given by the solution of*

$$\begin{vmatrix} x & y & 1 \\ n\bar{x} & n\bar{y} & n \\ \sum_{k=1}^n x_k^2 & \sum_{k=1}^n x_k y_k & n\bar{x} \end{vmatrix} = 0. \tag{8.18}$$

Proof. The aim is to find the values of $\boldsymbol{x} = {}^t(\alpha, \beta) \in \mathbb{R}^2$ for which

$$f(\boldsymbol{x}) = f(\alpha, \beta) = \sum_{k=1}^n (\alpha x_k + \beta - y_k)^2 \tag{8.19}$$

is the minimal. We find that

$$\nabla f = 2\left(\left(\sum_{k=1}^n x_k^2\right)\alpha + n\bar{x}\beta - \sum_{k=1}^n x_k y_k, n\bar{x}\alpha + n\beta - n\bar{y}\right) \tag{8.20}$$

whence the stationary point is the solution of the system of equations $B\boldsymbol{x} = \boldsymbol{b}$, where

$$B = \begin{pmatrix} \sum_{k=1}^n x_k^2 & n\bar{x} \\ n\bar{x} & n \end{pmatrix}, \boldsymbol{b} = \begin{pmatrix} \sum_{k=1}^n x_k y_k \\ n\bar{y} \end{pmatrix}, \boldsymbol{x} = \begin{pmatrix} \alpha \\ \beta \end{pmatrix}, \tag{8.21}$$

the determinant of B is

$$|B| = \left(\sum_{k=1}^n x_k^2\right)\left(\sum_{k=1}^n 1^2\right) - \left(\sum_{k=1}^n x_k\right)^2 = \frac{1}{2}\sum_{k,l=1}^n (x_k - x_l)^2 > 0 \tag{8.22}$$

by the Lagrange formula. Hence there is a unique solution $\boldsymbol{x}_0$ at which $\nabla f(\boldsymbol{x}_0) = \boldsymbol{o}$. Since $a = f_{xx}(\boldsymbol{x}_0) = 2\sum_{k=1}^n x_k^2 > 0$, $b = f_{xy}(\boldsymbol{x}_0) = 2n\bar{x}$, $c = f_{yy}(\boldsymbol{x}_0) = 2n$, the Hess matrix A is $2B$, so that $|A| > 0$ and the minimality conditions are satisfied. Hence f take on its local minimum at $\boldsymbol{x}_0$. Without resorting to any theorem we may check that the second differential, i.e. the degree two term $\frac{1}{2!}A[\boldsymbol{h}]$ in (8.13) is, writing $\boldsymbol{h} = {}^t(h, k)$ and omitting the constant,

$$ah^2 + 2bhk + ck^2 = a\left(h + \frac{b}{a}k\right)^2 - \frac{D}{a}, \tag{8.23}$$

where $D = b^2 - ac = -4|B| < 0$. Hence $\frac{1}{2!}A[\boldsymbol{h}] > 0$, so that f take on its local minimum at $\boldsymbol{x}_0$. □

Note that

$$|B| = n^2 s_x^2, \begin{vmatrix} \sum_{k=1}^n x_k y_k & n\bar{x} \\ n\bar{y} & n \end{vmatrix} = n^2 s_{xy}, \begin{vmatrix} \sum_{k=1}^n x_k^2 & n\bar{x} \\ \sum_{k=1}^n x_k y_k & n\bar{y} \end{vmatrix} = n^2(\bar{y}s_x^2 - \bar{x}s_{xy}), \tag{8.24}$$

so that

$$\boldsymbol{x}_0 = {}^t\left(\frac{s_{xy}}{s_x^2}, \frac{\bar{y}s_x^2 - \bar{x}s_{xy}}{s_x^2}\right) \tag{8.25}$$

which is a solution to the equation (8.18). We note that if $f(\boldsymbol{x})$ is a quadratic form $A[\boldsymbol{x}]$, then the Hess matrix is $2A$.

Elementary solution I. In (8.19) consider the increment $\boldsymbol{h} = {}^t(h, k) \to \boldsymbol{o}$.

$$f(\boldsymbol{x} + \boldsymbol{h}) = f(\alpha + h, \beta + k) = \sum_{i=1}^{n} (x_i\alpha + \beta - y_i + x_ih + k)^2 \tag{8.26}$$

$$= \sum_{i=1}^{n} (x_i\alpha + \beta - y_i)^2 + \sum_{i=1}^{n} (x_ih + k)^2 + 2\sum_{i=1}^{n} (x_i\alpha + \beta - y_i)(x_ih + k),$$

whence we have (8.20). Hence the stationary point is (8.25). At this point, (8.26) reads

$$f(\boldsymbol{x}_0 + \boldsymbol{h}) = f(\boldsymbol{x}_0) + \sum_{i=1}^{n} (x_ih + k)^2 \geq f(\boldsymbol{x}_0). \tag{8.27}$$

Hence $f(\boldsymbol{x}_0)$ is the local minimum.

Elementary solution II. We rewrite (8.19) as

$$\begin{aligned} f(\alpha, \beta) &= \sum_{k=1}^{n} (\alpha(x_k - \bar{x}) - (y_k - \bar{y}) + \alpha\bar{x} + \beta - \bar{y})^2 \\ &= \sum_{k=1}^{n} (\alpha(x_k - \bar{x}) - (y_k - \bar{y}))^2 + n(\alpha\bar{x} + \beta - \bar{y})^2 \\ &= ns_x^2\alpha^2 - 2ns_{xy}\alpha + ns_y^2 + n(\alpha\bar{x} + \beta - \bar{y})^2 \\ &= ns_x^2\left(\alpha - \frac{s_{xy}}{s_x}\right)^2 - n\frac{s_{xy}^2}{s_x^2} + ns_y^2 + n(\alpha\bar{x} + \beta - \bar{y})^2. \end{aligned} \tag{8.28}$$

Hence f has the local minimum at $\alpha = \frac{s_{xy}}{s_x}$ and $\beta = \bar{y} - \alpha\bar{x}$ as in (8.25).

Lemma 8.1. *Suppose $R \subset \mathbb{R}^2$ is a compact domain with boundary ∂R and $f \in C^1(\mathbb{R}^2)$. If for some $\boldsymbol{y}_0 \in R - \partial R = \operatorname{Int} R$, $f(\boldsymbol{y}_0) > f(\boldsymbol{x})$ $[f(\boldsymbol{y}_0) < f(\boldsymbol{x})]$ for all $\boldsymbol{x} \in \partial R$, then there exists a $\boldsymbol{x}_0 \in R - \partial R$ such that $f(\boldsymbol{x}_0)$ is the (global) maximum [minimum].*

Proof. Since R is compact, $f(\boldsymbol{x})$ has maximum and minimum by Weierstrass's theorem, Corollary 4.2. Consider the case of maximum $f(\boldsymbol{x}_0)$, where $\boldsymbol{x}_0 = {}^t(x_0, y_0)$. Then as $h \to 0$

$$\frac{f(x_0+h, y_0) - f(x_0, y_0)}{h} \begin{cases} \geq 0 & (h > 0) \\ \leq 0 & (h < 0) \end{cases}. \tag{8.29}$$

Hence as with (7.2)

$$\begin{aligned} 0 \leq \lim_{h \to 0+} \frac{f(x_0+h, y_0) - f(x_0, y_0)}{h} &= \frac{\partial}{\partial x}(\boldsymbol{x}_0) \\ = \lim_{h \to 0-} \frac{f(x_0+h, y_0) - f(x_0, y_0)}{h} &\leq 0, \end{aligned} \tag{8.30}$$

so that $\frac{\partial}{\partial x}(\boldsymbol{x}_0) = 0$. □

Since Lemma 8.1 applies (noting that $f(\alpha, \beta) \to \infty$ as $\alpha, \beta \to \infty$), the local minimum in Theorem 8.2 is the minimum.

Exercise 68. Prove Rolle's theorem, Theorem 7.1, and Lemma 8.1 by the method of proof of Lemma 7.1.

Example 8.1. We deduce the secular determinant for the molecular orbital Ψ consisting of n atomic orbitals:

$$\Psi = \sum_{k=1}^{n} c_k \phi_k, \tag{8.31}$$

where ϕ_k are atomic orbitals and c_k are (complex) coefficients. Let H denote the Hamiltonian of the molecule and let

$$E = \frac{\int_{\mathbb{R}^n} \Psi H \Psi \, \mathrm{d}\tau}{\int_{\mathbb{R}^n} \Psi^2 \, \mathrm{d}\tau}, \tag{8.32}$$

where in general, Ψ is to be treated as a complex vector, in which case $\Psi H \Psi$ resp. Ψ^2 are to be regarded as $\bar{\Psi} H \Psi$ resp. $|\Psi|^2$ and the integrals are over $\mathbb{C}^n$. We write

$$H_{ij} = H_{ji} = \int_{\mathbb{R}^n} \phi_i H \phi_j \, \mathrm{d}\tau, \quad S_{ij} = S_{ji} = \int_{\mathbb{R}^n} \phi_i \phi_j \, \mathrm{d}\tau. \tag{8.33}$$

Then

$$E = \frac{\sum_{i,j=1}^{n} H_{ij} c_i c_j}{\sum_{i,j=1}^{n} S_{ij} c_i c_j} = \frac{\sum_{i=1}^{n} \left\{ H_{ii} c_i^2 + 2c_i \sum_{\substack{k=1 \\ k \neq i}}^{n} H_{ki} c_k \right\}}{\sum_{i=1}^{n} \left\{ S_{ii} c_i^2 + 2c_i \sum_{\substack{k=1 \\ k \neq i}}^{n} S_{ki} c_k \right\}}. \tag{8.34}$$

Applying the differentiation rule for the quotient in the form

$$\left(\frac{f}{g}\right)' = \frac{f'}{g} - \frac{f}{g}\frac{g'}{g},$$

we deduce that

$$\frac{\partial E}{\partial c_i} = \frac{2H_{ii}c_i + 2\sum_{\substack{k=1\\k\neq i}}^{n} H_{ki}c_k}{\sum_{j=1}^{n}\left\{S_{jj}c_j^2 + 2c_j\sum_{\substack{k=1\\k\neq j}}^{n} S_{kj}c_k\right\}} - E\frac{2S_{ii}c_i + 2\sum_{\substack{k=1\\k\neq i}}^{n} S_{ki}c_k}{\sum_{j=1}^{n}\left\{S_{jj}c_j^2 + 2c_j\sum_{\substack{k=1\\k\neq j}}^{n} S_{kj}c_k\right\}}, \tag{8.35}$$

whence

$$2H_{ii}c_i + 2\sum_{k\neq i} H_{ki}c_k - 2ES_{ii}c_i - 2E\sum_{k\neq i} S_{ki}c_k = 0,$$

i.e. the system of linear equations

$$(H_{ii} - S_{ii}E)\,c_i + \sum_{k\neq i}(H_{ki} - ES_{ki})\,c_k = 0, \quad 1 \le i \le n. \tag{8.36}$$

For (8.36) to have a non-trivial solution c_i, the coefficient matrix must be singular, so that

$$\begin{vmatrix} H_{11} - S_{11}E & H_{12} - S_{12}E & \cdots & H_{1n} - S_{1n}E \\ H_{21} - S_{21}E & H_{22} - S_{22}E & \cdots & H_{2n} - S_{2n}E \\ \cdots & & & \\ H_{n1} - S_{n1}E & H_{n2} - S_{n2}E & \cdots & H_{nn} - S_{nn}E \end{vmatrix} = 0. \tag{8.37}$$

For a normalized molecule, we may suppose that $S_{ij} = \delta_{ij}$, in which case (8.37) reduces to

$$\begin{vmatrix} H_{11} - E & H_{12} & \cdots & H_{1n} \\ H_{21} & H_{22} - E & \cdots & H_{1n} \\ \cdots & & & \\ H_{n1} & H_{n2} & \cdots & H_{nn} - E \end{vmatrix} = 0, \tag{8.38}$$

which is the secular determinant for Ψ.

Exercise 69. Find the extremal value of the function

$$f(\boldsymbol{x}) = \frac{1}{2}A[\boldsymbol{x}] + \boldsymbol{b}\boldsymbol{x} = x^2 + y^2 + z^2 + xy + yz + zx + x + y + z, \tag{8.39}$$

where

$$A = \begin{pmatrix} 2 & 1 & 1 \\ 1 & 2 & 1 \\ 1 & 1 & 2 \end{pmatrix}, \quad \boldsymbol{b} = (1, 1, 1). \tag{8.40}$$

Solution. The stationary point is $\boldsymbol{x}_0 = {}^t\left(-\frac{1}{4}, -\frac{1}{4}, -\frac{1}{4}\right)$. Since the Hess matrix A is a circulant with $\gamma = (2, 1, 1)$ and the representor $p_\gamma(z) =$

$2 + z + z^2$, its eigenvalues are $4, 1, 1$. Hence the Hess matric is positive definite and $\boldsymbol{x}_0$ is a local minimal point with the function value $-\frac{3}{8}$.

Exercise 70. Find the extremal value of the function

$$f(\boldsymbol{x}) = x^4 + y^4 + z^4 - 4xyz. \tag{8.41}$$

Solution. Since

$$\nabla f = \left(4x^3 - 4yz, 4y^3 - 4xz, 4z^3 - 4xy\right), \tag{8.42}$$

the stationary point is $\boldsymbol{x}_0 = {}^t(0,0,0)$, $\boldsymbol{x}_1 = {}^t(1,1,1)$, $\boldsymbol{x}_2 = {}^t(1,-1,-1)$, $\boldsymbol{x}_3 = {}^t(-1,1,-1)$, $\boldsymbol{x}_4 = {}^t(-1,-1,1)$. The Hess matrix A is

$$A = \begin{pmatrix} 12x^2 & -4z & -4y \\ -4z & 12y^2 & -4x \\ -4y & -4x & 12z^2 \end{pmatrix}, \tag{8.43}$$

so that

$$A|_{\boldsymbol{x}_1} = 4\begin{pmatrix} 3 & -1 & -1 \\ -1 & 3 & -1 \\ -1 & -1 & 3 \end{pmatrix}, \quad A|_{\boldsymbol{x}_2} = 4\begin{pmatrix} 3 & 1 & 1 \\ 1 & 3 & -1 \\ 1 & -1 & 3 \end{pmatrix}. \tag{8.44}$$

We can check that $\boldsymbol{x}_0 = (0,0,0)$ is not an extremal point. Since $f_A(\lambda) = 4^3(\lambda-2)^2(\lambda-5)$, its eigenvalues are $2, 2, 5$. Hence this is a local minimal point with the value -1, so that it is positive definite and the function attains its local minimum at $\boldsymbol{x}_i$ $(i = 1, 2, 3, 4)$.

As one can perceive, it suffices to consider only the positive definite case, the negative definite case being the -1 times the positive definite. E.g. in the above exercises, $-A$ will give rise to the negative definite cases.

Exercise 71. Find the extremal value of the functions
(1) $f(\boldsymbol{x}) = 4xyz - x^4 - y^4 - z^4$.
(2) $f(\boldsymbol{x}) = x^2 + y^2 + z^2 - xy - x - z$.
(3) $f(\boldsymbol{x}) = 3x^2 + y^2 + z^2 - xy - xz + yz$.

8.1 Inverse functions, implicit functions

Extremal values occur at extremal points which are stationary points which in turn are singular points. We shall show some important results for differentiable functions at regular points, i.e. at those points for which the Jacobian does not vanish.

Lemma 8.2. *Let* $K = V_r(\boldsymbol{x}_0) \subset \mathbb{R}^n$ *and* $\boldsymbol{f} : \bar{K} \to \mathbb{R}^m$ *is continuous. Suppose that* $\boldsymbol{f}$ *is* $1:1$ *on* K *and at every point* $\boldsymbol{x} \in K$ *all partial derivatives exist and the Jacobian* $J_{\boldsymbol{f}}(\boldsymbol{x}) \neq 0$. *Then* $\boldsymbol{f}(K)$ *contains a neighborhood of* $\boldsymbol{f}(\boldsymbol{x}_0)$.

Lemma 8.3. *Suppose* $K \subset \mathbb{R}^n$ *is compact,* $\boldsymbol{f} : K \to \mathbb{R}^m$ *is continuous and that* $\boldsymbol{f}$ *is* $1:1$ *on* K, *so that the inverse function* $\boldsymbol{f}^{-1}$ *exists. Then* $\boldsymbol{f}^{-1}$ *is continuous on* $\boldsymbol{f}(K)$.

Proof. For an accumulation point $\boldsymbol{y} \in \boldsymbol{f}(K)$ by Proposition 3.1, we may take a sequence $\{\boldsymbol{y}_n\} \subset \boldsymbol{f}(K)$, $\boldsymbol{y}_n \neq \boldsymbol{y}$ being all distinct such that $\lim_{n\to\infty} \boldsymbol{y}_n = \boldsymbol{y}$. It suffices to show that $\lim_{n\to\infty} \boldsymbol{x}_n = \boldsymbol{f}^{-1}(\boldsymbol{y})$, where $\boldsymbol{x}_n = \boldsymbol{f}^{-1}(\boldsymbol{y}_n) \in K$. Since $\boldsymbol{x}_n$ are all distinct and K is bounded, it follows that the $\{\boldsymbol{x}_n\} \subset K$ is a bounded infinite set, so that by the Bolzano-Weierstrass axiom R_7, it has an accumulation point, say $\boldsymbol{x}$. Then similarly as above there exists a sequence $\{\boldsymbol{x}_{n_k}\} \subset \{\boldsymbol{x}_n\}$ such that $\lim_{k\to\infty} \boldsymbol{x}_{n_k} = \boldsymbol{x}$. But $\{\boldsymbol{f}(\boldsymbol{x}_{n_k})\}$ is a subsequence of $\{\boldsymbol{y}_n\}$, so that $\boldsymbol{f}(\boldsymbol{x}) = \lim_{k\to\infty} \boldsymbol{f}(\boldsymbol{x}_{n_k}) = \lim_{n\to\infty} \boldsymbol{f}(\boldsymbol{x}_n) = \boldsymbol{y}$ and so $\boldsymbol{f}(\boldsymbol{x}) = \boldsymbol{y}$. Hence the accumulation point of $\{\boldsymbol{x}_n\}$ is the only one $\boldsymbol{x} = \boldsymbol{f}^{-1}(\boldsymbol{y})$. Hence $\lim_{n\to\infty} \boldsymbol{x}_n = \boldsymbol{x} = \boldsymbol{f}^{-1}(\boldsymbol{y})$. □

The following theorem is the several variables generalization of the inverse function theorem in Theorem 6.2.

Theorem 8.3. (Inverse function theorem) *Let* $\Omega \subset \mathbb{R}^n$ *be an open set,* $\boldsymbol{f} = (f_1, \cdots, f_n) \in C^1(\Omega)$ *and that at* $\boldsymbol{x}_0$, *the Jacobian* $J_{\boldsymbol{f}}(\boldsymbol{x}_0) \neq 0$. *Then there exist an open set* $X \subset \Omega$ *and a unique inverse function* $\boldsymbol{g} : Y \to X$ *of* $\boldsymbol{f}$, *where* $Y = \boldsymbol{f}(X)$ *is an open set in* $\boldsymbol{f}(\Omega)$ *such that*
(i) $\boldsymbol{x}_0 \in X$, $\boldsymbol{f}(\boldsymbol{x}_0) \in Y$
(ii) $\boldsymbol{f}$ *is* $1:1$ *on* X
(iii) $\boldsymbol{g}(Y) = X$, $\boldsymbol{g}(\boldsymbol{f}(\boldsymbol{x})) = \boldsymbol{x}$ *for* $\forall \boldsymbol{x} \in X$
(iv) $\boldsymbol{g} \in C^1(Y)$.

Proof. 1°. First we shall show that in a neighborhood $V(\boldsymbol{x}_0)$, $\boldsymbol{f}$ is $1:1$. Let

$$\boldsymbol{z}_i = {}^t(z_i^{(1)}, \cdots, z_i^{(n)}) \in \mathbb{R}^n, \quad 1 \le i \le n$$

and

$$\boldsymbol{z} = {}^t(\boldsymbol{z}_1, \cdots, \boldsymbol{z}_n) \in \mathbb{R}^{n^2}.$$

Define the function $h : \mathbb{R}^{n^2} \to \mathbb{R}$ by

$$h(\boldsymbol{z}) = f(\boldsymbol{z}_1, \cdots, \boldsymbol{z}_n) = \det \begin{pmatrix} \nabla f_1(\boldsymbol{z}_1) \\ \vdots \\ \nabla f_n(\boldsymbol{z}_n) \end{pmatrix} = \det \left(\frac{\partial}{\partial z_i^{(j)}} f_i(\boldsymbol{z}_i) \right)_{1 \le i,j \le n}. \tag{8.45}$$

Since h is a polynomial in n^2 continuous entries, it is continuous on $\Omega^n \subset \mathbb{R}^{n^2}$. For the special argument $\boldsymbol{z}_{\boldsymbol{x}} := {}^t(\boldsymbol{x}, \cdots, \boldsymbol{x})$, h is the Jacobian $J_{\boldsymbol{f}(\boldsymbol{x})}$. Hence for $\boldsymbol{z}_0 := \boldsymbol{z}_{\boldsymbol{x}_0} = {}^t(\boldsymbol{x}_0, \cdots, \boldsymbol{x}_0)$, $h(\boldsymbol{z}_0) \neq 0$, so that by continuity in a neighborhood of $\boldsymbol{z}_0$, $h(\boldsymbol{z}) \neq 0$. Since neighborhoods may be taken in the form $V(\boldsymbol{x}_0) \times \cdots \times V(\boldsymbol{x}_0)$, where $V(\boldsymbol{x}_0) = V_r(\boldsymbol{x}_0)$, the open ball with radius $r > 0$ and center at $\boldsymbol{x}_0$, it follows that there exists a neighborhood $V(\boldsymbol{x}_0) \subset \mathbb{R}^n$ such that if $\boldsymbol{z}_i \in V(\boldsymbol{x}_0)$, $1 \le i \le n$, then $h(\boldsymbol{z}) \neq 0$. Since $V_r(\boldsymbol{x}_0)$ is convex, for any $\boldsymbol{x} \neq \boldsymbol{y} \in V_r(\boldsymbol{x}_0)$, the line segment (8.5) joining $\boldsymbol{x}, \boldsymbol{y}$ lies in it: $L(\boldsymbol{x}, \boldsymbol{y}) \subset V_r(\boldsymbol{x}_0)$. Hence by the mean value theorem, Corollary 8.1, there exist $\boldsymbol{z}_i$, $1 \le i \le n$ such that

$$f_i(\boldsymbol{y}) - f_i(\boldsymbol{x}) = \nabla f_i(\boldsymbol{z}_i)(\boldsymbol{y} - \boldsymbol{x}) \tag{8.46}$$

or

$$\begin{pmatrix} \nabla f_1(\boldsymbol{z}_1) \\ \vdots \\ \nabla f_n(\boldsymbol{z}_n) \end{pmatrix} (\boldsymbol{y} - \boldsymbol{x}) = \boldsymbol{f}(\boldsymbol{y}) - \boldsymbol{f}(\boldsymbol{x}). \tag{8.47}$$

Since the determinant of the coefficient matrix $= h(\boldsymbol{z}) \neq 0$, it follows by Cramer's formula, Theorem A.3, that $\boldsymbol{f}(\boldsymbol{y}) = \boldsymbol{f}(\boldsymbol{x})$ implies $\boldsymbol{y} - \boldsymbol{x}$, i.e. injectivity follows.

2°. On the basis of 1°, we choose a closed ball $\bar{K} \subset V(\boldsymbol{x}_0)$ on which $\boldsymbol{f}$ is $1:1$ and the open ball $K \ni \boldsymbol{x}_0$. By Lemma 8.2, there exists a neighborhood $Y \subset \boldsymbol{f}(K)$. Let $X = \boldsymbol{f}^{-1}(Y) \cap K$. Then this is an open set of Ω by Theorem 3.2, (ii). By Lemma 8.3, the inverse function exists such that $\boldsymbol{g} = \boldsymbol{f}^{-1} : \boldsymbol{f}(\bar{K}) \to \bar{K}$ and $\boldsymbol{g}(\boldsymbol{f})(\boldsymbol{x}) = \boldsymbol{x}$. Since $X \subset K$ and $Y \subset \boldsymbol{f}(K)$ and $\boldsymbol{f}$ and $\boldsymbol{g}$ are $1:1$ on $\bar{K} \supset K$, we have $\boldsymbol{g}(Y) = X$ and $\boldsymbol{f}(X) = Y$.

To prove (iv) we use the function h in (8.45) in the setting of 1°. We write the inverse function as $\boldsymbol{g} = (g_1, \cdots, g_n)$ and show that each $g_k \in C^1(\Omega)$. In (8.47) we choose $\boldsymbol{x}$ to be any $\boldsymbol{y} \in Y$ and $\boldsymbol{y}$ to be $\boldsymbol{y} + \lambda \boldsymbol{e}_l \in Y$, where λ is small enough so that $\boldsymbol{y} + \lambda \boldsymbol{e}_l \in Y$ for $1 \le l \le n$. Then $\boldsymbol{x} = \boldsymbol{g}(\boldsymbol{y})$

and $\boldsymbol{x}' = \boldsymbol{g}(\boldsymbol{y} + \lambda \boldsymbol{e}_l)$. Then in view of $\boldsymbol{f}(\boldsymbol{x}') - \boldsymbol{f}(\boldsymbol{x}) = \lambda \boldsymbol{e}_l$, (8.47) reads

$$\begin{pmatrix} \nabla f_1(\boldsymbol{z}_1) \\ \vdots \\ \nabla f_n(\boldsymbol{z}_n) \end{pmatrix} (\boldsymbol{x}' - \boldsymbol{x}) = \lambda \boldsymbol{e}_l \tag{8.48}$$

which is solvable and the solution is given (by dividing both sides by λ) by

$$\frac{x_k' - x_k}{\lambda} = \frac{1}{h(\boldsymbol{z})} \det \left(\nabla f_1(\boldsymbol{z}), \cdots, \overset{\overset{k}{\vee}}{\boldsymbol{e}_l}, \cdots, \nabla f_n(\boldsymbol{z}) \right), \tag{8.49}$$

where $\boldsymbol{e}_l$ lies in the kth column. Since $\boldsymbol{g}$ is continuous, we have $\boldsymbol{x}' = \boldsymbol{g}(\boldsymbol{y} + \lambda \boldsymbol{e}_l) \to \boldsymbol{g}(\boldsymbol{y}) = \boldsymbol{x}$ as $\lambda \to 0$, so that $L(\boldsymbol{x}', \boldsymbol{x}) \to \boldsymbol{x}$ and so do $\boldsymbol{z}_i$. Hence $\boldsymbol{z}_i \to \boldsymbol{x}$ and so $h(\boldsymbol{z}) \to J_{\boldsymbol{f}}(\boldsymbol{x}) \neq o$ as $\lambda \to 0$. This entails that the limit of the left-hand side of (8.49) exists and thence that

$$\lim_{\lambda \to 0} \frac{x_k' - x_k}{\lambda} = \frac{1}{h(\boldsymbol{x})} \det \left(\nabla f_1(\boldsymbol{x}), \cdots, \overset{\overset{k}{\vee}}{\boldsymbol{e}_l}, \cdots, \nabla f_n(\boldsymbol{x}) \right) = \frac{1}{h(\boldsymbol{x})} \tilde{a}_{lk}, \tag{8.50}$$

where $\tilde{a}_{lk} = (-1)^{l+k} D_{lk}$ is the (l,k)-cofactor of $J_{\boldsymbol{f}}(\boldsymbol{x})$ (cf. (A.10)). Since the left-hand side of (8.50) is

$$\lim_{\lambda \to 0} \frac{g_k(\boldsymbol{y} + \lambda \boldsymbol{e}_l) - g_k(\boldsymbol{y})}{\lambda} = \frac{\partial}{\partial y_l} g_k(\boldsymbol{y})$$

and the right-hand side of (8.50) is continuous, it follows that $\frac{\partial g_k}{\partial y_l}$ are continuous and $\boldsymbol{g} \in C^1(Y)$. □

Exercise 72. Assume Theorem 4.3 and prove the inverse function theorem, Theorem 6.2, for one variable by restricting the above proof to $n = 1$.

For $\boldsymbol{x} = {}^t(x_1, \cdots, x_n), \boldsymbol{t} = {}^t(t_1, \cdots, t_m)$, we view $\begin{pmatrix} \boldsymbol{x} \\ \boldsymbol{t} \end{pmatrix}$ as a point in $\mathbb{R}^{n+m}$ and write $\boldsymbol{f}\begin{pmatrix} \boldsymbol{x} \\ \boldsymbol{t} \end{pmatrix}$ for $\boldsymbol{f}\left(\begin{pmatrix} \boldsymbol{x} \\ \boldsymbol{t} \end{pmatrix}\right)$.

Theorem 8.4. (Implicit function theorem) *Let*

$$\boldsymbol{f} = {}^t(f_1, \cdots, f_n) : \Omega \subset \mathbb{R}^{n+m} \to \mathbb{R}^n$$

be $C^1(\Omega)$ such that at the point $\begin{pmatrix} \boldsymbol{x}_0 \\ \boldsymbol{t}_0 \end{pmatrix} \in \Omega$, $\boldsymbol{f}\begin{pmatrix} \boldsymbol{x}_0 \\ \boldsymbol{t}_0 \end{pmatrix} = \boldsymbol{o}$ and $J_{\boldsymbol{f}}\begin{pmatrix} \boldsymbol{x}_0 \\ \boldsymbol{t}_0 \end{pmatrix} = \det\left(\frac{\partial}{\partial x_j} f_i \begin{pmatrix} \boldsymbol{x}_0 \\ \boldsymbol{t}_0 \end{pmatrix} \right) \neq 0$. Then there exist a $T_0 = V(\boldsymbol{t}_0) \subset \mathbb{R}^m$ and a unique function $\boldsymbol{g} : T \to \mathbb{R}^n$ satisfying the conditions.

(i) $\boldsymbol{g} \in C^1(T_0)$ (ii) $\boldsymbol{g}(\boldsymbol{t}_0) = \boldsymbol{x}_0$ (iii) *for any $\boldsymbol{t} \in T_0$, $\boldsymbol{f}\begin{pmatrix} \boldsymbol{g}(\boldsymbol{t}) \\ \boldsymbol{t} \end{pmatrix} = \boldsymbol{o}$.*

Proof. We apply the inverse function theorem to the function

$$ {}^t(F_1, \cdots, F_n, F_{n+1}, \cdots, F_{n+m}) = \boldsymbol{F} = {}^t(\boldsymbol{f}; \boldsymbol{I}) = {}^t(\boldsymbol{f}^t(\boldsymbol{x}, \boldsymbol{t}); \boldsymbol{I}), \tag{8.51} $$

i.e.

$$ F_i \begin{pmatrix} \boldsymbol{x} \\ \boldsymbol{t} \end{pmatrix} = f_i \begin{pmatrix} \boldsymbol{x} \\ \boldsymbol{t} \end{pmatrix}, 1 \le i \le n, \quad F_{n+i} \begin{pmatrix} \boldsymbol{x} \\ \boldsymbol{t} \end{pmatrix} = t_i, 1 \le i \le m. \tag{8.52} $$

Hence $\frac{\partial}{\partial t_j} F_{n+i} \begin{pmatrix} \boldsymbol{x} \\ \boldsymbol{t} \end{pmatrix} = \delta_{ij}$ and $\frac{\partial}{\partial x_j} F_{n+i} \begin{pmatrix} \boldsymbol{x} \\ \boldsymbol{t} \end{pmatrix} = 0$. Hence

$$ J_{\boldsymbol{F}} \begin{pmatrix} \boldsymbol{x} \\ \boldsymbol{t} \end{pmatrix} = \det \begin{pmatrix} \frac{\partial}{\partial x_j} f_i \begin{pmatrix} \boldsymbol{x} \\ \boldsymbol{t} \end{pmatrix} & O \\ O & E \end{pmatrix} = \det \left(\frac{\partial}{\partial x_j} f_i \begin{pmatrix} \boldsymbol{x} \\ \boldsymbol{t} \end{pmatrix} \right). \tag{8.53} $$

Hence

$$ \boldsymbol{F} \begin{pmatrix} \boldsymbol{x}_0 \\ \boldsymbol{t}_0 \end{pmatrix} = {}^t \left(\boldsymbol{f} \begin{pmatrix} \boldsymbol{x}_0 \\ \boldsymbol{t}_0 \end{pmatrix}; \boldsymbol{t}_0 \right) = \begin{pmatrix} O & O \\ O & \begin{pmatrix} t_1 & & O \\ & \ddots & \\ O & & t_n \end{pmatrix} \end{pmatrix} \tag{8.54} $$

and $J_{\boldsymbol{F}} \begin{pmatrix} \boldsymbol{x}_0 \\ \boldsymbol{t}_0 \end{pmatrix} \neq 0$, so that Theorem 8.3 implies the existence of the following objects. Open sets $X \subset \Omega$ and $Y \subset \boldsymbol{F}(\Omega)$ such that $\begin{pmatrix} \boldsymbol{x}_0 \\ \boldsymbol{t}_0 \end{pmatrix} \in X$, $\begin{pmatrix} \boldsymbol{o} \\ \boldsymbol{t}_0 \end{pmatrix} \in Y$. $\boldsymbol{F}$ is $1:1$ on X, $X = \boldsymbol{F}^{-1}(Y)$. The unique inverse function $\boldsymbol{G}: Y \to X$ exists such that

$$ \boldsymbol{G} \left(\boldsymbol{F} \begin{pmatrix} \boldsymbol{x} \\ \boldsymbol{t} \end{pmatrix} \right) = \begin{pmatrix} \boldsymbol{x} \\ \boldsymbol{t} \end{pmatrix}. $$

We express $\boldsymbol{G}$ similarly to $\boldsymbol{F}$ in (8.51) as ${}^t(\boldsymbol{v}; \boldsymbol{w})$. Then

$$ \boldsymbol{v} \left(\boldsymbol{F} \begin{pmatrix} \boldsymbol{x} \\ \boldsymbol{t} \end{pmatrix} \right) = \boldsymbol{x}, \quad \boldsymbol{w} \left(\boldsymbol{F} \begin{pmatrix} \boldsymbol{x} \\ \boldsymbol{t} \end{pmatrix} \right) = \boldsymbol{t}. $$

For any $\begin{pmatrix} \boldsymbol{y} \\ \boldsymbol{t} \end{pmatrix} \in Y = \boldsymbol{F}(X)$, there exists a unique $\begin{pmatrix} \boldsymbol{x}_1 \\ \boldsymbol{t}_1 \end{pmatrix} \in X$ such that $\begin{pmatrix} \boldsymbol{y} \\ \boldsymbol{t} \end{pmatrix} = \boldsymbol{F} \begin{pmatrix} \boldsymbol{x}_1 \\ \boldsymbol{t}_1 \end{pmatrix} = \begin{pmatrix} \boldsymbol{f} \begin{pmatrix} \boldsymbol{x}_1 \\ \boldsymbol{t}_1 \end{pmatrix} \\ \boldsymbol{t}_1 \end{pmatrix}$, so that $\boldsymbol{t}_1 = \boldsymbol{t}$. Hence

$$ \boldsymbol{v} \begin{pmatrix} \boldsymbol{y} \\ \boldsymbol{t} \end{pmatrix} = \boldsymbol{v} \left(\boldsymbol{F} \begin{pmatrix} \boldsymbol{x}_1 \\ \boldsymbol{t} \end{pmatrix} \right) = \boldsymbol{x}_1, \quad \boldsymbol{w} \begin{pmatrix} \boldsymbol{y} \\ \boldsymbol{t} \end{pmatrix} = \boldsymbol{w} \left(\boldsymbol{F} \begin{pmatrix} \boldsymbol{x}_1 \\ \boldsymbol{t} \end{pmatrix} \right) = \boldsymbol{t} $$

or

$$G\begin{pmatrix} y \\ t \end{pmatrix} = \begin{pmatrix} x_1 \\ t \end{pmatrix},$$

where $x_1 \in \mathbb{R}^n$ satisfies $\begin{pmatrix} y \\ t \end{pmatrix} = F\begin{pmatrix} x_1 \\ t \end{pmatrix}$. We have therefore

$$F\begin{pmatrix} v\begin{pmatrix} y \\ t \end{pmatrix} \\ t \end{pmatrix} = F\begin{pmatrix} x_1 \\ t \end{pmatrix} = \begin{pmatrix} y \\ t \end{pmatrix}. \tag{8.55}$$

Let

$$T_0 = \left\{ t \in \mathbb{R}^m \middle| \begin{pmatrix} o \\ t \end{pmatrix} \in Y \right\}, \quad Y_0 = \left\{ \begin{pmatrix} o \\ t \end{pmatrix} \in \mathbb{R}^{m+n} \middle| \begin{pmatrix} o \\ t \end{pmatrix} \in Y \right\}, \tag{8.56}$$

which turns to be an open set. Define for $t \in T_0$

$$g(t) = v\begin{pmatrix} o \\ t \end{pmatrix}. \tag{8.57}$$

Since $Y_0 \subset Y$, $G \in C^1(Y)$ and g is formed from the components of G, it follows that $g \in C^1(T_0)$, i.e. (i).

Recalling (8.54), we have

$$g(t_0) = v\begin{pmatrix} o \\ t_0 \end{pmatrix} = v\left(F\begin{pmatrix} x_0 \\ t_0 \end{pmatrix}\right) = x_0$$

i.e. (ii). To check (iii) we note by (8.55) that for any $\begin{pmatrix} y \\ t \end{pmatrix} \in Y$

$$\begin{pmatrix} y \\ t \end{pmatrix} = F\begin{pmatrix} v\begin{pmatrix} y \\ t \end{pmatrix} \\ t \end{pmatrix} = \begin{pmatrix} f\begin{pmatrix} v\begin{pmatrix} y \\ t \end{pmatrix} \\ t \end{pmatrix} \\ t \end{pmatrix},$$

whence

$$y = f\begin{pmatrix} v\begin{pmatrix} y \\ t \end{pmatrix} \\ t \end{pmatrix}. \tag{8.58}$$

Putting $y = o$, we have $f\begin{pmatrix} v\begin{pmatrix} o \\ t \end{pmatrix} \\ t \end{pmatrix} = o$. Combining this with (8.57), we conclude that

$$f\begin{pmatrix} g(t) \\ t \end{pmatrix} = f\begin{pmatrix} v\begin{pmatrix} o \\ t \end{pmatrix} \\ t \end{pmatrix} = o. \tag{8.59}$$

Finally, uniqueness of $\boldsymbol{g}$ follows from the remark that if there is another $\boldsymbol{h}$ satisfying (iii), then we have

$$\boldsymbol{f}\begin{pmatrix}\boldsymbol{g}(\boldsymbol{t})\\ \boldsymbol{t}\end{pmatrix} = \boldsymbol{o} = \boldsymbol{f}\begin{pmatrix}\boldsymbol{h}(\boldsymbol{t})\\ \boldsymbol{t}\end{pmatrix}$$

entails, $\boldsymbol{f}$ being $1:1$, that the arguments must coincide, and *a fortiori*, $\boldsymbol{g}(\boldsymbol{t}) = \boldsymbol{h}(\boldsymbol{t})$ for all $\boldsymbol{t} \in T$. This complete the proof. □

8.2 Lagrange multiplier method

Theorem 8.5. (Lagrange (undetermined) multiplier method) *Let $X \subset \mathbb{R}^n$ be an open set, $f : X \to \mathbb{R}$, $f \in C^1(X)$ and let $\boldsymbol{g} = (g_1, \cdots, g_m) \in C^1(X)$ with $m < n$. Further let*

$$X_0 = \{\boldsymbol{x} \in X | \boldsymbol{g}(\boldsymbol{x}) = \boldsymbol{o}\},$$

and suppose that at $\boldsymbol{x}_0 \in X_0$, there exists a neighborhood $V(\boldsymbol{x}_0)$ such that $f(\boldsymbol{x}) \le f(\boldsymbol{x}_0)$ or $f(\boldsymbol{x}) \ge f(\boldsymbol{x}_0)$ for all $\boldsymbol{x} \in X_0 \cap V(\boldsymbol{x}_0)$ and

$$\det\left(\frac{\partial}{\partial x_j}\boldsymbol{g}(\boldsymbol{x}_0)\right)_{1\le i,j\le m} \ne 0. \tag{8.60}$$

Then putting

$$L = L(\boldsymbol{x}, \lambda_k) = f(\boldsymbol{x}) + \sum_{k=1}^{m} \lambda_k g_k(\boldsymbol{x}), \tag{8.61}$$

we have the multipliers $\{\lambda_k\}$ such that

$$\nabla L = \nabla f(\boldsymbol{x}_0) + \sum_{k=1}^{m} \lambda_k \nabla g_k(\boldsymbol{x}_0) = \boldsymbol{o} \tag{8.62}$$

or in component form

$$\frac{\partial}{\partial x_j} f(\boldsymbol{x}_0) + \sum_{k=1}^{m} \lambda_k \frac{\partial}{\partial x_j} g_k(\boldsymbol{x}_0) = 0 \quad (1 \le j \le n). \tag{8.63}$$

Proof. We view the first m equations in (8.63) as the system of equations in m unknowns λ_k, $1 \le k \le m$:

$$\sum_{k=1}^{m} \lambda_k \frac{\partial}{\partial x_j} g_k(\boldsymbol{x}) = -\frac{\partial}{\partial x_j} f(\boldsymbol{x}) \quad (1 \le j \le m). \tag{8.64}$$

Since the determinant of the coefficient matrix is the one (8.60), so that by Cramér's rule, (8.64) has a unique solution λ_k, $1 \le k \le m$. It suffices to prove that this satisfies the remaining $n-m$ equations in (8.63).

We apply the implicit functions theorem in the following setting.

$$\boldsymbol{x} = {}^t(x_1, \cdots, x_n) \longleftrightarrow \boldsymbol{x}' = {}^t(x_1, \cdots, x_m),$$
$$\boldsymbol{t} = {}^t(t_1, \cdots, t_m) \longleftrightarrow \boldsymbol{t} = {}^t(x_{m+1}, \cdots, x_n),$$
$$\begin{pmatrix} \boldsymbol{x} \\ \boldsymbol{t} \end{pmatrix} \longleftrightarrow \boldsymbol{x} = \begin{pmatrix} \boldsymbol{x}' \\ \boldsymbol{t} \end{pmatrix}$$
$$\boldsymbol{f} = {}^t(f_1, \cdots, f_n) : \Omega \subset \mathbb{R}^{n+m} \to \mathbb{R}^n \longleftrightarrow \boldsymbol{g} = {}^t(g_1, \cdots, g_m) : X \subset \mathbb{R}^n \to \mathbb{R}^m$$
$$\boldsymbol{f}\begin{pmatrix} \boldsymbol{x}_0 \\ \boldsymbol{t}_0 \end{pmatrix} = \boldsymbol{o} \quad \left(\begin{pmatrix} \boldsymbol{x}_0 \\ \boldsymbol{t}_0 \end{pmatrix} \in \Omega\right) \longleftrightarrow \boldsymbol{g}\begin{pmatrix} \boldsymbol{x}_0' \\ \boldsymbol{t}_0 \end{pmatrix} = \boldsymbol{g}(\boldsymbol{x}_0) = \boldsymbol{o} \quad \left(\begin{pmatrix} \boldsymbol{x}_0' \\ \boldsymbol{t}_0 \end{pmatrix} = \boldsymbol{x}_0 \in X_0\right)$$
$$\det\left(\frac{\partial}{\partial x_j} f_i \begin{pmatrix} \boldsymbol{x}_0 \\ \boldsymbol{t}_0 \end{pmatrix}\right)_{1 \le i,j \le n} \ne 0 \longleftrightarrow \det\left(\frac{\partial}{\partial x_j} g_i \begin{pmatrix} \boldsymbol{x}_0' \\ \boldsymbol{t}_0 \end{pmatrix}\right)_{1 \le i,j \le m} \ne 0.$$

Hence Theorem 8.4 implies that there exists an open set $T_0 = V(\boldsymbol{t}_0) \subset \mathbb{R}^{n-m}$ and a unique function $\boldsymbol{h} : T_0 \to \mathbb{R}^m$ such that $\boldsymbol{h} \in C^1(T_0)$, $\boldsymbol{h}(\boldsymbol{t}_0) = \boldsymbol{x}_0'$ and $\boldsymbol{g}\begin{pmatrix} \boldsymbol{h}(\boldsymbol{t}) \\ \boldsymbol{t} \end{pmatrix} = \boldsymbol{o}$ for all $\boldsymbol{t} \in T_0$.

Putting

$$\boldsymbol{H}(\boldsymbol{t}) = \begin{pmatrix} \boldsymbol{h}(\boldsymbol{t}) \\ \boldsymbol{t} \end{pmatrix} = {}^t(h_1(\boldsymbol{t}), \cdots, h_m(\boldsymbol{t}), x_{m+1}, \cdots, x_n) \tag{8.65}$$

for $\boldsymbol{t} \in T_0$, we introduce new functions

$$F(\boldsymbol{t}) = f(\boldsymbol{H}(\boldsymbol{t})), \quad \boldsymbol{G}(\boldsymbol{t}) = \boldsymbol{g}\begin{pmatrix} \boldsymbol{h}(\boldsymbol{t}) \\ \boldsymbol{t} \end{pmatrix} = \boldsymbol{g}(\boldsymbol{H}(\boldsymbol{t})).$$

As stated above $\boldsymbol{G}(\boldsymbol{t}) = \boldsymbol{o}$ for $\boldsymbol{t} \in T_0$, and so *a fortiori* $\nabla \boldsymbol{G}(\boldsymbol{t}) = O$. By the chain rule, Theorem 6.1,

$$O = \nabla \boldsymbol{G}(\boldsymbol{t}_0) = \nabla \boldsymbol{g}(\boldsymbol{H}(\boldsymbol{t}_0)) \nabla \boldsymbol{H}(\boldsymbol{t}_0) = \nabla \boldsymbol{g}(\boldsymbol{x}_0) \nabla \boldsymbol{H}(\boldsymbol{t}_0) \tag{8.66}$$

since

$$\boldsymbol{H}(\boldsymbol{t}_0) = \begin{pmatrix} \boldsymbol{h}(\boldsymbol{t}_0) \\ \boldsymbol{t}_0 \end{pmatrix} = \begin{pmatrix} \boldsymbol{x}_0' \\ \boldsymbol{t}_0 \end{pmatrix} = \boldsymbol{x}_0. \tag{8.67}$$

By (8.65), we have

$$\nabla \boldsymbol{H}(\boldsymbol{t}) = \begin{pmatrix} B \\ E_{n-m} \end{pmatrix}, \quad B = \begin{pmatrix} \frac{\partial h_1}{\partial x_{m+1}} & \cdots & \frac{\partial h_1}{\partial x_n} \\ \vdots & & \vdots \\ \frac{\partial h_m}{\partial x_{m+1}} & \cdots & \frac{\partial h_m}{\partial x_n} \end{pmatrix}. \tag{8.68}$$

Putting

$$\nabla \boldsymbol{g}(\boldsymbol{x}) = \begin{pmatrix} \nabla g_1 \\ \vdots \\ \nabla g_m \end{pmatrix} = (A_1, C_1), \tag{8.69}$$

as in Exercise 73, we deduce that

$$(\lambda_1, \cdots, \lambda_m)\nabla \boldsymbol{g}(\boldsymbol{x}_0) = \left(\sum_{k=1}^m \lambda_k \nabla g_k(\boldsymbol{x}_0)\right) = (A, C), \tag{8.70}$$

where

$$A = \left(\sum_{k=1}^m \lambda_k \frac{\partial g_k}{\partial x_1}(\boldsymbol{x}_0), \cdots, \sum_{k=1}^m \lambda_k \frac{\partial g_k}{\partial x_m}(\boldsymbol{x}_0)\right), \tag{8.71}$$
$$C = \left(\sum_{k=1}^m \lambda_k \frac{\partial g_k}{\partial x_{m+1}}(\boldsymbol{x}_0), \cdots, \sum_{k=1}^m \lambda_k \frac{\partial g_k}{\partial x_n}(\boldsymbol{x}_0)\right).$$

Hence (8.66) multiplied by $(\lambda_1, \cdots, \lambda_m)$ from the left amounts to

$$\boldsymbol{o} = (\lambda_1, \cdots, \lambda_m)\nabla \boldsymbol{g}(\boldsymbol{x}_0)\nabla \boldsymbol{H}(\boldsymbol{t}_0) = (AB,\ C). \tag{8.72}$$

On the other hand, since $\boldsymbol{h}$ is continuous on T_0 and *a fortiori* at $\boldsymbol{t}_0$ and so is $\boldsymbol{H}$. Hence by (8.67) there exists a $U(\boldsymbol{t}_0) \subset T_0$ such that $\boldsymbol{H}(\boldsymbol{t}) \in X_0$ for all $\boldsymbol{t} \in U(\boldsymbol{t}_0)$. It follows that $\boldsymbol{H}(\boldsymbol{t}) \in X_0 \cap V(\boldsymbol{x}_0)$ on which either $F(\boldsymbol{t}_0) \ge F(\boldsymbol{t})$ or $F(\boldsymbol{t}_0) \le F(\boldsymbol{t})$. Hence $F(\boldsymbol{t})$ must attain its extremum at $\boldsymbol{t}_0$ and so analogously to (8.66) we have

$$\boldsymbol{o} = \nabla F(\boldsymbol{t}_0) = \nabla f(\boldsymbol{x}_0)\nabla \boldsymbol{H}(\boldsymbol{t}_0). \tag{8.73}$$

Noting that

$$\nabla f(\boldsymbol{x}_0) = (A', C'), \quad A' = \left(\frac{\partial f}{\partial x_1}(\boldsymbol{x}_0), \cdots, \frac{\partial f}{\partial x_m}(\boldsymbol{x}_0)\right), \tag{8.74}$$
$$C' = \left(\frac{\partial f}{\partial x_{m+1}}(\boldsymbol{x}_0), \cdots, \frac{\partial f}{\partial x_n}(\boldsymbol{x}_0)\right)$$

and invoking (8.68), we conclude that

$$\boldsymbol{o} = \nabla f(\boldsymbol{x}_0)\nabla \boldsymbol{H}(\boldsymbol{t}_0) = (A'B,\ C'). \tag{8.75}$$

Now summing (8.72) and (8.75), we obtain

$$\boldsymbol{o} = ((A + A')B,\ C + C'). \tag{8.76}$$

$A + A' = \boldsymbol{o}$ by (8.64) and $C + C' = \boldsymbol{o}$ means that the remaining $(n - m)$ equations of (8.63) are satisfied. □

Exercise 73. Use (8.69) to express (8.66) explicitly.

Solution. We have evidently

$$A_1 = \begin{pmatrix} \frac{\partial g_1}{\partial x_1} & \cdots & \frac{\partial g_1}{\partial x_m} \\ \vdots & & \vdots \\ \frac{\partial g_m}{\partial x_1} & \cdots & \frac{\partial g_m}{\partial x_m} \end{pmatrix}, \quad C_1 = \begin{pmatrix} \frac{\partial g_1}{\partial x_{m+1}} & \cdots & \frac{\partial g_1}{\partial x_n} \\ \vdots & & \vdots \\ \frac{\partial g_m}{\partial x_{m+1}} & \cdots & \frac{\partial g_m}{\partial x_n} \end{pmatrix}.$$

Hence (8.66) amounts to

$$\boldsymbol{o} = \nabla \boldsymbol{g}(\boldsymbol{x}_0)\nabla \boldsymbol{H}(\boldsymbol{t}_0) = (A_1 B,\ C_1), \tag{8.77}$$

evaluated at $\boldsymbol{x}_0$.

Corollary 8.3. *Let $X \subset \mathbb{R}^n$ be an open set, $f : X \to \mathbb{R}$, $f \in C^1(X)$ and let $g \in C^1(X)$. Further let*

$$X_0 = \{\boldsymbol{x} \in X \mid g(\boldsymbol{x}) = 0\},$$

and suppose that in a neighborhood $V(\boldsymbol{x}_0)$ of $\boldsymbol{x}_0 \in X_0$, $f(\boldsymbol{x})$ takes its local extremum and $\nabla g(\boldsymbol{x}_0) \neq \boldsymbol{o}$ (i.e. $g_{x_1}^2 + \cdots + g_{x_n}^2 \neq 0$).

Then putting

$$L = L(\boldsymbol{x}, \lambda) = f(\boldsymbol{x}) + \lambda g(\boldsymbol{x}), \tag{8.78}$$

we have

$$\nabla L = \nabla f(\boldsymbol{x}) + \lambda \nabla g(\boldsymbol{x}) = \boldsymbol{o} \tag{8.79}$$

or in component form

$$\frac{\partial}{\partial x_j} f(\boldsymbol{x}) + \lambda \frac{\partial}{\partial x_j} g(\boldsymbol{x}) = \boldsymbol{o} \quad (1 \le j \le n). \tag{8.80}$$

We explain Corollary 8.3 in the case of two variables x, y. Then (8.80) reads

$$\begin{cases} \frac{\partial}{\partial x} f(x,y) + \lambda \frac{\partial}{\partial x} g(x,y) = 0 \\ \frac{\partial}{\partial y} f(x,y) + \lambda \frac{\partial}{\partial y} g(x,y) = 0. \end{cases} \tag{8.81}$$

Hence Corollary asserts that the extremal points of $f(x,y)$ on the curve $g(x,y) = 0$ are contained in (8.81) and the constraint $g(x,y) = 0$. Since $g_x^2 + g_y^2 \neq 0$, we may suppose that $g_y \neq 0$. Then by the implicit function theorem, y can be solved in x and g becomes a function $v(x) = g(x, y(x))$ in X, i.e. $g(x, y(x)) = 0$. By the chain rule

$$0 = \frac{\mathrm{d}v}{\mathrm{d}x} = \frac{\partial g}{\partial x} + \frac{\partial g}{\partial y}\frac{\mathrm{d}y}{\mathrm{d}x}, \tag{8.82}$$

whence

$$\frac{\mathrm{d}y}{\mathrm{d}x} = -\frac{\frac{\partial g}{\partial x}}{\frac{\partial g}{\partial y}}. \tag{8.83}$$

Then $f(x, y) = f(x, y(x)) = u = u(x)$, say and

$$\frac{\mathrm{d}u}{\mathrm{d}x} = f_x + f_y \frac{\mathrm{d}y}{\mathrm{d}x} = f_x - f_y \frac{g_x}{g_y} \tag{8.84}$$

by (8.83). (8.84) is equivalent to

$$\frac{\partial(f, g)}{\partial(x, y)} = f_x g_y - f_y g_x = 0, \tag{8.85}$$

where $\frac{\partial(f,g)}{\partial(x,y)} = \det \nabla \boldsymbol{f}$ is the Jacobian, and where $\boldsymbol{f} = (f, g)$. Hence the extremal points are contained in the system of equations

$$\frac{\partial(f, g)}{\partial(x, y)} = 0, \quad g(x, y) = 0. \tag{8.86}$$

In the case $g_x \neq 0$ we lead to (8.83). In the process, we need to distinguish which variable is independent, which results in cumbersomeness. The **Lagrange (undetermined) multiplier method** eliminates this complexity by introducing the multiplier λ: Consider

$$L = L(x, y) = f(x, y) + \lambda g(x, y) \tag{8.87}$$

and viewing both are independent variables, we have

$$\frac{\partial L}{\partial x} = f_x + \lambda g_x = 0, \quad \frac{\partial L}{\partial y} = f_y + \lambda g_y = 0 \tag{8.88}$$

one of which is solvable in λ. Substituting the value of λ in the other, we obtain (8.86). Hence (8.86) amounts to (8.88) or (8.81).

Example 8.2. Given a probability distribution of an information system, $\{p_1, \cdots, p_n\}$, $0 < p_k < 1$,

$$\sum_{k=1}^{n} p_k = 1, \tag{8.89}$$

we find the values of p_k for which the **information entropy**

$$S = S(p_1, \cdots, p_n) = -\sum_{k=1}^{n} p_k \log p_k \tag{8.90}$$

attains its maximum. Here we apply the **principle of the entropy increase.** We apply the Lagrange (undetermined) multiplier method. Putting

$$L = L(p_1, \cdots, p_n, \lambda) = S(p_1, \cdots, p_n) + \lambda \left(\sum_{k=1}^{n} p_k - 1 \right), \tag{8.91}$$

where λ is a parameter, we may find the extremal points from the equation $\nabla L = \boldsymbol{o}$: Since

$$\begin{aligned} \nabla L &= \left(\frac{\partial L}{\partial p_1}, \cdots, \frac{\partial L}{\partial p_n}, \frac{\partial L}{\partial \lambda} \right) \\ &= \left(-\log p_1 - 1 + \lambda, \cdots, -\log p_n - 1 + \lambda, \sum_{k=1}^{n} p_k - 1 \right), \end{aligned} \tag{8.92}$$

it follows that $-\log p_k - 1 + \lambda = 0$, i.e. $\log p_k = \lambda - 1$, $p_k = e^{\lambda-1}$. Substituting this in (8.89), we conclude that $ne^{\lambda-1} = 1$, or $e^{\lambda-1} = \frac{1}{n}$, whence that

$$p_1 = \cdots = p_n = \frac{1}{n}. \tag{8.93}$$

Equation (8.93) is in conformity with our intuition that the entropy becomes the maximum when all the variables have the same value. Consider, e.g. the casting of a dice.

Example 8.3. We find extremal values of the function

$$f_1(\boldsymbol{x}) = \prod_{k=1}^{n} x_k^2, \quad f_2(\boldsymbol{x}) = |\boldsymbol{x}|^2 \tag{8.94}$$

under the constraints

$$g(\boldsymbol{x}) = A[\boldsymbol{x}] - 1 = 0, \tag{8.95}$$

where $A = E$ for $f_2(\boldsymbol{x})$ and A is a positive definite real symmetric matrix for $f_2(\boldsymbol{x})$. We assume no entry of f_1 is 0.

Proof. The extremal points are the solutions of the vector equation

$$\nabla f_j(\boldsymbol{x}) + \lambda_j \nabla g(\boldsymbol{x}) = \boldsymbol{o}, \quad j = 1, 2, \tag{8.96}$$

where

$$\nabla f_1(\boldsymbol{x}) = 2 f_1(\boldsymbol{x}) \left(\frac{1}{x_1}, \cdots, \frac{1}{x_n} \right), \quad \nabla f_2(\boldsymbol{x}) = 2^t \boldsymbol{x} \tag{8.97}$$

and

$$\nabla g(\boldsymbol{x}) = \nabla A[\boldsymbol{x}] = 2^t \boldsymbol{x} A. \tag{8.98}$$

Hence the scalar product of (8.96) by ${}^t\boldsymbol{x}$ leads to

$$0 = a_j f_j(\boldsymbol{x}) + \lambda_j A[\boldsymbol{x}] = a_j f_j(\boldsymbol{x}) + \lambda_j \tag{8.99}$$

because of the constraint (8.95) where $a_1 = n$ and $a_2 = 1$. Hence

$$\lambda_j = -a_j f_j(\boldsymbol{x}). \tag{8.100}$$

From here we must digress. Substituting this in (8.96), we deduce that

$$f_1(\boldsymbol{x})\left(\frac{1}{x_1}, \cdots, \frac{1}{x_n}\right) - n f_1(\boldsymbol{x})^t\boldsymbol{x} = \boldsymbol{o} \tag{8.101}$$

or

$$x_k^2 = \frac{1}{n}, \quad 1 \le k \le n, \tag{8.102}$$

so that $\boldsymbol{x} = {}^t\left(\frac{1}{\sqrt{n}}, \cdots, \frac{1}{\sqrt{n}}\right)$ at which

$$f_1(\boldsymbol{x}) = \frac{1}{n^n}. \tag{8.103}$$

This must be the maximum since $f_1(\boldsymbol{x})$ can be smaller. Hence $\max f(\boldsymbol{x}) = \frac{1}{n^n}$ as long as $x_k = \frac{1}{\sqrt{n}}$, $1 \le k \le n$. Then choosing $x_k = \frac{a_k}{|\boldsymbol{a}|}$, the constraint conditions is satisfied and we have

$$\prod_{k=1}^n \frac{a_k^2}{|\boldsymbol{a}|^2} = \frac{1}{n^n},$$

or

$$\prod_{k=1}^n a_k^2 \le \frac{1}{n^n}|\boldsymbol{a}|^{2n}, \quad \left(\prod_{k=1}^n a_k^2\right)^{1/n} \le \frac{\sum_{k=1}^n a_k^2}{n} \tag{8.104}$$

i.e., the **arithmetic-geometric mean inequality**. For an elementary proof cf. Exercise 6.

With (8.100), (8.96) reads (on putting $t = |\boldsymbol{x}|^{-2}$)

$$(tE - A)\boldsymbol{x} = \boldsymbol{o}. \tag{8.105}$$

Since this has a non-trivial solution, t is a solution of the eigen equation $|tE - A| = 0$. Since A is positive definite, all the eigenvalues are positive and $\min f_2(\boldsymbol{x})$ is $\frac{1}{\alpha}$ with α the maximal root of the equation. $\sqrt{\alpha}$ gives the length of the chord from the origin to the nearest point on the centered quadratic curve. □

8.3 Power series

A **power series** is like a polynomial of infinite degree as given by Definition 1.3 and is of the form

$$\sum_{n=0}^{\infty} a_n z^n,$$

a_n being called the nth coefficient. The coefficients are uniquely determined, i.e. if $\sum_{n=0}^{\infty} a_n z^n = \sum_{n=0}^{\infty} b_n z^n$ in some domain, then $a_n = b_n$. Recall that the geometric series (1.27)

$$\sum_{n=0}^{\infty} z^n = \frac{-1}{z-1}$$

is absolutely and uniformly convergent in $|z| < 1$, divergent in $|z| > 1$ by Theorem 8.6. On the circle $|z| = 1$ it is divergent but Abel summable except for $z = 1$ while the series $\sum_{n=1}^{\infty} \frac{z^n}{n}$ is (conditionally) convergent on the unit circle except for $z = 1$, which is a singular point of the limit function. We can see this because on the unit circle, we have $z = e^{2\pi i x}$, $x \in \mathbb{R}$, which is 1 if and only if $x \in \mathbb{Z}$. It turns out that the domain of convergence of power series is always a circle (finite or infinite) and the threshold circle as above is called the **circle of convergence**. Inside the circle of convergence power series are absolutely and uniformly convergent, so that they are analytic functions there. The radius r is called the **radius of convergence** and can be given in a closed form due to Cauchy and can be most easily determined by the D'Alembert test in the following:

Theorem 8.6. *The radius of convergence is given*

$$\frac{1}{\limsup_{n\to\infty} |a_n|^{1/n}} \tag{8.106}$$

and most conveniently by

$$r = \lim_{n\to\infty} \left|\frac{a_n}{a_{n+1}}\right|, \tag{8.107}$$

provided that $a_n \neq 0$ and the limit exists.

- Within the circle of convergence, the power series behave exactly like ordinary polynomials, i.e. we may perform addition, subtraction, multiplication and division (provided that the denominator $\neq 0$) among them. They can be differentiated termwise any number of times and can be integrated termwise along any path lying in the circle of convergence.

- The multiplication is performed as with polynomials, i.e. by forming the Cauchy product (1.40): $\sum_{m+n=l} a_m b_n$ is the lth coefficient of the product $(\sum_{m=0}^{\infty} a_m z^m)(\sum_{n=0}^{\infty} b_n z^n)$, which is sometimes referred to as the **Abel convolution** as opposed to the Dirichlet convolution.
- The division of a power series by another one gives rise to a meromorphic function. The division can also be performed formally as with polynomials.

Example 8.4. The radius of convergence of the power series

$$\sum_{n=0}^{\infty} \frac{z^n}{n!} \tag{8.108}$$

is ∞, i.e. it is absolutely and uniformly convergent over the whole complex plane and defines an analytic function. This can be checked by Theorem 8.6. By the Cauchy formula coupled with (1.2) also works:

$$\left(\sqrt{2\pi n}\left(\frac{n}{e}\right)^n\right)^{\frac{1}{n}} = (2\pi n)^{\frac{1}{2n}}\frac{n}{e} \to \infty. \tag{8.109}$$

For $z = x \in \mathbb{R}$, (8.108) coincides with the Taylor expansion for the exponential function e^x, so that (8.108) is the only way of continuing e^x to an analytic function (the principle of analytic continuation). Thus this gives a good motivation to denote (8.108) by e^z (or often $\exp z$) and call it the (complex) **exponential function**:

$$e^z = \sum_{n=0}^{\infty} \frac{z^n}{n!}. \tag{8.110}$$

Similarly, we can define the complex **sine function** and the **cosine function**, respectively by

$$\sin z = \sum_{n=0}^{\infty} \frac{(-1)^n z^{2n+1}}{(2n+1)!}, \quad \cos z = \sum_{n=0}^{\infty} \frac{(-1)^n z^{2n}}{(2n)!}. \tag{8.111}$$

Remark 8.1. To determine the Taylor coefficients we may apply the **method of undetermined coefficients**. E.g. supposing the differentiation formula $(e^z)' = e^z$ and the value $e^0 = 1$, we may determine the coefficients a_n for $e^z = \sum_{n=0}^{\infty} a_n z^n$ as follows. First the y-intercept is $a_0 = e^0 = 1$. Differentiating k-times, we get $e^z = \sum_{n=k}^{\infty} n(n-1)\cdots(n-k+1)a_k z^{n-k} = \sum_{n=0}^{\infty} \frac{(n+k)!}{n!} a_{n+k} z^n$, whence $a_n = \frac{1}{n!}$. The same method applies to the sine and cosine functions, but for them, it is far simpler to appeal to Euler's identity below.

Example 8.5. Since the power series (8.110) is absolutely convergent, we may form the Cauchy product for e^{z_1} and e^{z_2} to deduce the **exponential law**:

$$e^{z_1}e^{z_2} = e^{z_1+z_2}. \tag{8.112}$$

We may say that "the binomial theorem implies the exponential law".

From (8.117) and (8.112) we may deduce a conventional definition of the complex exponential function ($z = x + iy \in \mathbb{C}$, $x, y \in \mathbb{R}$)

$$e^z = e^x e^{iy} = e^x(\cos y + i \sin y). \tag{8.113}$$

We note that adopting (8.113) as the definition of the exponential function is problematic because we are left with the same type of problem as to how we define the function e^{iy}, and we are convinced that the above is the only proper way of introducing the exponential function. We can also see that

$$e^{z+2\pi in} = e^z, \quad n \in \mathbb{Z}, \tag{8.114}$$

i.e. the exponential function is a periodic function of period 2π. This periodicity of e^z of period $2\pi i$ is reflected on the multi-valuedness of its inverse function, $\log z$.

Exercise 74. Prove (8.112) and in particular the exponential law for

$$e^{i\alpha} \cdot e^{i\beta} = e^{i(\alpha+\beta)}. \tag{8.115}$$

Solution. It suffices to prove (8.112).

$$e^{z_1}e^{z_2} = \sum_{n=0}^{\infty} \frac{1}{n!} \sum_{l+m=n} \frac{n!}{\ell!m!} {z_1}^{\ell} {z_2}^{m} = \sum_{n=0}^{\infty} \frac{1}{n!}(z_1 + z_2)^n = e^{z_1+z_2}.$$

8.4 Euler's identity

In the expansion formula (8.110) for e^z, we substitute ix and classify the terms into 4 classes according to the values of i^n to obtain

$$e^{ix} = \sum_{n=0}^{\infty} \frac{(-1)^n}{(2n)!} x^{2n} + i \sum_{n=0}^{\infty} \frac{(-1)^n}{(2n+1)!} x^{2n+1} = \cos x + i \sin x, \tag{8.116}$$

which is called **Euler's identity**. For $z \in \mathbb{C}$ we obtain the (general) **Euler's identity**:

$$e^{iz} = \cos z + i \sin z. \tag{8.117}$$

In the scope of real analysis, exponential functions and trigonometric functions have nothing to do with one another, but in complex analysis, they are almost the same. Indeed, we have $\cos(-z) = \sum_{n=0}^{\infty} \frac{(-1)^n}{(2n)!}(-z)^{2n} = \cos z$ and $\sin(-z) = -\sin z$, so that $e^{-iz} = \cos z - i \sin z$. Hence

$$\cos z = \frac{1}{2}\left(e^{iz} + e^{-iz}\right), \quad \sin z = \frac{1}{2i}\left(e^{iz} - e^{-iz}\right), \tag{8.118}$$

i.e. $\cos z$, $\sin z$ are expressed in terms of e^{iz}, $e^{-iz} = \frac{1}{e^{iz}}$.

Remark 8.2. To find power series expansions (8.111) for the sine and cosine functions, it is the easiest to use (8.118). It is quite instructive to draw the figure of (8.118) when $z = \theta$ indicates an angle. Indeed, the vector $\boldsymbol{z} = e^{i\theta}$ starts from the origin reaching to the point on the unit circle on the complex plane. The vector $e^{-i\theta}$ is the reflection of $\boldsymbol{z}$ so that their sum divided by 2 is exactly the vector on the real axis pointing to $\cos\theta$. The case of the sine function is more complicated. Express it in the form $\frac{1}{2}\left(e^{i\theta} - e^{-i\theta}\right)e^{-\frac{\pi}{2}i}$. Then we may understand the first difference is the vector on the imaginary axis pointing to $\sin\theta$. Then the factor $e^{-\frac{\pi}{2}i}$ means the clockwise rotation by $\frac{\pi}{2}$ and it amounts to viewing the vector on the imaginary axis on the real axis, thus giving $\sin\theta$.

This clockwise rotation by $\frac{\pi}{2}$ has a more far-reaching effect. Modular functions are defined for the upper half-plane $\mathcal{H} = \{\tau | \operatorname{Im}\tau > 0\}$ and rotating it by $\frac{\pi}{2}$ is given by $s = -i\tau$. Then the variable s lies in the right half-plane

$$\mathcal{RHP} = \{s | \operatorname{Re} s > 0\}. \tag{8.119}$$

We have the correspondence

$$\mathcal{RHP} \leftrightarrow \mathcal{H} \leftrightarrow 0 < |q| < 1. \tag{8.120}$$

Cf. [Chakraborty *et al.* (2016), Remark 3.2, p. 20] for more details.

In view of (8.118), we may introduce the **hyperbolic functions** by

$$\cosh x = \cos ix = \frac{e^x + e^{-x}}{2}, \quad \sinh x = -i \sin ix = \frac{e^x - e^{-x}}{2}, \tag{8.121}$$

$$\tanh x = \frac{\sinh x}{\cosh x} = \frac{e^x - e^{-x}}{e^x + e^{-x}}.$$

These functions will be used quite often without notice in what follows.

Example 8.6. (A) By (8.118) and the binomial theorem,

$$\cos^3 x = \left(\frac{e^{ix} + e^{-ix}}{2}\right)^3 = \frac{1}{4}\left(\frac{e^{3ix} + e^{-3ix}}{2} + 3\frac{e^{ix} + e^{-ix}}{2}\right),$$

whence $\cos^3 x = \frac{1}{4}(\cos 3x + 3\cos x)$.

(B) Similarly,

$$\sin^3 x = \left(\frac{e^{ix} - e^{-ix}}{2i}\right)^3 = -\frac{1}{4}\left(\frac{e^{3ix} - e^{-3ix}}{2i} - 3\frac{e^{ix} - e^{-ix}}{2i}\right),$$

whence $\sin^3 x = \frac{1}{4}(-\sin 3x + 3\sin x)$. Of course, these can be easily proved by the triplication formulas.

Exercise 75. (A) Use (8.118) to deduce the following formulas.
(i) $\cos^4 x = \frac{1}{8}(\cos 4x + 4\cos 2x + 3)$, $\sin^4 x = \frac{1}{8}(\cos 4x - 4\cos 2x + 3)$.
(ii)

$$\cos^5 x = \frac{1}{16}(\cos 5x + 5\cos 3x + 10\cos x),$$
$$\sin^5 x = \frac{1}{16}(\sin 5x - 5\sin 3x + 10\sin x).$$

(iii)

$$\cos^6 x = \frac{1}{32}(\cos 6x + 6\cos 4x + 15\cos 2x + 10),$$
$$\sin^6 x = -\frac{1}{32}(\cos 6x - 6\cos 4x + 15\cos 2x - 10).$$

(iv)

$$\cos^7 x = \frac{1}{64}(\cos 7x + 7\cos 5x + 21\cos 3x + 35\cos x),$$
$$\sin^7 x = -\frac{1}{64}(\sin 7x - 7\sin 5x + 21\sin 3x - 35\sin x).$$

(v)

$$\cos^8 x = \frac{1}{128}(\cos 8x + 8\cos 6x + 28\cos 4x + 56\cos 2x + 35),$$
$$\sin^8 x = \frac{1}{128}(\cos 8x - 8\cos 6x + 28\cos 4x - 56\cos 2x + 35).$$

(B) Evaluate the following definite integrals on the basis of (A). Also evaluate them by recurrences.
(i) $\int_0^{\frac{\pi}{6}} \sin^6 x\,dx$ (ii) $\int_0^{\frac{\pi}{6}} \cos^6 x\,dx$ (iii) $\int_0^{\frac{\pi}{4}} \sin^6 x\,dx$ (iv) $\int_0^{\frac{\pi}{3}} \sin^5 x\,dx$
(v) $\int_0^{\frac{\pi}{3}} \cos^5 x\,dx$ (vi) $\int_0^{\frac{\pi}{4}} \sin^5 x\,dx$ (vii) $\int_0^{\frac{\pi}{2}} \sin^6 x\,dx$ (viii) $\int_0^{\frac{\pi}{2}} \cos^6 x\,dx$
(ix) $\int_0^{\frac{\pi}{6}} \cos^4 x\,dx$ (x) $\int_0^{\frac{\pi}{6}} \sin^4 x\,dx$ (xi) $\int_0^{\frac{\pi}{4}} \sin^4 x\,dx$.

Exercise 76. Use (8.118) to prove the following equalities.
(i)

$$\cos^{2m} x = \frac{1}{4^m}\left(\binom{2m}{m} + 2\sum_{k=0}^{m-1}\binom{2m}{k}\cos 2(m-k)x\right)$$

(ii)

$$\sin^{2m} x = \frac{1}{4^m}\left(\binom{2m}{m} + 2\sum_{k=0}^{m-1}(-1)^{m-k}\binom{2m}{k}\cos 2\left(m-k\right)x\right)$$

(iii)

$$\cos^{2m+1} x = \frac{1}{4^m}\left(\sum_{k=0}^{m}\binom{2m+1}{k}\cos\left(2\left(m-k\right)+1\right)x\right)$$

(iv)

$$\sin^{2m+1} x = \frac{1}{4^m}\left(\sum_{k=0}^{m}\binom{2m+1}{k}(-1)^{m-k}\sin(2(m-k)+1)x\right).$$

Proposition 8.1. *The addition theorem*

$$\begin{cases}\sin(\alpha+\beta) = \sin\alpha\cos\beta + \cos\alpha\sin\beta, \\ \cos\left(\alpha+\beta\right) = \cos\alpha\cos\beta - \sin\alpha\sin\beta,\end{cases} \tag{8.122}$$

for trigonometric functions amounts to the exponential law (8.115) *for the complex exponential function (under Euler's identity)*

$$e^{i\alpha}\cdot e^{i\beta} = e^{i(\alpha+\beta)}.$$

Proof. From Euler's identity,

$$\begin{aligned} e^{i\alpha}\cdot e^{i\beta} &= \left(\cos\alpha + i\sin\alpha\right)\left(\cos\beta + i\sin\beta\right) \\ &= \cos\alpha\cos\beta - \sin\alpha\sin\beta + i\left(\sin\alpha\cos\beta + \cos\alpha\sin\beta\right). \end{aligned}$$

By (8.115) this is equal to

$$e^{i(\alpha+\beta)} = \cos\left(\alpha+\beta\right) + i\sin\left(\alpha+\beta\right)$$

by Euler's identity. Hence comparing the real and imaginary parts, we prove our proposition. □

For another proof, cf. [Chakraborty *et al.* (2016), p. 23].

The polar from (1.73) can be concisely expressed in view of Euler's identity as

$$z = re^{i\theta}, \quad r = |z|, \quad \theta = \arg z. \tag{8.123}$$

Especially, for $n \in \mathbb{Z}$, de Moivre's formula in Exercise 8 is nothing but the exponential law

$$(e^{i\theta})^n = e^{in\theta}. \tag{8.124}$$

Exercise 77. For $\mathbb{R} \ni x \neq 2n\pi$ $(n \in \mathbb{Z})$ prove the formulas
(i)

$$\sum_{k=1}^{n} e^{ikx} = e^{ix}\frac{1-e^{inx}}{1-e^{ix}} = \frac{\sin\left({}^{nx}\!/_{2}\right)}{\sin\left({}^{x}\!/_{2}\right)} e^{i(n+1)x/2}.$$

(ii)

$$\sum_{k=1}^{n} \sin kx = \sin\frac{nx}{2}\sin\left(n+1\right)\frac{x}{2}\Big/\sin\frac{x}{2},$$

$$\sum_{k=1}^{n} \cos kx = \sin\frac{nx}{2}\cos\left(n+1\right)\frac{x}{2}\Big/\sin\frac{x}{2}.$$

(iii)

$$\sum_{k=1}^{n} e^{i(2k-1)x} = e^{-ix}\sum_{k=1}^{n} e^{2ikx} = \frac{\sin nx}{\sin x}e^{inx}.$$

Solution. (i) The first equality follows from the sum formula (1.22) for a geometric sequence and the second follows by factoring out $e^{\frac{n+2}{2}xi}$ and $e^{\frac{1}{2}xi}$ from the numerator and the denominator, respectively and then applying (8.118).

Exercise 78. Deduce the quintuplicate formula

$$\sin 5\theta = 16\sin^5\theta - 20\sin^3\theta + 5\sin\theta,$$

$$\cos 5\theta = 16\cos^5\theta - 20\cos^3\theta + 5\cos\theta.$$

Using the first to find the values $\sin\frac{\pi}{5}$, $\cos\frac{\pi}{5}$.

Solution. In the same way as above, we compare the 5th powers of $e^{i\theta}$ and $\cos 5\theta + i\sin 5\theta$. To find the value of $\sin\theta$ with $\theta = \frac{\pi}{5}$, we note that $\sin 5\theta = 0$, whence we obtain the quartic equation

$$16\sin^4\theta - 20\sin^2\theta + 5 = 0$$

whose solution satisfying the restriction $0 < \sin\theta < \sin\frac{\pi}{4} = \frac{1}{\sqrt{2}}$ is $\sin\theta = \left(\frac{10-2\sqrt{5}}{16}\right)^{1/2}$. More interesting is the value of the cosine:

$$2\cos\frac{\pi}{5} = \tau = \frac{1+\sqrt{5}}{2} = 1.618\cdots \tag{8.125}$$

known as the **golden ratio**, cf. e.g. [Chakraborty *et al.* (2009)].

Fig. 8.1: Dirichlet

Chapter 9

Algorithms

In this section we present some well-known algorithms. The first is the *g*-**adic expansion algorithm**. In the situation of Definition 4.3, we assume for simplicity $0 < x \in \mathbb{R}$. Recall the integral part $[x]$ of x from Corollary 4.3: the fractional part (which is sometimes denoted $\{x\}$)

$$x_1 = x - [x] \tag{9.1}$$

satisfies the inequality

$$0 \le x_1 < 1, \tag{9.2}$$

whence

$$x_2 = gx_1 - [gx_1] = gx_1 - c_1 \tag{9.3}$$

lies in $[0, 1)$. Continuing this process, we define

$$x_{n+1} = gx_n - c_n, \quad c_n = [gx_n] \tag{9.4}$$

to obtain a sequence $\{x_n\} \subset [0, 1)$. It follows that

$$0 \le c_n < g. \tag{9.5}$$

From the definition we have successively

$$x = [x] + x_1, x_1 = \frac{c_1 + x_2}{g}, \cdots, x_n = \frac{c_n + x_{n+1}}{g}, \tag{9.6}$$

whence

$$x = [x] + \frac{c_1}{g} + \cdots + \frac{c_n}{g^n} + \frac{x_{n+1}}{g^n} := a_n + \frac{x_{n+1}}{g^n}, \tag{9.7}$$

say. Since

$$\lim_{n\to\infty} \frac{x_{n+1}}{g^n} = 0, \tag{9.8}$$

it follows that $\lim_{n\to\infty} a_n = x$. This limit is called the g-adic expansion of x and denoted

$$x = [x] + \frac{c_1}{g} + \cdots + \frac{c_n}{g^n} + \cdots = [x].c_1c_2\cdots \tag{9.9}$$

and c_n is called the n-digit number. We note from (9.7) that

$$\begin{aligned} 0 \le x - a_n &= \frac{c_{n+1}}{g^{n+1}} + \frac{c_{n+2}}{g^{n+2}} + \cdots \\ &\le \sum_{k=n+1}^{\infty} \frac{g-1}{g^k} = \frac{1}{g^n} \end{aligned} \tag{9.10}$$

by the geometric series (Exercise 11). Hence it follows that $x = a_n + \frac{1}{g^n}$ is equivalent to $c_{n+1} = c_{n+2} = \cdots = g-1$. Hence

$$x = [x] + \frac{c_1}{g} + \cdots + \frac{c_n}{g^n} + \frac{1}{g^n}, \tag{9.11}$$

which is of the form $\frac{x'}{g^n}$ with $x' \in \mathbb{Z}$.

Put $m = \min\{m \in \mathbb{N} \mid n \ge m \Longrightarrow c_n = g-1\}$. Then m cannot be 1. For if so, then $x = [x] + \frac{g-1+1}{g} = [x]+1$, a contradiction. Hence $c_n - 1 = g - 1$ $(n \ge m)$ and so $c_{m-1} \le g-2$, whence $c'_{m-1} := c_{m-1}+1$ is an $(m-1)$-digit number. Hence (9.11) with $n = m$ reads

$$x = [x] + \frac{c_1}{g} + \cdots + \frac{c'_{m-1}}{g^{m-1}} + \frac{0}{g^m} + \cdots, \tag{9.12}$$

which is another expression for $x = \frac{x'}{g^n}$ different from (9.9).

Conversely, if $x = \frac{x'}{g^n}$, then we may expand $x' \in \mathbb{Z}$ into the g-adic expansion $c_\ell g^\ell + \cdots + c_0$ with digits $0 \le c_k \le g-1$. For let $c_\ell = \left[\frac{x'}{g^\ell}\right]$. Then $0 \le r_1 = x' - c_\ell g^\ell < g^\ell$. Then put $c_{\ell-1} = \left[\frac{r_1}{g^{\ell-1}}\right]$ and so on. This leads to two different expressions for x.

On the other hand, if x is not of the form $\frac{x'}{g^n}$, then some of (and therefore infinitely many) c_n's are not equal to $g-1$. Hence (9.10) leads to $0 \le x - a_n < g^{-n}$. Multiplying this by g^n and noting that $g^n a_n \in \mathbb{Z}$, we deduce that $0 \le g^n x - g^n a_n < 1$, whence

$$[g^n x] = g^n a_n. \tag{9.13}$$

Noting that $a_n - a_{n-1} = \frac{c_n}{g^n}$, we have $c_n = g^n a_n - g(g^{n-1}a_{n-1})$ and so by (9.13)

$$c_n = [g^n x] - g[g^{n-1}x] \tag{9.14}$$

and x has a unique expression (9.9) and there appear infinitely many c_n's different from $g-1$.

Definition 9.1. For any $a \in \mathbb{R}\backslash\mathbb{Q}$, the following process is known as the **continued fraction expansion.**

(1) $a_0 = [a] \in \mathbb{Z}, \quad 0 < b_0 := a - a_0 < 1, \quad a = a_0 + b_0$

(2) $a_1 = \left[\frac{1}{b_0}\right] \in \mathbb{N}, \quad 0 < b_1 = \frac{1}{b_0} - a_1 < 1, \quad b_0 = \frac{1}{a_1+b_1}$

$$\vdots$$

(n) $a_n = \left[\frac{1}{b_{n-1}}\right] \in \mathbb{N}, \quad 0 < b_n = \frac{1}{b_{n-1}} - a_n < 1, \quad b_{n-1} = \frac{1}{a_n+b_n}.$

$$a = a_0 + \frac{1}{a_1+b_1} = a_0 + \frac{1}{a_1 + \frac{1}{a_2+b_2}} = a_0 + \frac{1}{a_1 + \frac{1}{a_2 + \frac{1}{\ddots + \frac{1}{a_n+b_n}}}} \tag{9.15}$$

$$= [a_0; a_1, a_2, \cdots, a_n + b_n],$$

say. We refer to (9.15) as the nth continued fraction expansion of a. The process terminates for $a \in \mathbb{Q}$.

Lemma 9.1. *If we write* (9.15) *with* $b_n = 0$ *as*

$$\alpha_n = a_0 + \frac{1}{a_1 + \frac{1}{a_2 + \frac{1}{\ddots + \frac{1}{a_n}}}} = [a_0; a_1, a_2, \cdots, a_n] = \frac{p_n}{q_n}, \tag{9.16}$$

the nth **convergent**, *then*

$$p_n = a_n p_{n-1} + p_{n-2}, \quad q_n = a_n q_{n-1} + q_{n-2}, \quad n \geq 3 \tag{9.17}$$

and

$$p_{n+1} q_n - p_n q_{n+1} = (-1)^n, \quad n \in \mathbb{N}. \tag{9.18}$$

Proof. First three values of convergents are as in the following table.

n	p_n	q_n
0	a_0	1
1	$a_0 a_1 + 1$	a_1
2	$a_0(a_1 a_2 + 1) + a_2$	$a_2 a_1 + 1$
2	$a_0 a_1 a_2 a_3 + a_0 a_1 + a_0 a_3 + a_2 a_3 + 1$	$a_1 a_2 a_3 + a_1 + a_3$

Table 2. Convergents

Proof of (9.17) is by induction. It is true for $n = 3$ by the table. Indeed, it is also true for $n = 2$. Assume it is true for n. Then α_{n+1} is obtained from α_n by replacing a_n by $a_n + \frac{1}{a_{n+1}}$. Hence

$$\frac{p_{n+1}}{q_{n+1}} = \alpha_{n+1} = \frac{\left(a_n + \frac{1}{a_{n+1}}\right) p_n + p_{n-1}}{\left(a_n + \frac{1}{a_{n+1}}\right) q_n + q_{n-1}} = \frac{a_{n+1}p_n + p_{n-1}}{a_{n+1}q_n + q_{n-1}}. \tag{9.19}$$

(9.18) follows from induction and (9.17). □

Theorem 9.1. *The sequence* $\{\alpha_n\}$ *in* (9.16) *converges to* a *in* (9.15).

Proof. By the definition (9.16), replacing a_n by $a_n + b_n$, we should recover a:

$$a = \frac{(a_n + b_n)p_{n-1} + p_{n-2}}{(a_n + b_n)q_{n-1} + q_{n-2}} = \frac{p'_n}{q'_n} \tag{9.20}$$

say. Hence

$$a - \alpha_n = \frac{p'_n q_n - p_n q'_n}{q_n q'_n} = \frac{b_n(p_{n-1}q_n - q_{n-1}p_n)}{q_n q'_n} = \frac{b_n(-1)^n}{q_n q'_n}$$

by (9.18). Since $0 < b_n < 1$ by Definition 9.1 and *a fortiori* $q_n < q'_n$, we infer that

$$|a - \alpha_n| < \frac{1}{q_n q'_n} < \frac{1}{q_n^2}. \tag{9.21}$$

Since $\{q_n\} \subset \mathbb{Z}$ is increasing and $q_n > 0$, we have $q_n \geq n$. Hence the inequality (9.21) leads to $|a - \alpha_n| < \frac{1}{n^2} \to 0$, whence the result. □

Example 9.1. The golden ratio $\tau = \frac{1+\sqrt{5}}{2}$ in (8.125) satisfies the quadratic equation

$$\tau^2 = \tau + 1$$

or

$$\tau = 1 + \frac{1}{\tau}.$$

Hence apparently,

$$\tau = [1; 1, 1, \cdots] = [1.\bar{1}].$$

Cf. [Sierpinski (1987)] for more details.

The following is the terminating continued fraction expansion of $\approx \frac{\log 3}{\log 2} - 1$.

$$\frac{1760913}{3010300} = \frac{1}{1+}\,\frac{1}{1+}\,\frac{1}{2+}\,\frac{1}{2+}\,\frac{1}{3+}\,\frac{1}{1+}\,\frac{1}{5+}\,\frac{1}{2+}\,\frac{1}{23+}\cdots = \cfrac{1}{1+\cfrac{1}{1+\cfrac{1}{2+\cfrac{1}{\ddots +}}}}. \tag{9.22}$$

The fourth convergent is

$$\frac{1}{1+}\,\frac{1}{1+}\,\frac{1}{2+}\,\frac{1}{2} = \cfrac{1}{1+\cfrac{1}{1+\frac{1}{2}+\frac{1}{\frac{1}{2}}}} = \frac{7}{12}. \tag{9.23}$$

This is the 12-note tempered scale.

The sixth convergent is

$$\frac{1}{1+}\,\frac{1}{1+}\,\frac{1}{2+}\,\frac{1}{2+}\,\frac{1}{3+1} = \frac{31}{53}. \tag{9.24}$$

This is the 53-note tempered scale due to Mercator.

Fig. 9.1: Euler

Chapter 10

Theory of Riemann-Stieltjes integration

There are many instances where the partial summation and integration by parts are regarded as different processes. However, from the general point of view of Stieltjes integration they are exactly the same and it allows one to treat the sum and integral in an essentially similar fashion. Cf. [Apostol (1957)], [Widder (1946), Chapter 1] and [Widder (1989)] for a rather complete theory of Stieltjes integrals.

Definition 10.1. For bounded functions f, g defined on the closed interval $I = [a, b]$ one introduces the (Riemann-)Stieltjes integral in almost verbatim to that of the Riemann integral:

$$\int_a^b f(x)\,\mathrm{d}g(x) = \int_a^b f\,\mathrm{d}g. \tag{10.1}$$

The only difference from the Riemann integral is that one uses $g(x_{j+1}) - g(x_j)$ for the difference $x_{j+1} - x_j$.

We consider a **division** (or sometimes partition) Δ of I into m disjoint subintervals:

$$\Delta : a = x_0 < x_1 < \ldots < x_m = b. \tag{10.2}$$

We understand Δ means the division of I into the union of subintervals $I = \bigcup_{k=1}^{m} I_k, I_k = (x_{k-1}, x_k]$ (here only $I_1 = [a, x_1]$) as well as the set of all division points $\{x_0, \cdots, x_m\}$. Adding more division points gives rise to a refinement of I. We denote the maximum of the width (1-dimensional measure) of m subintervals $\mu(I_k) = x_k - x_{k-1}$ by $|\Delta|$ and call it the **size** of Δ:

$$|\,\Delta\,| = \max\{\mu(I_k) \mid 1 \le k \le m\}.$$

From each I_k choose $\forall \xi_k \in I_k$ and let $\Xi = (\xi_1, \ldots, \xi_m)$. Then form the finite sum

$$S(f) = S(f, \Delta) = S(f, \Xi, \Delta) := \sum_{k=1}^{m} f(\xi_k)(g(x_k) - g(x_{k-1})) \tag{10.3}$$

and call it the **Riemann-Stieltjes sum.**

If $S(f, \Xi, \Delta)$ approaches a value S say, independently of the choice of ξ_k as we make the size $|\Delta|$ of the division smaller (refining and making the size smaller), i.e.

$$\forall \varepsilon > 0,\ \exists \delta = \delta(\varepsilon) > 0\ \forall \Delta\ \forall \Xi\ (\ |\Delta| < \delta\ \Rightarrow\ |S(f, \Xi, \Delta) - S| < \varepsilon),$$

we say that f is **integrable** (in the Riemann-Stieltjes sense) on I, and call the value S the **(Riemann-)Stieltjes integral** (10.1) of f with respect to g (with a and b lower and upper limit, respectively).

Definition 10.2. If a function f on $[a, b]$ satisfies the condition that the sum of all differences is bounded:

$$\sum_{k=1}^{m} |f(x_k) - f(x_{k-1}))| \ = \ O(1) \tag{10.4}$$

for any division (10.1), then f is said to be **of bounded variation.**

$$V_f = V_f([a, b]) = \sup\{\sum_{k=1}^{m} |f(x_k) - f(x_{k-1})||\Delta\}, \tag{10.5}$$

where the supremum is over all divisions Δ of $[a, b]$, is called the **total variation** of f on $[a, b]$. $V_f \geq 0$ is finite.

Lemma 10.1. (Additivity with respect to intervals) *Let f be of bounded variation on $[a, b]$ and let $c \in (a, b)$. Then f is of bounded variation on $[a, c]$ and $[c, b]$ and we have*

$$V_f(a, b) = V_f(a, c) + V_f(c, b). \tag{10.6}$$

Proof. Let Δ_1, Δ_2 denote the division of $[a, c]$ and $[c, d]$, respectively. Then $\Delta_0 = \Delta_1 \cup \Delta_2$ is a division of $[a, b]$. Let $S(\Delta_j) = \sum_{k=1}^{m} |f(x_k) - f(x_{k-1})|$ be the sum corresponding to the division Δ_j. Then

$$S(\Delta_1) + S(\Delta_2) = S(\Delta_0) \leq V_f(a, b),$$

whence it follows that f is of bounded variation on $[a, c]$ and $[c.b]$ and that $V_f(a, c) + V_f(c, b) \leq V_f(a, b)$ in view of (2.29).

To prove the reverse inequality, let Δ be a division (10.2) and consider a possibly new division $\Delta' = \Delta \cup \{c\}$ and suppose $x_{k-1} < c \le x_k$. Then $\Delta_1 : a = x_0 < x_1 < \ldots < x_{k-1} < c$ is a division of $[a, c]$ and $\Delta_2 : c < x_k < \ldots < x_m - b$ is a division of $[c, b]$. Since

$$|f(x_k) - f(x_{k-1})| \le |f(x_k) - f(c)| + |f(c) - f(x_{k-1})|$$

it follows that

$$S(\Delta) \le S(\Delta') = S(\Delta_1) + S(\Delta_2) \le V_f(a, c) + V_f(c, b),$$

whence (10.6) follows. □

Lemma 10.2. *Let f be of bounded variation on $[a, b]$ and let $V_f(a, b)$ be the total variation. Let $V(x) = V_f(a, x)$ for $a < x \le b$ and $V(a) = 0$. Then V and $D := V - f$ are increasing functions on $[a, b]$.*

Proof. If $a < x < y \le b$, then by Lemma 10.1, $V_f(a, y) = V_f(a, x) + V_f(x, y)$ which amounts to $V(y) - V(x) = V_f(x, y) \ge 0$.

On the other hand, note that $D(y) - D(x) = V(y) - V(x) - (f(y) - f(x)) = V_f(x, y) - (f(y) - f(x)) \ge 0$. □

Theorem 10.1. *Any function of bounded variation may be expressed as a difference of two monotone increasing [decreasing] functions.*

Denote $\{I_j | 1 \le j \le m\}$ by λ. Another division $\lambda' = \{I'_k | 1 \le k \le n\}$ $(m \le n)$ is said to be a refinement of λ if for any $I'_k \in \lambda'$ there is an $I_j \in \lambda$ such that $I'_k \subset I_j$, denoted $\lambda \prec \lambda'$. The set Λ of all divisions is a directed set with respect to this ordering: $(\Lambda, \prec)$. For it is enough to check condition (iv) in Definition 16.12. For $\lambda_j = \{I^j_k\}$, $j = 1, 2$ let $I^3_{ij} = I^1_i \cap I^2_j$ and form $\lambda_3 = \{I^3_{ij}\}$ as long as $I^3_{ij} \ne \emptyset$. (This means that division points in I^3_{ij} are the union of $\lambda_j = \{I^j_k\}$ while the intervals are a superposition.) Then $\lambda_1, \lambda_2 \prec \lambda_3$. For directed sets cf. §16.8.

The main result is Theorem 10.2, (i), where f is continuous and g is of bounded variation. By Theorem 10.1, it suffices to consider an increasing function g. Correspondingly to (10.3) we form two generalized sequences for $\lambda \in \Lambda$

$$\bar{S}(f, \lambda) = \sum_{k=1}^{m} \sup f(I_k)(g(x_k) - g(x_{k-1})), \tag{10.7}$$

$$\underline{S}(f, \lambda) = \sum_{k=1}^{m} \inf f(I_k)(g(x_k) - g(x_{k-1})),$$

called the upper and the lower Riemann (-Stieltjes) sum associated to the division λ. Then define the upper and the lower integrals by

$$\overline{\int}_{[a,b]} f = \inf_{\lambda\in\Lambda} \bar{S}(f,\lambda), \qquad \underline{\int}_{[a,b]} f = \sup_{\lambda\in\Lambda} \underline{S}(f,\lambda). \tag{10.8}$$

Now we express the definition of the Riemann-Stieltjes integral in Definition 10.1 by the following.

We say that f is integrable with respect to g if

$$\overline{\int}_{[a,b]} f = \underline{\int}_{[a,b]} f \tag{10.9}$$

and denote the common value by (10.1).

Lemma 10.3. *The generalized sequence* $\bar{S}(f,\lambda)$ *is non-increasing and* $\underline{S}(f,\lambda)$ *is non-decreasing, so that*

$$\lim_{\lambda\in\Lambda} \bar{S}(f,\lambda) = \overline{\int}_{[a,b]} f, \qquad \lim_{\lambda\in\Lambda} \underline{S}(f,\lambda) = \underline{\int}_{[a,b]} f. \tag{10.10}$$

Proof. Suppose $\lambda = \{I_j | 1 \le j \le m\}$ and $\lambda' = \{I'_k | 1 \le k \le n\}$ $(m \le n)$ and that $\lambda \prec \lambda'$. We prove the first assertion. For any $I'_k \in \lambda'$ there is an $I_j \in \lambda$ such that $I'_k \subset I_j$. From this and the meaning of a division it follows that

$$I_j = \bigcup_{h=1}^{l} I_{j'+h}, \quad g(x_j) - g(x_{j-1}) = \sum_{h=1}^{l} \left(g(x'_{j'+h}) - g(x'_{j'+h-1})\right) \tag{10.11}$$

with $j \le j' + l \le n$. Since $I'_k \subset I_j$, we have $\sup f(I_j) \ge \sup f(I'_k)$, so that $\bar{S}(f,\lambda) \ge \bar{S}(f,\lambda')$. □

Theorem 10.2.

(i) *The Stieltjes integral* $\int_a^b f\,\mathrm{d}g$ *exists if* f *is continuous and* g *is of bounded variation and linear in* f *and* g*. The role can be changed in view of Item* (ii)*. It holds that*

$$\int_a^b \mathrm{d}g(x) = g(b) - g(a). \tag{10.12}$$

(ii) *The formula for integration by parts holds true:*

$$\int_a^b f(x)\,\mathrm{d}g(x) = [f(x)g(x)]_a^b - \int_a^b g(x)\,\mathrm{d}f(x), \tag{10.13}$$

provided that f is continuous and g is of bounded variation or g is continuous and f is of bounded variation.

(iii) *If g is a step function with jumps a_n at x_n, the Stieltjes integral reduces to the sum:*

$$\int_a^b f(x)\,\mathrm{d}g(x) = \sum_{a<x_n\le b} f(x_n)a_n. \tag{10.14}$$

(iv) *If f is continuous and g is differentiable, then the Stieltjes integral reduces to the Riemann integral:*

$$\int_a^b f(x)\,\mathrm{d}g(x) = \int_a^b f(x)g'(x)\,\mathrm{d}x. \tag{10.15}$$

(v) *The Stieltjes integral $\int_a^b f\,\mathrm{d}g$ shares the properties stated in Corollary 11.1 for Riemann integrals, i.e. linearity in the integrator as well as the integrand, additivity with respect to subintervals, and the monotonicity holds when g is increasing:*

$$f_1(x) \le f_2(x) \quad (x \in I) \Longrightarrow \int_a^b f_1\,\mathrm{d}g \le \int_a^b f_2\,\mathrm{d}g.$$

Proof. (i) We assume g is increasing. By Theorem 4.4, f is uniformly continuous, i.e. given $\varepsilon > 0$ there exists a $\delta = \delta(\varepsilon) > 0$ such that for any $x_1, x_2 \in [a,b]$ satisfying $|x_1 - x_2| < \delta$ we have $|f(x_1) - f(x_2)| < \varepsilon$. Choose λ_ε such that $|\lambda_\varepsilon| < \delta$. Then for $\lambda_\varepsilon \prec \lambda$ we have

$$\begin{aligned}
\bar{S}(f,\lambda) - \underline{S}(f,\lambda) &= \sum_{k=1}^m \sup f(I_k)(g(x_k) - g(x_{k-1})) \qquad (10.16)\\
&- \sum_{k=1}^m \inf f(I_k)(g(x_k) - g(x_{k-1}))\\
&= \sum_{k=1}^m (\sup f(I_k) - \inf f(I_k))\,(g(x_k) - g(x_{k-1}))\\
&\le \varepsilon \sum_{k=1}^m (g(x_k) - g(x_{k-1})) \le \varepsilon(g(b) - g(a)).
\end{aligned}$$

Since the left-hand side member is $\geq \overline{\int} f - \underline{\int} f \geq 0$, it follows that

$$0 \leq \overline{\int}_{[a,b]} f - \underline{\int}_{[a,b]} f \leq \varepsilon(g(b) - g(a)).$$

Since $\varepsilon > 0$ is arbitrary, this leads to (10.9).

Since the integral $\int_a^b f \, \mathrm{d}g$ is linear in g:

$$\int_a^b f \, \mathrm{d}(c_1 g_1 + c_2 g_2) = c_1 \int_a^b f \, \mathrm{d}g + c_2 \int_a^b f \, \mathrm{d}g$$

and *a fortiori* for $g_1 - g_2$, we conclude the assertion (i) for g of bounded variation. □

Remark 10.1. If in (10.14), f is differentiable, then applying (10.13) to it, we deduce that

$$\sum_{a < x_n \leq b} f(x_n) a_n = \int_a^b f(x) \, \mathrm{d}g(x) = [f(x)g(x)]_a^b - \int_a^b g(x) f'(x) \, \mathrm{d}x, \quad (10.17)$$

where $g(x) = A(x) = \sum_{x_k \leq x} a_k$. This is a formula for **partial summation**. Below we give some variants and specializations of this for ease of applications. In many of them, by shifting the term $-A(a)f(a)$ to the left, the right-hand side appears in the form $A(b)f(b) - \int_a^b A(x) f'(x) \, \mathrm{d}x$.

When f and g are differentiable of class C^1, say, we combine (10.13) and (10.15) to deduce

$$\int_a^b f(x) g'(x) \, \mathrm{d}x = [f(x)g(x)]_a^b - \int_a^b g(x) f'(x) \, \mathrm{d}x, \quad (10.18)$$

which is the formula for integration by parts. Cf. (1.20).

If in (10.14), f is not differentiable, then

$$\sum_{a < x_n \leq b} f(x_n) a_n = \int_a^b f(x) \, \mathrm{d}g(x) = [f(x)g(x)]_a^b - \int_a^b g(x) \, \mathrm{d}f(x). \quad (10.19)$$

Now specify $a = 1, b = n$, $x_k = k$, $f(k) = a_k$, $g(x) = B(x) = \sum_{k \leq x} b_k$. Then

$$\sum_{1 < k \leq n} a_k b_k = a_n B(n) - a_1 b_1 - \sum_{k=1}^{n-1} B(k) \int_k^{k+1} \mathrm{d}f(x). \quad (10.20)$$

By definition, $\int_k^{k+1} \mathrm{d}f(x) = a_{k+1} - a_k$, so that (10.20) implies

Corollary 10.1. (Partial summation) *Let $B_n = \sum_{k=1}^{n} b_k$ be the nth partial sum of $\{b_n\}$. Then*

$$\sum_{1\le k\le n} a_k b_k = a_n B_n - \sum_{k=1}^{n-1} B_k(a_{k+1} - a_k). \tag{10.21}$$

We give another direct proof. By differencing $b_k = B_k - B_{k-1}$, and so we find that

$$\begin{aligned}\sum_{1<k\le n} a_k b_k &= \sum_{2\le k\le n} a_k(B_k - B_{k-1}) \qquad (10.22)\\ &= a_n B_n - \sum_{2\le k\le n-1} (a_{k+1} - a_k)B_k - a_1 b_1,\end{aligned}$$

whence the result.

Remark 10.2. Recalling the difference operator, Corollary 10.1 may be stated schematically as

$$\Delta^{-1} a_k \Delta B_k = [a_k B_k] - \Delta^{-1} b_k \Delta a_k, \tag{10.23}$$

a prototype of integration by parts.

We derive (2.8) as a special case. Putting $B_k = 1$, we have

$$0 = \sum_{m\le k\le n} a_k(B_k - B_{k-1}) = [a_k B_k]_m^n - \sum_{k=m}^{n-1} (a_{k+1} - a_k) \tag{10.24}$$

whence (2.8).

Exercise 79. [Prachar (1957), Satz 1.4, p. 371] Let $\{\lambda_n\} \subset \mathbb{R}$ be an increasing sequence with $\lambda_1 > 0$. Then for any complex sequence $\{a_n\}$ and $x > 0$ we write

$$A(x) = \sum_{\lambda_1\le \lambda_n\le x} a_n \tag{10.25}$$

with empty sum being 0. Suppose $g(t)$ is of class C^1 on $[\lambda_1, \infty)$. Then prove the partial summation formula

$$\sum_{\lambda_1\le \lambda_k\le x} a_k g(\lambda_k) = A(x)g(x) - \int_{\lambda_1}^{x} A(t)g'(t)\,\mathrm{d}t. \tag{10.26}$$

Solution. Correspondingly to (10.22) we have with $\lambda_n \le x < \lambda_{n+1}$

$$\sum_{\lambda_1<\lambda_k\le x} a_k g(\lambda_k) = A(\lambda_n)g(\lambda_n) - \sum_{\lambda_1\le\lambda_k\le\lambda_{n-1}} A(\lambda_k)\int_{\lambda_k}^{\lambda_{k+1}} g'(t)\,\mathrm{d}t - a_1 g(\lambda_1) \tag{10.27}$$

$$= A(\lambda_n)g(\lambda_n) - \int_{\lambda_1}^{\lambda_n} A(t)g'(t)\,\mathrm{d}t - a_1 g(\lambda_1).$$

The integral $-\int_{\lambda_1}^{\lambda_n} A(t)g'(t)\,\mathrm{d}t$ amounts to $A(x)g(x) - A(x)g(\lambda_n) - \int_{\lambda_1}^{x} A(t)g'(t)\,\mathrm{d}t$ for $\lambda_1 \le x < \lambda_{n+1}$ and the result follows.
Another solution.

$$A(x)g(x) - \sum_{\lambda_k\le x} a_k g(\lambda_k) = \sum_{\lambda_k\le x} a_k(g(x) - g(\lambda_k)) \tag{10.28}$$

$$= \sum_{\lambda_k\le x}\int_{\lambda_k}^{x} a_k g'(u)\,\mathrm{d}u = \int_{\lambda_1}^{x}\sum_{\lambda_k\le u} a_k g'(u)\,\mathrm{d}u,$$

which is the last term of the right-hand side of (10.26) with minus sign. We note the last step in (10.28), which is usually regarded as a coincidence of the exchange of the sum and the integral but what we really made is the change of order of repeated the (Stieltjes) integral. Cf. [Widder (1946)].

Exercise 80. Let $\{\lambda_n\} \subset \mathbb{R}$ be an increasing sequence with $\lambda_1 > 0$. Then for any complex sequence $\{a_n\}$ and $x > 0$ we write

$$A_m(x) = \sum_{\lambda_{m+1}\le\lambda_n\le x} a_n$$

with empty sum being 0. Suppose $g(t)$ is of class C^1 on $[\lambda_1, \infty)$. Then prove a variant of the partial summation formula

$$\sum_{\lambda_{m+1}\le\lambda_k\le x} a_k g(\lambda_k) = A_m(x)g(x) - \int_{\lambda_{m+1}}^{x} A_m(t)g'(t)\,\mathrm{d}t. \tag{10.29}$$

Solution. Correspondingly to (10.27) we have with $\lambda_n \le x < \lambda_{n+1}$, and $A_m(\lambda_m) = 0$,

$$\sum_{\lambda_{m+1}\le\lambda_k\le x} a_k g(\lambda_k) = A_m(\lambda_n)g(\lambda_n) - \sum_{\lambda_m\le\lambda_k\le\lambda_{n-1}} A_m(\lambda_k)\int_{\lambda_k}^{\lambda_{k+1}} g'(t)\,\mathrm{d}t$$

$$- A_m(\lambda_m)g(\lambda_m) = A_m(\lambda_n)g(\lambda_n) - \int_{\lambda_{m+1}}^{\lambda_n} A_m(t)g'(t)\,\mathrm{d}t. \tag{10.30}$$

The integral $-\int_{\lambda_{m+1}}^{\lambda_n} A_m(t)g'(t)\,\mathrm{d}t$ amounts to $A(x)g(x) - A(x)g(\lambda_n) - \int_{\lambda_{m+1}}^{x} A_m(t)g'(t)\,\mathrm{d}t$ and the result follows. This also follows from Exercise 79 by subtracting the sum $A(\lambda_m)$ in (10.25).

In literature [Hardy and Riesz (1972), pp. 2–4], [Serre (1973), pp. 65–66], this lemma is stated and used in the form of the sum, Lemma 10.1:

Lemma 10.4. *For* $m \in \mathbb{N}$

$$\sum_{m\le k\le x} a_k g(k) = A_m(x)g([x]) - \sum_{k=m}^{[x]-1} A_m(k)\Delta g, \tag{10.31}$$

coupled with

Lemma 10.5. *If* $\Re s = \sigma \neq 0$, *then*

$$|\Delta e^{-\lambda_n s}| \le \frac{|s|}{\sigma}\Delta e^{-\lambda_n \sigma}. \tag{10.32}$$

Proof follows from

$$|\Delta e^{-\lambda_n s}| = \left|\int_{\lambda_n}^{\lambda_{n+1}} s e^{-su}\mathrm{d}u\right| \le |s|\int_{\lambda_n}^{\lambda_{n+1}} e^{-\sigma u}\,\mathrm{d}u.$$

Corollary 10.2. (Euler's summation formula) *Suppose* $f \in C^1$. *Then*

$$\sum_{a<n\le b} f(n) = \int_a^b f(x)\,\mathrm{d}x - [\bar{B}_1(x)f(x)]_a^b + \int_a^b \bar{B}_1(x)f'(x)\,\mathrm{d}x, \tag{10.33}$$

where $\bar{B}_1(x) = B_1(x-[x]) = x-[x]-\frac{1}{2}$ *is the periodic Bernoulli polynomial in* (13.79).

Proof. We apply (10.13) with $g(x) = x - [x] = \bar{B}_1(x) + \frac{1}{2}$ to obtain

$$\int_a^b f(x)\mathrm{d}x - \int_a^b f(x)\mathrm{d}[x] = \left[f(x)\left(\bar{B}_1(x)+\frac{1}{2}\right)\right]_a^b - \int_a^b \left(\bar{B}_1(x)+\frac{1}{2}\right)f'(x)\mathrm{d}x. \tag{10.34}$$

Since the second integral on the left is, by (10.14),

$$\int_a^b f(x)\,\mathrm{d}[x] = \sum_{a<n\le b} f(n), \tag{10.35}$$

we conclude the assertion on noting cancellation of terms. □

Definition 10.3. Given two functions A and B defined for $x \geq 1$ of bounded variation on each bounded interval, we define the **Stieltjes resultant** C of A and B on the basis of local-global principle (11.5) by

$$C(x)=(A\times B)(x)=\sum_{n\le x}(A*B)(n)=\sum_{mn\le x}A(m)B(n)=\iint_{uv\le x}\mathrm{d}A(u)\mathrm{d}B(v). \tag{10.36}$$

Corollary 10.3. *If $B(x)$ is a step function, then*

$$\int_{u\le x}\mathrm{d}B(u)=B(x). \tag{10.37}$$

We may also express the Stieltjes resultant C of A and B as

$$C(x)=(A\times B)(x)=\int_{u\le x}\int_{v\le x/u}\mathrm{d}A(v)\,\mathrm{d}B(u)=\int_a^x A(x/u)\,\mathrm{d}B(u), \tag{10.38}$$

whenever the integral exists and for all x, $C(x)$ lies between the limits $\lim_{h\to+0}C(x\pm h)$.

We may deduce a recurrence formula for the **sum of kth powers**

$$Z(n,k)=\sum_{j=1}^{n} j^k \tag{10.39}$$

from Theorem 10.2:

Corollary 10.4.

$$(k+1)Z(n,k)=n(n+1)^k-\sum_{r=0}^{k-2}\binom{k}{r}Z(n,r+1), \tag{10.40}$$

where an empty sum is to be interpreted to mean 0.

Proof. We evaluate the integral

$$I_{n+1}=I_{n+1}(k):=\int_0^{n+1}x^k\,\mathrm{d}[x] \tag{10.41}$$

in two ways. First by Theorem 10.2, (iii)

$$I_{n+1}=\sum_{j=1}^{n+1}j^k=Z(n+1,k)=Z(n,k)+(n+1)^k. \tag{10.42}$$

On the other hand, by Theorem 10.2, (ii) and (iv),

$$I_{n+1}=x^k[x]\Big|_0^{n+1}-\int_0^{n+1}[x]\,\mathrm{d}x^k=(n+1)^{k+1}-\sum_{j=0}^{n}j\int_j^{j+1}kx^{k-1}\,\mathrm{d}x, \tag{10.43}$$

so that

$$I_{n+1} = (n+1)^{k+1} - \sum_{j=1}^{n} jx^k\Big|_j^{j+1} = (n+1)^{k+1} - \sum_{j=1}^{n} j((j+1)^k - j^k) \tag{10.44}$$

$$= (n+1)^{k+1} - \sum_{j=1}^{n} j \sum_{r=0}^{k-1} \binom{k}{r} j^r = (n+1)^{k+1} - \sum_{r=0}^{k-1} \binom{k}{r} Z(n, r+1).$$

Hence by equating (10.42) and (10.44) and shifting the term with $r = k-1$ to the left it follows that

$$(k+1)Z(n,k) + (n+1)^k = (n+1)^{k+1} - \sum_{r=0}^{k-2} \binom{k}{r} Z(n, r+1)$$

or (10.40). □

It is customary to express the sum of kth powers by Bernoulli polynomials (cf. §13.7).

Theorem 10.3. *For the sum* (10.39) *we have*

$$Z(n,k) = \frac{1}{k+1}\left(B_{k+1}(n+1) - B_{k+1}(1)\right). \tag{10.45}$$

Proof. First by Exercise 81

$$(n+1)^{k+1} - 1 = \sum_{l=0}^{k} \binom{k+1}{l} Z(n,l). \tag{10.46}$$

Assuming (10.45) with $k = l$ for $l = 0, 1, 2, \cdots, k-1$ and substituting in (10.46), we deduce that

$$(n+1)^{k+1} - 1 = S_{k-1} + (k+1)Z(n,k), \tag{10.47}$$

where

$$S_{k-1} = S_{k-1}(n) = \sum_{l=0}^{k-1} \binom{k+1}{l} \frac{1}{l+1}\left(B_{l+1}(n+1) - B_{l+1}(1)\right). \tag{10.48}$$

Note that

$$\binom{k+1}{l}\frac{1}{l+1} = \frac{1}{k+2}\binom{k+2}{l+1},$$

whence that

$$S_{k-1} = \frac{1}{k+2}\sum_{l=0}^{k-1} \binom{k+2}{l+1} \left(B_{l+1}(n+1) - B_{l+1}(1)\right)$$

$$= \frac{1}{k+2}\sum_{l=0}^{k-1} \binom{k+2}{l+1} \left(B_{l+1}(n+1) - B_{l+1}(0)\right) - B_1(1) + B_1(0)$$

by replacing $B_{l+1}(1)$ by $B_{l+1}(0)$ with modification for $l+1=1$. We rewrite this further as

$$S_{k-1} = \frac{1}{k+2}\sum_{r=1}^{k}\binom{k+2}{r}\left(B_r(n+1)-B_r(0)\right)-1 \tag{10.49}$$
$$= \frac{1}{k+2}\sum_{r=0}^{k+2}\binom{k+2}{r}B_r(n+1) - \frac{1}{k+2}\sum_{r=0}^{k+2}\binom{k+2}{r}B_r(0)$$
$$-\left(B_{k+1}(n+1)-B_{k+1}(0)\right)-\frac{1}{k+2}\left(B_{k+2}(n+1)-B_{k+2}(0)\right)-1.$$

Applying the addition formula [Kanemitsu and Tsukada (2007), p. 3]

$$B_n(x+y)=\sum_{r=0}^{n}\binom{n}{r}B_r(x)y^{n-r} \tag{10.50}$$

to the two sums on the right-hand side of (10.49), we see that (10.49) becomes

$$S_{k-1} = \frac{1}{k+2}\left(B_{k+2}(n+2)-B_{k+2}(1)\right) \tag{10.51}$$
$$-\left(B_{k+1}(n+1)-B_{k+1}(0)\right)-\frac{1}{k+2}\left(B_{k+2}(n+1)-B_{k+2}(0)\right)-1.$$

Hence, substituting (10.51) in (10.47), we conclude that

$$(n+1)^{k+1}-1 = \frac{1}{k+2}\left(B_{k+2}(n+2)-B_{k+2}(n+1)\right) \tag{10.52}$$
$$-\left(B_{k+1}(n+1)-B_{k+1}\right)-1+(k+1)Z(n,k).$$

Applying the basic difference equation

$$\frac{1}{k+2}\left(B_{k+2}(n+2)-B_{k+2}(n+1)\right)=(n+1)^{k+1}, \tag{10.53}$$

we conclude (10.45), completing the proof. □

Exercise 81. Deduce (10.46) from (10.40).

Solution. We may rewrite (10.40) as

$$(k+1)Z(n,k)=n(n+1)^k-\sum_{l=1}^{k-1}\binom{k}{l-1}Z(n,l)$$

or

$$(n+1)^{k+1}-(n+1)^k=\sum_{l=1}^{k}\binom{k+1}{l}Z(n,l)-\sum_{l=1}^{k-1}\left(\binom{k+1}{l}-\binom{k}{l-1}\right)Z(n,l)$$
$$=\sum_{l=1}^{k}\binom{k+1}{l}Z(n,l)-\sum_{l=1}^{k-1}\binom{k}{l}Z(n,l)$$
$$=\sum_{l=0}^{k}\binom{k+1}{l}Z(n,l)-\sum_{l=0}^{k-1}\binom{k}{l}Z(n,l). \tag{10.54}$$

Hence by induction, we prove (10.46).

Exercise 82. By the method of proof of Corollary 10.4, deduce the following

$$\sum_{j=1}^{n} j = \frac{1}{2}n(n+1), \quad \sum_{j=1}^{n} j^2 = \frac{1}{6}n(n+1)(2n+1). \tag{10.55}$$

Solution. We follow the above argument verbatim. For (10.42), we have

$$I_{n+1}(1) = \sum_{j=1}^{n+1} j = Z(n,1) + n + 1, \tag{10.56}$$

$$I_{n+1}(2) = \sum_{j=1}^{n+1} j^2 = Z(n,2) + (n+1)^2.$$

On the other hand, (10.43) for $k = 1$ reads

$$I_{n+1}(1) = x[x]\Big|_0^{n+1} - \int_0^{n+1} [x]\,\mathrm{d}x = (n+1)^2 - \sum_{j=0}^{n} j \int_j^{j+1} \mathrm{d}x \tag{10.57}$$
$$= (n+1)^2 - Z(n,1),$$

whence the first equality of (10.55).

Now (10.43) for $k = 2$ reads

$$I_{n+1}(2) = x^2[x]\Big|_0^{n+1} - \int_0^{n+1} [x]\,\mathrm{d}x^2 = (n+1)^3 - \sum_{j=0}^{n} j \int_j^{j+1} 2x\,\mathrm{d}x \tag{10.58}$$

$$= (n+1)^3 - \sum_{j=1}^{n} jx^2\Big|_j^{j+1} = (n+1)^3 - \sum_{j=1}^{n} j(2j+1)$$
$$= (n+1)^3 - 2Z(n,2) - Z(n,1),$$

whence the second equality of (10.55).

Exercise 83. Let n be a natural number. Prove that

$$I := \int_0^n (x^2+1)\,\mathrm{d}[x] = \frac{1}{6}n(2n^2+3n+7). \tag{10.59}$$

Solution. We use Theorem 10.2, (ii) and (iv).

$$I = (x^2+1)[x]\Big|_0^n - \int_0^n [x]\,\mathrm{d}(x^2+1) = (n^2+1)n - 2\sum_{j=0}^{n-1} \int_j^{j+1} [x]x\,\mathrm{d}x \tag{10.60}$$

$$= n(n^2+1) - 2\sum_{j=0}^{n-1} j\frac{1}{2}x^2\Big|_j^{j+1} = n(n^2+1) - \sum_{j=1}^{n-1}(2j^2+j)$$

whence (10.59) follows after simplification.

The following theorem specifies to the first mean value theorem for Riemann integrals in Theorem 11.4 and its corollary.

Theorem 10.4. (The first mean value theorem for Stieltjes integrals) *Suppose f is continuous and g is monotone increasing on $I = [a, b]$. Then there exists a $\xi \in I$ such that*

$$\int_a^b f(x)\,\mathrm{d}g(x) = f(\xi) \int_a^b \mathrm{d}g(x). \tag{10.61}$$

Proof. If $g(b) = g(a)$, then both sides are 0. We may suppose $g(b) > g(a)$. Proof is almost verbatim to that of Theorem 11.4 below. Only (11.22) is to be replaced by

$$m(g(b) - g(a)) \le \underline{S}(f, \lambda) \le \bar{S}\,(f, \lambda) \le M(g(b) - g(a)), \tag{10.62}$$

where lower and upper Riemann sums are defined in (10.7). The integral must lie in the same range, so that the λ in (11.21) is to be replaced by

$$\lambda = \frac{\int_a^b f\,(x)\,\mathrm{d}g(x)}{g(b) - g(a)}, \tag{10.63}$$

which lies in $[m, M]$ and the intermediate value theorem concludes the assertion. □

We may prove the counterpart of Theorem 11.5, the second mean value theorem for Stieltjes integrals.

Theorem 10.5. (Second mean value theorem for integrals) *If g is continuous and f is monotone increasing on $I = [a, b]$, then there exists a $\xi \in I$ such that*

$$\int_a^b f\,(x)\,\mathrm{d}g(x) = f\,(a) \int_a^\xi \mathrm{d}g(x) + f\,(b) \int_\xi^b \mathrm{d}g(x). \tag{10.64}$$

Proof. Proof follows on applying Theorem 10.4 to (10.13) in the form

$$\int_a^b f(x)\,\mathrm{d}g(x) = f(b)g(b) - f(a)g(a) - \int_a^b g(x)\,\mathrm{d}f(x),$$

the integral on the right being $g(\xi)(f(b) - f(a))$. We may apply (10.12). □

Chapter 11

Theory of Riemann integration

In this section we specify the theory of Riemann-Stieltjes integration to that of Riemann integration. It was developed in Riemann's Haibilitationsschrift [Riemann (1854)].

If a function $f(x)$ is defined (and bounded) on an interval $I = [a, b]$, the area (with sign) of the figure formed by the graph of $y = f(x)$ and the lines $x = a$, $x = b$, $y = 0$ is called the **definite integral** with lower limit a, upper limit b (or the integral of f from a to b) and denoted by

$$\int_a^b f(x)\,\mathrm{d}x.$$

Mathematically, it is defined by the division of the interval as in Definition 10.1.

Definition 11.1. Let $I = [a, b]$ be a bounded closed interval and let $f : I \to \mathbb{R}$ be bounded (i.e. $\exists M > 0\ \forall x \in I\,(|f(x)| < M)$). We consider a **division** Δ of I into m disjoint subintervals:

$$\Delta : a = x_0 < x_1 < \ldots < x_m = b.$$

We understand Δ means the division of I into the union of subintervals $I = \bigcup_{k=1}^{m} I_k$ ($I_1 = [x_0, x_1], I_k = (x_{k-1}, x_k], k \geq 2$) as well as the set of all division points $\{x_0, \ldots, x_m\}$. Adding more division points gives rise to a refinement of I. We denote the maximum of the width (1-dimensional measure) of m subintervals $\mu(I_k) = x_k - x_{k-1}$ by $|\Delta|$ and call it the **size**, mesh or norm of Δ:

$$|\,\Delta\,| = \max\{\mu(I_k)\mid 1 \leq k \leq m\}.$$

From each I_k choose $\forall \xi_k \in I_k$ and let $\Xi = (\xi_1, \ldots, \xi_m)$. Then form the finite sum

$$S(f) = S(f, \Delta) = S(f, \Xi\Delta) := \sum_{k=1}^{m} f(\xi_k)\,\mu(I_k) \tag{11.1}$$

and call it the **Riemann sum**.

If $S(\Delta, f, \Xi)$ approaches a value S say, independently of the choice of ξ_k as we make the size $|\Delta|$ of the division smaller (refining and making the size smaller), i.e.

$$\forall \varepsilon > 0,\ \exists \delta = \delta(\varepsilon) > 0\ \forall \Delta\ \forall \Xi\ \left(\, |\Delta| < \delta\ \Rightarrow\ |S(\Delta, f, \Xi) - S| < \varepsilon\right),$$

we say that f is **integrable** (in the Riemann sense) on I, and call the value S the **definite integral** of f (with a and b lower and upper limits, respectively), denoted

$$\int_a^b f(x)\,\mathrm{d}x \quad \text{or} \quad \int_I f(x)\,\mathrm{d}x.$$

The value S may be interpreted to mean the area with sign of the figure formed by the graph of f and the x-axis (counting the area of that part which lies above the axis as plus and one under minus). In contrast to multiple integrals to be introduced presently, these are sometimes called single integrals.

We write

$$\mathbb{I} = [a_1, b_1] \times \cdots \times [a_n, b_n] \tag{11.2}$$

and call it an n-dimensional closed interval (or a rectangle). Similarly to single integrals, for a bounded function $z = f(x, y)$ in two variables defined on the 2-dimensional bounded closed interval $\mathbb{I} \subset \mathbb{R}^2$ we make a division into subintervals to define the **double integral** of f

$$\int_{\mathbb{I}} f\,\mathrm{d}\boldsymbol{x} = \iint_{\mathbb{I}} f\,\mathrm{d}x\mathrm{d}y$$

which signifies the volume of the figure formed by the graphic surface $z = f(x, y)$ and the xy-plane.

In general we introduce the **n-ple integral**

$$\int_{\mathbb{I}} f\,\mathrm{d}\boldsymbol{x} = \int\!\cdots\!\int_{\mathbb{I}} f\,\mathrm{d}x_1\cdots\mathrm{d}x_n$$

of a function $f(x_1, \cdots, x_n)$ in n variables defined on a finite closed interval (11.2) by making the division of the interval.

Let $R \subset \mathbb{R}^n$ be a bounded closed (i.e. compact) set and suppose f is defined and bounded on R and that R is contained in an n-dimensional interval $\mathbb{I}$. We define a new function $\bar{f}$ called **zero-extension** of f by

$$\bar{f}(x) = \begin{cases} f(x) & \text{if} \quad x \in R \\ 0 & \text{if} \quad x \in \mathbb{I} - R \end{cases}. \tag{11.3}$$

Then we define the n-ple integral

$$\int \cdots \int_R f \, \mathrm{d}x_1 \cdots \mathrm{d}x_n = \int \cdots \int_{\mathbb{I}} \bar{f} \, \mathrm{d}x_1 \cdots \mathrm{d}x_n. \tag{11.4}$$

What we encounter as a domain R is an ordinate set (or its slight generalization) as in Definition 11.4.

As can be seen from the definition by division into subintervals, the definite integral depends on the **local-global principle**:

$$\mathrm{d}A = f(\boldsymbol{x}) \, \mathrm{d}\boldsymbol{x} \Longrightarrow \int \mathrm{d}A = \int_R f(\boldsymbol{x}) \, \mathrm{d}\boldsymbol{x}, \tag{11.5}$$

where R is a certain compact domain in $\mathbb{R}^n$, say, i.e. *collecting local data gives rise to the whole*, a reminiscent of the principle of molecular biology "Genotype determines phenotype". The principle manifests typically in the form of the change of variable formula ((14.11)). It applies also to the definition of integrals over a curve or a surface, where there is no Jacobian but there is the length or the area element. This implies implicitly that the case $f(x) \cong 1$ gives rise to the volume $\mu(\Omega)$ of Ω, which is to be non-negative.

Subsequently, we sometimes provide an intuitive view and as soon as we may express the data of infinitesimal part in the form

$$f(\boldsymbol{x}) \, \mathrm{d}\boldsymbol{x}$$

we immediately refer to the local-global principle to conclude the results. More rigorous treatment is possible and can be found in many standard textbooks. In this book we assume the following theorem. What appears is the cases $n = 1, 2, 3$.

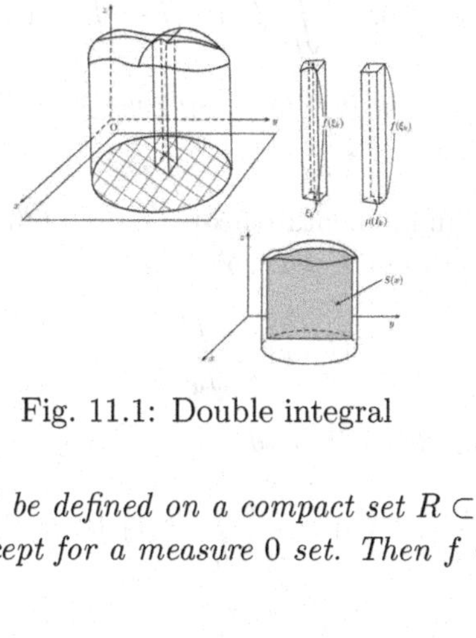

Fig. 11.1: Double integral

Theorem 11.1. *Let f be defined on a compact set $R \subset \mathbb{R}^n$ and suppose f is continuous on R except for a measure 0 set. Then f is integrable on R.*

Corollary 11.1. *Let $R \subset \mathbb{R}^n$ be a compact set and suppose $f, g : \mathbb{I} \to \mathbb{R}$ are continuous except for a measure 0 set E. Then*
(i) *(linearity) For $c_1, c_2 \in \mathbb{R}$,*

$$\int_R (c_1 f + c_2 g)\, \mathrm{d}\boldsymbol{x} = c_1 \int_R f\, \mathrm{d}\boldsymbol{x} + c_2 \int_R g\, \mathrm{d}\boldsymbol{x}$$

(ii) *(monotonicity) $f(\boldsymbol{x}) \leq g(\boldsymbol{x}) \quad (\forall \boldsymbol{x} \in R)$*

$$\Rightarrow \int_R f(\boldsymbol{x})\, \mathrm{d}\boldsymbol{x} \leq \int_R g(\boldsymbol{x})\, \mathrm{d}\boldsymbol{x}$$

(iii) *The absolute value $|f| : R \to \mathbb{R}_{\geq 0}$ of f is also integrable and*

$$\left| \int_R f(\boldsymbol{x})\, \mathrm{d}\boldsymbol{x} \right| \leq \int_R |f|(\boldsymbol{x})\, \mathrm{d}\boldsymbol{x}$$

where $|f|(\boldsymbol{x}) = |f(\boldsymbol{x})|$.
(iv) (*additivity with respect to intervals in* 1*-dimensional case) If $\int_a^b f(x)\, \mathrm{d}x$ exists for $a < b$, we define*

$$\int_b^a f(x)\, \mathrm{d}x = -\int_a^b f(x)\, \mathrm{d}x.$$

Also we define

$$\int_a^a f(x)\, \mathrm{d}x = 0.$$

Then for $a, b, c \in \mathbb{R}$, we have

$$\int_a^b f(x)\, \mathrm{d}x + \int_b^c f(x)\, \mathrm{d}x + \int_c^a f(x)\, \mathrm{d}x = 0$$

in the sense that if two of the three integrals exist then the third exists and the equality holds.

Corollary 11.2. (Fundamental theorem for infinitesimal calculus) *If $f : I = [a, b] \to \mathbb{R}$ is continuous, then putting*

$$F(x) = \int_a^x f(t)\, \mathrm{d}t$$

$(a \leq x \leq b)$, F is of class C^1, and

$$\frac{\mathrm{d}}{\mathrm{d}x} F(x) = \frac{\mathrm{d}}{\mathrm{d}x} \int_a^x f(t)\, \mathrm{d}t = f(x). \tag{11.6}$$

Proof. Fix a point z_0 and consider the indefinite integral

$$F(z) = \int_{z_0}^{z} f(w)\,\mathrm{d}w. \tag{11.7}$$

Since $F(z)$ does not depend on the choice of the path connecting z_0 and z (by the Cauchy integral theorem), it is uniquely defined. Since $f(z)$ is continuous on I, for any $z \in I$ and any $\varepsilon > 0$ there exists a $\delta >$ such that

$$|f(z+h) - f(z)| < \varepsilon \quad \text{for } |h| < \delta,\ z+h \in I.$$

By Corollary 11.1,

$$\frac{F(z+h) - F(z)}{h} = \frac{1}{h}\int_{z}^{z+h} f(w)\,\mathrm{d}w.$$

Substituting $f(z) = \frac{f(z)}{h}\int_z^{z+h} \mathrm{d}w$ from this, we deduce that

$$\left|\frac{F(z+h) - F(z)}{h} - f(z)\right| \le \frac{1}{|h|}\int_{z}^{z+h} |f(w) - f(z)|\,\mathrm{d}w < \frac{1}{|h|}\varepsilon|h| = \varepsilon. \tag{11.8}$$

Hence $F'(z) = f(z)$ and $F(z)$ is analytic in I. □

11.1 Evaluation of definite integrals

Definition 11.2. Given a function $f(x)$, any differentiable function $F(x)$ is called a **primitive function** or an **anti-derivative** of $f(x)$ if

$$F'(x) = f(x), \tag{11.9}$$

or denoting the differential operator by D, $Df = f'$, then $D^{-1}F = f$. I.e. a primitive function of f is "any differentiable function giving f when differentiated".

By Corollary 11.2, $\int_a^x f(t)\,\mathrm{d}t$ is a primitive function of $f(x)$, and so from the Newton-Leibniz rule (Corollary 7.1), it follows that

$$F(x) = \int_a^x f(t)\,\mathrm{d}t + C \tag{11.10}$$

(C being an integral constant). It is natural to denote the set of all primitive functions of f by and call it the indefinite integral:

$$\int f(x)\,\mathrm{d}x = F(x) + C, \quad F'(x) = f(x). \tag{11.11}$$

Together with (11.6), we may write

$$\int Df(x)\,\mathrm{d}x = D^{-1}Df(x) = f(x) + C, \quad \frac{\mathrm{d}}{\mathrm{d}x}\int_a^x f(t)\,\mathrm{d}t = f(x). \quad (11.12)$$

For computation of definite integrals in calculus, recourse is made to the following formula and one gets a feeling that one can always find a primitive function. But as we have seen, most primitive functions cannot be expressed in terms of elementary functions.

Corollary 11.3. (Computation of definite integrals) *Suppose $f : I \to \mathbb{R}$ is continuous on the open interval I and that for $\forall\,[a,b] \subset I$, F is a* **primitive function** *of f (on I) (i.e. a function such that $F'(x) = f(x)$). Then we may compute the value as*

$$\int_a^b f(x)\,\mathrm{d}x = F(b) - F(a). \quad (11.13)$$

We often denote the right-hand side by $[F(x)]_a^b$ or $F(x)|_a^b$.

Proof. Proof follows from (11.10). □

Theorem 11.2. (Formula for change of variable) *Let J be an interval with end-points α, β and suppose that φ has a continuous derivative. Let $\varphi(\alpha) = a, \varphi(\beta) = b$ and f is continuous on $\varphi(J)$. Then*

$$\int_a^b f(x)\,\mathrm{d}x = \int_\alpha^\beta f(\varphi(t))\varphi'(t)\,\mathrm{d}t. \quad (11.14)$$

Proof. More generally we prove the following. Let

$$F(u) = \int_a^u f(x)\,\mathrm{d}x, \quad (11.15)$$

for $u \in \varphi(J)$. Then $G(t) = \int_\alpha^t f(\varphi(s))\varphi'(s)\,\mathrm{d}s$ exists and we prove that it is equal to $F(\varphi(t))$.

By Corollary 11.2,

$$G'(t) = f(\varphi(t))\varphi'(t).$$

By the same corollary and the chain rule, Corollary 6.2,

$$(F(\varphi(t)))' = f(\varphi(t))\varphi'(t).$$

It follows that $(F(\varphi(t)))' - G'(t) = 0$ on the interval with end points α, β. By the Newton-Leibniz rule, Corollary 7.2, we conclude that $F(\varphi(t)) - G(t) = c$, c being a constant. Putting $t = \alpha$, we have $G(\alpha) = 0$ and $F(\varphi(\alpha)) = F(a) = 0$, so that $c = 0$, whence $G(t) = F(\varphi(t))$. In particular $G(\beta) = F(\varphi(\beta))$ which is the assertion. □

Corollary 11.4. *If $\varphi'(t) \neq 0$ on J, then*

$$\int_J f(x)\,\mathrm{d}x = \int_{\varphi^{-1}(J)} f(\varphi(t))|\varphi'(t)|\,\mathrm{d}t. \tag{11.16}$$

Exercise 84. For $m, n \in \mathbb{N}$ prove the following formulas, where δ_{mn} is defined by (1.84).

(i) $\dfrac{1}{\pi}\displaystyle\int_{-\pi}^{\pi} \cos mt \cos nt\,\mathrm{d}t = \delta_{mn}$

(ii) $\dfrac{1}{\pi}\displaystyle\int_{-\pi}^{\pi} \sin mt \sin nt\,\mathrm{d}t = \delta_{mn}$

(iii) $\displaystyle\int_{-\pi}^{\pi} \cos mt \sin nt\,\mathrm{d}t = 0.$

11.2 Taylor expansions again and mean value theorems

Theorem 11.3. (nth Taylor expansion) *Suppose in a neighborhood of x_0 $C^n(V(x_0))$ f is n times continuously differentiable. Then*

$$f(x) = \sum_{k=0}^{n-1} \frac{f^{(k)}(x_0)}{k!}(x - x_0)^k + R_n(x),$$

where the remainder term $R_n = R_n(x)$ is given by

$$R_n = R_n(x) = \frac{1}{(n-1)!}\int_{x_0}^{x} (x-t)^{n-1} f^{(n)}(t)\,\mathrm{d}t. \tag{11.17}$$

Proof. By integration by parts,

$$\begin{aligned} &R_n(x) &(11.18)\\ &= \frac{1}{(n-1)!}\left[(x-t)^{n-1} f^{(n-1)}(t)\right]_{x_0}^{x} + \frac{1}{(n-2)!}\int_{x_0}^{x} (x-t)^{n-2} f^{(n-1)}(t)\,dt \\ &= -\frac{f^{(n-1)}(x_0)}{(n-1)!}(x-x_0)^{n-1} + R_{n-1}(x). \end{aligned}$$

Hence by the telescoping series technique (2.8),

$$R_n(x) = -\sum_{k=1}^{n-1} \frac{f^{(k)}(x_0)}{k!}(x-x_0)^k + R_1(x).$$

Substituting

$$R_1(x) = \int_{x_0}^{x} f'(t)\,\mathrm{d}t = f(x) - f(x_0)$$

and moving the terms to the other side, the conclusion follows. □

Corollary 11.5. *If* $f(x) \in C^{\infty}$ *and*

$$\exists M > 0 \ s.t.\ f^{(n)}(x) = O(M^n)$$

holds in an interval I*, then* $f(x)$ *is* **real analytic** *in that interval (can be expanded into a power series):*

$$f(x) = \sum_{n=0}^{\infty} \frac{f^{(n)}(x_0)}{n!}(x - x_0)^n,$$

which is called the **Taylor expansion** *around* x_0*. In particular, the Taylor expansion around* $x_0 = 0$ *is called the* **Maclaurin expansion**.

Proof. For any two points x_0, x on I, the remainder term of the nth Taylor expansion is

$$\begin{aligned} |R_n| &= \left| \frac{1}{(n-1)!} \int_{x_0}^{x} (x-t)^{n-1} f^{(n)}(t)\, \mathrm{d}t \right| \qquad (11.19) \\ &\leq \frac{1}{(n-1)!} \int_{x_0}^{x} \left| (x-t)^{n-1} f^{(n)}(t) \right| \mathrm{d}t \\ &= O\left(\frac{M^n}{(n-1)!} \int_{x_0}^{x} |x-t|^{n-1}\, \mathrm{d}t \right) = O\left(\frac{M^n}{n!} |x - x_0|^n \right). \end{aligned}$$

Hence, noting Exercise 65 and letting $n \to \infty$, we conclude the assertion. □

Exercise 85. Prove that if $f^{(n)}$ is monotone, then the remainder term (11.17) amounts to the Lagrange remainder (7.22).

Exercise 86. Use the first mean value theorem for integrals, Theorem 11.4, to prove that the remainder term in the nth Taylor expansion in Theorem 11.3 may be written in the form of (7.32). Review the expansions (7.33) and (7.34) for $e^x, \sin x, \cos x$, respectively.

Theorem 11.4. (The first mean value theorem for integrals) *Suppose* $f(x), g(x) \in C$ *and that* $g(x)$ *is of constant sign (always positive or negative) on the interval* $I = [a, b]$*. Then*

$$\exists \xi \in I \ s.t.\ \int_a^b f(x) g(x)\, \mathrm{d}x = f(\xi) \int_a^b g(x)\, \mathrm{d}x. \qquad (11.20)$$

Proof. We may suppose that $g(x) > 0$. Let M, m be the maximum and minimum of $f(x)$ on I which exist by Corollary 4.2. Hence we have $mg(x) \leq f(x)g(x) \leq Mg(x)$, $x \in I$, so that by monotonicity

$$m \int_a^b g(x)\, \mathrm{d}x \leq \int_a^b f(x) g(x)\, \mathrm{d}x \leq M \int_a^b g(x)\, \mathrm{d}x.$$

Hence putting

$$\lambda = \frac{\int_a^b f(x)\, g(x)\ \mathrm{d}x}{\int_a^b g(x)\ \mathrm{d}x}, \tag{11.21}$$

we have

$$\int_a^b f(x)\, g(x)\ \mathrm{d}x = \lambda \int_a^b g(x)\ \mathrm{d}x. \tag{11.22}$$

Since $\lambda \in [m, M] = f(I)$, it follows from Theorem 4.1 that

$$\exists \xi \in I \text{ s.t. } f(\xi) = \lambda,$$

i.e. (11.20). □

Corollary 11.6. *A special case of Theorem 11.4 reads*

$$\int_a^b f(x)\,\mathrm{d}x = f(\xi)(b-a). \tag{11.23}$$

Exercise 87. Prove that (11.23) implies (11.6) and that (11.6) and Theorem 7.3 imply (11.23).

Solution. Dividing (11.23) with b replaced by x by $x-a$ and taking the limit as $x \to a$, we deduce that

$$\lim_{x\to a} \frac{1}{x-a}\int_a^x f(t)\,\mathrm{d}t = \lim_{x\to a} f(\xi) = f(x)$$

since f is continuous at $x \in [a, b]$, whence (11.6).

Conversely, (11.6) implies (11.13). By Theorem 7.3, $F(b) - F(a) = F'(\xi)(b-a)$, which is $f(\xi)(b-a)$, by (11.6) and (11.23) follows.

Theorem 11.5. (Second mean value theorem for integrals) *If $f, g(x) \in L^1$ and f is monotone on $I = [a, b]$, then*

$$\exists \xi \in I \ s.t. \int_a^b f(x)\, g(x)\,\mathrm{d}x = f(a)\int_a^\xi g(x)\,\mathrm{d}x + f(b)\int_\xi^b g(x)\,\mathrm{d}x.$$

Corollary 11.7. (Bonnet's form) *If $f(x)$ is monotone increasing function,*

$$g(x) \in C,\ f(x) \ge 0 \ on\, I = [a, b]$$

then there exists a $\xi \in I$ such that

$$\int_a^b f(x)g(x)\,\mathrm{d}x = f(b)\int_\xi^b g(x)\,\mathrm{d}x. \tag{11.24}$$

If f is decreasing, then (11.24) *should read*

$$\int_a^b f(x)g(x)\,\mathrm{d}x = f(a)\int_a^\xi g(x)\,\mathrm{d}x. \tag{11.25}$$

Proof. For a division

$$\Delta : a = x_0 < x_1 < \ldots < x_n = b, \tag{11.26}$$

for every $\varepsilon > 0$ there exists a $\delta > 0$ such that $|\Delta| = \max\{x_i - x_{i-1}\} < \delta$ implies that the Riemann sums approximate the integrals:

$$\left|\sum_{i=1}^{m} g(x_{i-1})(x_i - x_{i-1}) - \int_a^{\xi_1} g(x)\,\mathrm{d}x\right| < \varepsilon, \tag{11.27}$$

$$\left|S_\Delta - \int_a^b f(x)g(x)\,\mathrm{d}x\right| < \varepsilon,$$

where $\xi_1 = x_m$ and

$$S_\Delta = \sum_{i=1}^{n} f(x_{i-1})g(x_{i-1})(x_i - x_{i-1}).$$

We treat the case of f being positive and decreasing. By partial summation (10.21) with $b_i = g(x_{i-1})(x_i - x_{i-1})$ and $B_m = \sum_{i=1}^m b_i$, we have

$$S_\Delta = \sum_{i=1}^{n-1} B_i(f(x_i) - f(x_{i+1})) + B_n f(x_n).$$

Replacing B_{i-1} by $\min B_m$ and $\max B_m$ $(1 \leq m \leq n)$ implies

$$f(a)\min B_m < S_\Delta < f(a)\max B_m.$$

By the first inequality in (11.27), the Riemann sums B_m can be approximated by the integral up to $\pm\varepsilon$. The integral $\int_a^{\xi_1} g(x)\,\mathrm{d}x$, being continuous, takes max U and min L, say. Hence

$$f(a)(L - \varepsilon) < S_\Delta < f(a)(U + \varepsilon).$$

By the second inequality in (11.27), S_Δ can approximate the integral up to $\pm\varepsilon$. Hence it follows that

$$f(a)(L - 2\varepsilon) < \int_a^b f(x)g(x)\,\mathrm{d}x < f(a)(U + 2\varepsilon), \tag{11.28}$$

whence

$$f(a)L \leq \int_a^b f(x)g(x)\,\mathrm{d}x \leq f(a)U. \tag{11.29}$$

The integral $f(a)\int_a^{\xi_1} g(x)\,\mathrm{d}x$, being continuous, can take any intermediate value between $[f(a)L, f(a)U]$, whence for some ξ, (11.25) holds true. □

Theorem 11.5 is sometimes referred to as Du Bois Raymond's form and Corollary 11.7 as Bonnet's form of the second mean value theorem for integrals. The above proof is much more involved than the corresponding Theorem 10.5 for Stieltjes integrals in view of the fact that the integration by parts in used in a twisted form as partial summation.

Exercise 88. Deduce Du Bois Raymond's form from Bonnet's form by taking $|f(x) - f(b)|$ for $f(x)$.

11.3 Elementary functions

So far we have used elementary functions freely without mentioning their definition. Here we mean by elementary function, the polynomial functions, exponential functions, trigonometric functions and their inverse functions. In this subsection we introduce some of them according to Klein's method of using integrals.

(i) Polynomial functions are the same as polynomials introduced in §1.3 because $\operatorname{char}\mathbb{R} = 0$. They can be differentiated and integrated freely with the rule $(x^n)' = nx^{n-1}$ and applying linearity. If degree is n, then after $n+1$ times differentiation, it becomes 0.

(ii) The **logarithm** function $\log x$ or $\ln x$ is defined by the definite integral for $x > 0$

$$\log x = \int_1^x \frac{1}{t}\,\mathrm{d}t. \tag{11.30}$$

It has the properties $(x, y > 0)$

(a) $\log 1 = 0$

(b)

$$(\log x)' = \frac{1}{x} > 0, \tag{11.31}$$

whence $\log x$ is a monotone increasing function.

(c)

$$\log xy = \log x + \log y. \tag{11.32}$$

Proof.

$$\frac{\mathrm{d}}{\mathrm{d}x}(\log xy) = \frac{y}{xy} = \frac{1}{x}. \tag{11.33}$$

Hence $\log xy = \log x + C$. Putting $x = 1$ gives $C = \log y$, whence the result. □

For $n \in \mathbb{N}$ we have

$$\log \prod_{k=1}^{n} a_k = \sum_{k=1}^{n} \log a_k, \quad \log a^n = n \log a.$$

(d)

$$\left(\log \frac{1}{x}\right) = -\log x.$$

For by (11.32), $0 = \log 1 = \log x\frac{1}{x} = \log x + \log\frac{1}{x}$.
(iii) The exponential function $e^x = \exp(x)$ is defined as the inverse function of the logarithm function

$$e^x = \log^{-1} x; \quad e^x = y \Longleftrightarrow x = \log y \quad (y > 0) \tag{11.34}$$

and it is a map from $\mathbb{R}$ to $(0, \infty)$. This exists and is differentiable on $\mathbb{R}$ in view of (11.31).
(a) $e^0 = 1$
(b)

$$\frac{\mathrm{d}}{\mathrm{d}x}(e^x) = e^x. \tag{11.35}$$

By the inverse function theorem

$$\frac{\mathrm{d}}{\mathrm{d}x}(e^x) = \frac{1}{\frac{\mathrm{d}}{\mathrm{d}x}\log y}\bigg|_{y=\exp(x)} = y\big|_{y=\exp(x)} = e^x.$$

(c) The exponential law holds true:

$$e^{x+y} = e^x e^y. \tag{11.36}$$

Proof. Let $\xi = e^x, \eta = e^y$. Then by (11.32),

$$\exp(x+y) = \exp(\log\xi + \log\eta) = \exp(\log\xi\eta) = \xi\eta = e^x e^y. \tag{11.37}$$

□

For $n \in \mathbb{N}$ we have

$$\exp\left(\sum_{k=1}^{n} \exp(a_k)\right) = \prod_{k=1}^{n} \exp(a_k), \quad \exp(a^n) = n\exp(a).$$

(d)

$$\frac{1}{e^x} = e^{-x}.$$

For by (11.36), $1 = e^0 = e^{x-x} = e^x e^{-x}$.
(e) By Remark 12.1 and Exercise 111, we contend that

$$e = \lim_{n\to\infty}\left(1 + \frac{1}{n}\right)^n = e^1 = \sum_{k=0}^{\infty}\frac{1}{k!}. \tag{11.38}$$

(iv) The **power function** x^u defined for $x > 0$ and $u \in \mathbb{R}$ and the definition is inherited in complex functions, where the logarithm is the complex logarithm and a branch is to be chosen.

$$x^u = e^{u\log x}. \tag{11.39}$$

Properties of this function in x or in u can be easily read off and we omit them. We only remark that

$$\frac{\mathrm{d}}{\mathrm{d}x}(x^u) = \frac{u}{x}x^u, \quad \frac{\mathrm{d}}{\mathrm{d}u}(x^u) = x^u \log x. \tag{11.40}$$

The first equality entails $\frac{\mathrm{d}}{\mathrm{d}x}(x^\alpha) = \alpha x^{\alpha-1}$ and the second $\frac{\mathrm{d}}{\mathrm{d}x}(a^x) = a^x \log a$.
(v) The trigonometric functions $e^x = \exp(x)$ are introduced as the definite integrals as in (i).
For $|x| < 1$

$$\arcsin x = \int_0^x \frac{1}{\sqrt{1-t^2}}\,\mathrm{d}t. \tag{11.41}$$

It has the properties $(-1 < x < 1)$
(a) $\arcsin 0 = 0$
(b)

$$(\arcsin x)' = \frac{1}{\sqrt{1-x^2}} > 0, \tag{11.42}$$

whence $\arcsin x$ is a monotone increasing function.

The sine function $\sin x$ is defined as the inverse function of the inverse sine function

$$\sin x = \arcsin^{-1} x; \quad \sin x = y \Longleftrightarrow x = \arcsin y \quad (|y| < 1) \tag{11.43}$$

for $-\frac{\pi}{2} \le x \le \frac{\pi}{2}$ in the first place. However, it is more convenient to introduce it through the cosine function in view of

$$\arcsin x = \frac{\pi}{2} - \arccos x. \tag{11.44}$$

We introduce $\arccos x$ in the same way as $\arcsin x$ is introduced above:

$$\arccos x = \frac{\pi}{2} - \int_0^x \frac{1}{\sqrt{1-t^2}}\,\mathrm{d}t. \tag{11.45}$$

(a) $\arccos 0 = \frac{\pi}{2}$
(b)

$$\frac{\mathrm{d}}{\mathrm{d}x}(\arccos x) = -\frac{1}{\sqrt{1-x^2}}. \tag{11.46}$$

It is a monotone decreasing function from $[-1,1]$ to $[0,\pi]$. Hence the inverse function $\cos x$ is defined on $[0,\pi]$ by

$$\cos x = \arccos^{-1} x; \quad \cos x = y \Longleftrightarrow x = \arccos y \quad (|y| < 1). \tag{11.47}$$

We extend this to $[-\pi, \pi]$ by the evenness condition $\cos(-x) = \cos x$. Then extend it to $\mathbb{R}$ by periodicity: $\cos(x + 2n\pi) = \cos x$. Thus we obtain a periodic function differentiable on $\mathbb{R}$. And it is a map from $\mathbb{R}$ to $[-1, 1]$.
(a) $\cos 0 = 1$
(b)

$$\frac{\mathrm{d}}{\mathrm{d}x}(\cos x) = -\sin x. \tag{11.48}$$

In view of (11.44), we define the sine function on $\mathbb{R}$ by

$$\sin x = \cos\left(\frac{\pi}{2} - x\right) \tag{11.49}$$

and we have $\frac{\mathrm{d}}{\mathrm{d}x}(\sin x) = \cos x$ on $\mathbb{R}$. Other trigonometric function are defined as usual, in particular $\tan x = \frac{\sin x}{\cos x}$ on $\left(-\frac{\pi}{2}, \frac{\pi}{2}\right)$ and then periodically.

Their Taylor expansions will be introduced in §11.4.

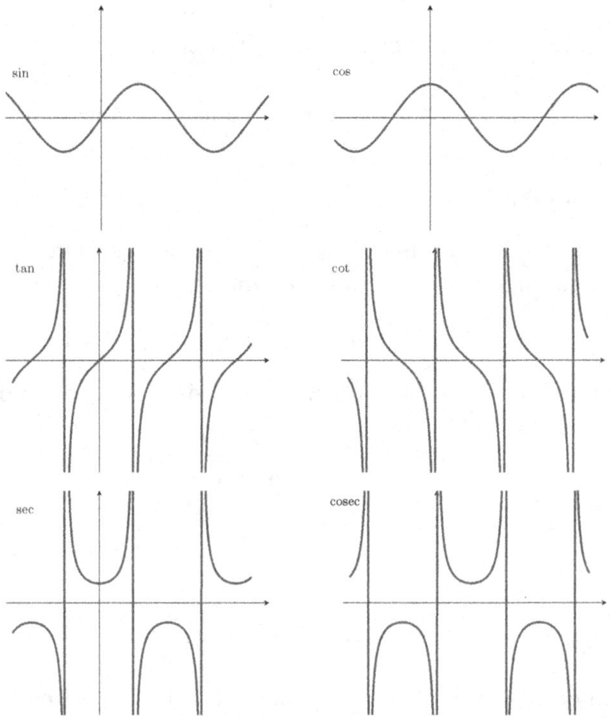

Fig. 11.2: Trigonometric functions

11.4 Binomial expansion and consequences

In this section we shall restrict to the Taylor expansion of the power function $(1+x)^\alpha$ and introduce elementary functions as its consequences. For $|x| < 1$ let

$$f(x) = (1+x)^\alpha = e^{\alpha \log(1+x)}. \tag{11.50}$$

Then

$$f^{(k)}(x) = k!\binom{\alpha}{k}(1+x)^{\alpha-k}, \tag{11.51}$$

where $\binom{\alpha}{k}$ is the generalized binomial coefficient introduced in (1.3). Hence if $\alpha = n \in \mathbb{N}$, then $f^{(k)}(x) = 0$ for $k > n = \alpha$. Applying Theorem 7.5 with the Lagrange remainder term

$$R_n(x) = \binom{\alpha}{n}(1+\theta x)^{\alpha-n}x^n \tag{11.52}$$

and noting that

$$\lim_{n\to\infty}\binom{\alpha}{n}|x|^n = 0, \tag{11.53}$$

we have

Theorem 11.6. *For $|x| < 1$ we have the general binomial expansion*

$$(1+x)^\alpha = \sum_{n=0}^{\infty}\binom{\alpha}{n}x^n, \tag{11.54}$$

which reduces to the binomial theorem (1.1) *in case $\alpha \in \mathbb{N}$.*

Corollary 11.8. (i) *For $|x| < 1$*

$$-\log(1-x) = \sum_{n=1}^{\infty}\frac{1}{n}x^n, \quad \log(1+x) = \sum_{n=1}^{\infty}\frac{(-1)^{n-1}}{n}x^n. \tag{11.55}$$

(ii) *For $|x| \le 1$*

$$\arctan x = \sum_{n=0}^{\infty}\frac{(-1)^n}{2n+1}x^{2n+1}. \tag{11.56}$$

For $|x| < 1$

$$\arcsin x = \sum_{n=0}^{\infty}\frac{(2n-1)!!}{(2n)!!(2n+1)}x^{2n+1} = \sum_{n=0}^{\infty}\frac{(2n)!}{2^{2n}(n!)^2(2n+1)}x^{2n+1}, \tag{11.57}$$

where $(2n)!! = (2n)(2n-2)\cdots 2$, $(2n-1)!! = (2n-1)(2n-3)\cdots 1$ *and* $0!! = (-1)!! = 1$.

Proof. Proof of (11.55) and (11.56) depends on the geometric series, which is a special case of (11.54).

Integrating the power series (1.27) over $[0, x]$ term by term we deduce the first equality in (11.55), whereby we perform the termwise integration on the right-side, which is legitimate in view of absolute convergence. Or integrating the nth Taylor expansion with the remainder $R_n = O((1-x)^{-1}x^n)$ and then letting $n \to \infty$ leads to the expansion (here by a direct application, we have the estimate $R_n = O((1-x)^{-1-n}x^n)$ but we may use the closed formula (1.22)).

(11.56) for $|x| < 1$ follows on integrating the geometric series

$$\frac{1}{1+x^2} = \sum_{n=0}^{\infty} (-1)^n x^{2n}. \tag{11.58}$$

Or we may appeal to the binomial expansion

$$\frac{1}{1+x^2} = (1+x^2)^{-1} = \sum_{n=0}^{\infty} \binom{-1}{n} x^{2n}, \tag{11.59}$$

which amounts to (11.58) in view of $\binom{-1}{n} = (-1)^n$. The cases $x = \pm 1$ will be treated separately below.

For the proof of (11.57) we may apply the binomial expansion

$$\frac{1}{\sqrt{1-x^2}} = (1-x^2)^{-\frac{1}{2}} = \sum_{n=0}^{\infty} \binom{-\frac{1}{2}}{n} (-1)^n x^{2n} = \sum_{n=0}^{\infty} \frac{(2n-1)!!}{(2n)!!} x^{2n}. \tag{11.60}$$

□

Exercise 89. Check (11.60):

$$\frac{1}{\sqrt{1-x}} = \sum_{n=0}^{\infty} \frac{(2n-1)!!}{(2n)!!} x^n \tag{11.61}$$

and deduce

$$\sqrt{1-x} = 1 - \sum_{n=1}^{\infty} \frac{(2n-3)!!}{(2n)!!} x^n. \tag{11.62}$$

Theorem 11.7. (Polynomial approximation of elementary functions)
(i)

$$e^x = \sum_{k=0}^{n-1} \frac{x^k}{k!} + O\left(e^{|x|} \frac{|x|^n}{n!}\right). \tag{11.63}$$

(ii)

$$\sin x = \sum_{k=0}^{n-1} \frac{(-1)^k}{(2k+1)!} x^{2k+1} + O\left(\frac{|x|^{2n}}{(2n)!}\right). \tag{11.64}$$

(iii)

$$\cos x = \sum_{k=0}^{n} \frac{(-1)^k}{(2k)!} x^{2k} + O\left(\frac{|x|^{2n+1}}{(2n+1)!}\right). \tag{11.65}$$

(iv) *For* $|x| < 1, \alpha \in \mathbb{R}$

$$(1+x)^\alpha = \sum_{k=0}^{n-1} \binom{\alpha}{k} x^k + R_n(x), \tag{11.66}$$

where

$$R_n(x) = \begin{cases} O(|x^n|) & \alpha \geq 1 \\ O((1-|x|)^{\alpha-1}|x|^n) & \alpha < 1 \end{cases}. \tag{11.67}$$

(iv) *For* $0 < x < 1$

$$\log(1+x) = \sum_{k=1}^{n-1} \frac{(-1)^{k-1}}{k} x^k + O(x^n), \quad -\log(1-x) = \sum_{n=1}^{\infty} \frac{1}{k} x^k + O\left(\frac{x^n}{1-x}\right). \tag{11.68}$$

(v) *For* $|x| < 1$

$$\arcsin x = \sum_{k=0}^{n-1} \frac{(2k-1)!!}{(2k)!!(2k+1)} x^{2k+1} + O\left(\frac{|x|^{2n+1}}{1-|x|}\right). \tag{11.69}$$

(vi) *For* $|x| \leq 1$

$$\arctan x = \sum_{k=1}^{n-1} \frac{(-1)^k}{2k+1} x^{2k+1} + O(|x|^{2n}). \tag{11.70}$$

Cf. Example 5.1 above.

11.5 Formulas for integrals

Exercise 90. Prove the formulas by the Taylor expansion

$$\int_a^b (x-a)(x-b)\,\mathrm{d}x = -\frac{1}{6}(b-a)^3, \quad \int_a^b (x-a)^2(x-b)\,\mathrm{d}x = -\frac{1}{12}(b-a)^4. \tag{11.71}$$

Exercise 91. Allowing the use of complex functions (assuming $x^2 - A \neq 0$, $A \neq 0$) prove the formulas

$$\int \frac{1}{x^2 - A}\,\mathrm{d}x = \frac{1}{\sqrt{-A}} \arctan\left(\frac{x}{\sqrt{-A}}\right) + C = \frac{1}{\sqrt{A}} \operatorname{arctanh}\left(\frac{x}{\sqrt{A}}\right) + C. \tag{11.72}$$

Note that (11.72) entails ($a > 0$)

$$\int \frac{1}{x^2 + a^2}\,\mathrm{d}x = \frac{1}{a} \arctan\left(\frac{x}{a}\right) + C. \tag{11.73}$$

Exercise 92. Prove the formula

$$\int \sqrt{1 + Ax^2}\,\mathrm{d}x = \frac{1}{2}\left(x\sqrt{1+Ax^2} + I\right), \tag{11.74}$$

where $I = \int \frac{1}{\sqrt{1+Ax^2}}\,\mathrm{d}x$. Allowing the use of complex functions (assuming $1 + Ax^2 > 0$) prove the formulas

$$\int \frac{1}{\sqrt{1 + Ax^2}}\,\mathrm{d}x = \frac{1}{\sqrt{-A}} \arcsin\left(\sqrt{-A}x\right) + C = -\frac{1}{\sqrt{A}} \operatorname{arcsinh}\left(\sqrt{A}x\right) + C. \tag{11.75}$$

Note that (11.75) entails (assuming $a > 0$, $1 + Ax^2 > 0$, $A \neq 0$)

$$\int \frac{1}{\sqrt{a^2 - x^2}}\,\mathrm{d}x = \arcsin\frac{x}{a} + C, \quad \int \frac{1}{\sqrt{A + x^2}}\,\mathrm{d}x = \log|x + \sqrt{A + x^2}| + C. \tag{11.76}$$

Exercise 93. Prove the following.

$$\int \frac{x^2}{\sqrt{A + x^2}}\,\mathrm{d}x = \frac{1}{2}\left(x\sqrt{A+x^2} - A\log|x + \sqrt{A + x^2}|\right) + C.$$

Exercise 94. Deduce the formula

$$\begin{aligned} I &= \int \csc x\,\mathrm{d}x = \int \frac{1}{\sin x}\,\mathrm{d}x \\ &= \log\left|\tan\frac{x}{2}\right| + C = -\log|\csc x + \cot x| + C. \end{aligned} \tag{11.77}$$

Solution. It amounts to the well-known transformation

$$\frac{1}{\sin t}=\frac{1}{2\sin\frac{t}{2}\cos\frac{t}{2}}=\frac{1}{2\tan\frac{t}{2}\cos^2\frac{t}{2}}. \tag{11.78}$$

Exercise 95. Deduce the formulas

$$I_1=\int\frac{1}{\sqrt{1-x^2}}\arcsin x\,\mathrm{d}x=\frac{1}{2}\arcsin^2 x+C, \tag{11.79}$$

$$I_2=\int\arcsin x\,\mathrm{d}x=x\arcsin x+\sqrt{1-x^2}+C, \tag{11.80}$$

$$I_3=\int\sqrt{1-x^2}\arcsin x\,\mathrm{d}x=\frac{1}{2}\left(x\sqrt{1-x^2}\arcsin x-\frac{1}{2}x^2+\frac{1}{2}\arcsin^2 x\right)+C, \tag{11.81}$$

$$I_4=\int\arcsin^2 x\,\mathrm{d}x=x\arcsin^2 x+2\sqrt{1-x^2}\arcsin x-2x+C, \tag{11.82}$$

$$\begin{aligned}I_5&=\int x\arcsin^2 x\,\mathrm{d}x \qquad (11.83)\\&=\frac{1}{4}\left(2x^2\arcsin^2 x-\arcsin^2 x+2x\sqrt{1-x^2}\arcsin x-x^2\right)+C.\end{aligned}$$

Solution. (11.79) follows from

$$\int f'f\,\mathrm{d}x=\frac{1}{2}f^2+C. \tag{11.84}$$

(11.80) follows by integration by parts.

To prove (11.81), we apply integration by parts using (11.80) to deduce that

$$I_3=\sqrt{1-x^2}\left(x\arcsin x+\sqrt{1-x^2}\right)-I_3+I_1+\frac{1}{2}x^2. \tag{11.85}$$

By (11.82), we have

$$I_5=\frac{1}{2}x\left(x\arcsin^2 x+2\sqrt{1-x^2}\arcsin x-2x\right)+\frac{1}{2}x^2-I_3.$$

Substituting (11.81), we deduce (11.83).

Exercise 96. Let for $m,n\in\mathbb{Z}$

$$I_{m,n}=\int\sin^m x\cos^n x\,\mathrm{d}x. \tag{11.86}$$

For $m,n\in\mathbb{N}\cup\{0\}, m+n\neq 0$ prove that

$$\begin{aligned}I_{m,n}&=\frac{1}{m+n}\sin^{m+1}x\cos^{n-1}x+\frac{n-1}{m+n}I_{m,n-2},\\I_{m,n}&=\frac{1}{m+n}(-\sin^{m-1}x\cos^{n+1}x)+\frac{m-1}{m+n}I_{m-2,n},\end{aligned} \tag{11.87}$$

and that for $m\geq 2$,

$$I_{m,-m}=\frac{1}{m-1}\tan^{m-1}x-I_{m-2,-(m-2)}.$$

Exercise 97. Let for $n \in \mathbb{Z}$

$$S_n = I_{n,0} = \int_0^{\frac{\pi}{2}} \sin^n x \, dx. \tag{11.88}$$

For $m, n \in \mathbb{N} \cup \{0\}$ check that

$$S_n = \begin{cases} \frac{(n-1)!!}{(n)!!} \frac{\pi}{2} & 2 \mid n \\ \frac{(n-1)!!}{(n)!!} & 2 \nmid n \end{cases}, \tag{11.89}$$

and prove that

$$\lim_{n\to\infty} \frac{S_{2n}}{S_{2n+1}} = 1.$$

Substituting S_n in this, prove Wallis' formula

$$\frac{2}{\pi} = \prod_{n=1}^{\infty} \left(1 - \frac{1}{4n^2}\right). \tag{11.90}$$

Prove that

$$\lim_{n\to\infty} \sqrt{n} S_{2n+1} = \frac{\sqrt{\pi}}{2}, \quad \lim_{n\to\infty} \frac{\sqrt{n}}{2^{2n}} \binom{2n}{n} = \frac{1}{\sqrt{\pi}}. \tag{11.91}$$

Exercise 98. By the same technique by which one deduces (11.77), deduce the series representation for the integral

$$2G := \int_0^{\frac{\pi}{2}} \frac{x}{\sin x} \, dx = \sum_{n=0}^{\infty} \frac{(-1)^n}{(2n+1)^2} = L(2, \chi_4), \tag{11.92}$$

where the far-right side member is the value of the Dirichlet L-function at $s = 2$ and is referred to as the **Catalan constant**.

Solution. By (11.78), we see that

$$2G = \frac{1}{2} \int_0^{\frac{\pi}{2}} \frac{t}{\tan\frac{t}{2} \cos^2\frac{t}{2}} \, dt, \tag{11.93}$$

whence by the change of variable $u = \tan\frac{t}{2}$, we find that

$$2G = \int_0^1 \frac{\arctan u}{u} du. \tag{11.94}$$

By (11.56) and termwise integration, (11.92) follows.

Exercise 99. Prove the formula

$$G_1 := \int_0^{\pi} \frac{\sin u}{u} du = \pi - \frac{\pi^3}{3!3} + \frac{\pi^5}{5!5} - \cdots, \tag{11.95}$$

where G_1 is often referred to as the **Gibbs constant**.

Solution. By the Maclaurin expansion (7.34) and termwise integration, we see that

$$G_1 = \sum_{n=1}^{\infty} \frac{(-1)^{n-1}\pi^{2n-1}}{(2n-1)!(2n-1)}, \tag{11.96}$$

which is (11.95).

Exercise 100. Deduce another expression for G:

$$2G = \frac{1}{2}\sum_{k=0}^{\infty} \frac{(2k)!}{2^{2k}(k!)^2(2k+1)} \frac{(k-1/2)\cdots\frac{1}{2}\cdot\pi}{k!} = \sum_{k=0}^{\infty} \frac{\{(2k)!\}^2\pi}{2^{4k}(k!)^4(2k+1)}. \tag{11.97}$$

Solution. Making the change of variable $u = \sin t$ in (11.92), we obtain

$$2G = \int_0^1 \frac{\arcsin u}{u\sqrt{1-u^2}}\, du. \tag{11.98}$$

Applying (11.57), we have

$$2G = \int_0^1 \sum_{k=0}^{\infty} \frac{(2k)!}{2^{2k}(k!)^2(2k+1)} u^{2k} \frac{1}{\sqrt{1-u^2}}\, du. \tag{11.99}$$

Making the change of variable $u^2 = v$ in (11.99) and integrating term by term, we have

$$2G = \frac{1}{2}\sum_{k=0}^{\infty} \frac{(2k)!}{2^{2k}(k!)^2(2k+1)} \int_0^1 v^{k-1/2}(1-v)^{-1/2} dv. \tag{11.100}$$

Since the inner integral is the beta-function

$$B(k+1/2, 1/2) = \frac{\Gamma(k+1/2)\Gamma(1/2)}{\Gamma(k+1)} = \frac{(k-1/2)\cdots\frac{1}{2}\cdot\pi}{k!},$$

where we used the value of the probability integral $\Gamma(1/2) = \sqrt{\pi}$, we deduce the formula.

Exercise 101. Prove the identity

$$\int_0^1 \frac{\sin 2\pi nt}{\cos \pi t} dt = \frac{4(-1)^{n+1}}{\pi} \sum_{k=0}^{n-1} \frac{(-1)^k}{2k+1} = \frac{4(-1)^n}{\pi} \sum_{k=1}^{\infty} \frac{(-1)^k}{2k-1}. \tag{11.101}$$

Solution. By Euler's identity and the factorization formula, we immediately see that

$$\frac{\sin 2\pi nt}{\cos \pi t} = -i\left(e^{\pi it} - e^{-\pi it}\right) \sum_{r=0}^{n-1} e^{2\pi i(n-1-2r)t} \tag{11.102}$$

$$= -i\left(\sum_{r=0}^{n-1} e^{\pi i(2n-1-4r)t} - \sum_{r=0}^{n-1} e^{\pi i(2n-3-4r)t}\right). \tag{11.103}$$

Hence, integrating (11.102) and transforming, we arrive at

$$\int_0^1 \frac{\sin 2\pi nt}{\cos \pi t}\mathrm{d}t = \frac{2}{\pi}\left(\sum_{r=0}^{n-1}\frac{1}{2n-1-4r} - \sum_{r=0}^{n-1}\frac{1}{2n-3-4r}\right). \tag{11.104}$$

Now dividing the sum over $0 \le r \le n-1$ into two $0 \le r \le [\frac{n-1}{2}]$, $[\frac{n-1}{2}] \le r \le n-1$ and writing $2n-1-4r = 2(n-1-2r)+1$ and $2n-3-4r = 2(n-1-2r-1)+1$, we find

$$\int_0^1 \frac{\sin 2\pi nt}{\cos \pi t}\mathrm{d}t = \frac{4}{\pi}\sum_{k=0}^{n-1}\frac{(-1)^k}{2(n-1-k)+1}, \tag{11.105}$$

which amounts to (11.101).

11.6 Improper integrals and integral transforms

For a real-valued function $f(x)$ defined and bounded on the finite interval $[a,b]$, the important quantity (1-dimensional) definite integral $\int_a^b f(x)\,\mathrm{d}x$ is defined in Definition 11.1. In this subsection we are going to generalize this notion of definite integrals to:
Case (i) the interval is infinite, e.g. of the type $[a,\infty)$
and
Case (ii) the function $f(x)$ is unbounded at an end-point of the interval, e.g. $\lim_{\varepsilon\to+0}|f(a+\varepsilon)| = \infty$
and show that in these cases we may also define a quantity corresponding to the area with sign of the figure formed by the graph of the function $y = f(x)$ and the x-axis.

Infinite integrals are analogous to infinite series and $I(b) = \int_0^b a(t)\,\mathrm{d}t$ corresponds to the partial sum $S_n = \sum_{m=0}^n a(m)$, so that $I(b)$ is what we call, the "partial integral". Indeed, from the point of view of Stieltjes integrals, both of them are special cases of Stieltjes integrals (cf. Theorem 10.2 and [Apostol (1957)]. Although b varies continuously $\to\infty$ and n discretely $\to\infty$, those theorems on infinite series have their counterparts in infinite integrals. By stating Case (i) rather in detail, we hope that the reader will be able to perceive the corresponding results on infinite series, and *a fortiori* the limits of a sequence since the limit of a sequence is the same as the limit of the corresponding telescoping series.

Case (i). We define the **infinite integral** or **improper integral of the first kind** $\int_a^\infty f(x)\,\mathrm{d}x$ as the limit $b\to\infty$ of the proper integral

$I(b) = \int_a^b f(x)\,\mathrm{d}x$:

$$\int_a^{\infty} f(x)\,\mathrm{d}x = \lim_{b\to\infty} \int_a^b f(x)\,\mathrm{d}x.$$

Similarly, we define

$$\int_{-\infty}^{b} f(x)\,\mathrm{d}x = \lim_{a\to-\infty} \int_a^b f(x)\,\mathrm{d}x.$$

If for some constant c, both $\int_{-\infty}^{c} f(x)\,\mathrm{d}x$ and $\int_c^{\infty} f(x)\,\mathrm{d}x$ exist, then we denote their sum by $\int_{-\infty}^{\infty} f(x)\,\mathrm{d}x$:

$$\begin{aligned}\int_{-\infty}^{\infty} f(x)\,\mathrm{d}x &= \int_{-\infty}^{c} f(x)\,\mathrm{d}x + \int_c^{\infty} f(x)\,\mathrm{d}x \\ &= \lim_{a\to\infty} \int_{-a}^{c} f(x)\,\mathrm{d}x + \lim_{b\to\infty} \int_c^b f(x)\,\mathrm{d}x,\end{aligned}$$

where it is important that a and b independently $a, b \to \infty$. Especially, if $a = b \to \infty$, i.e. $\lim_{a\to\infty} \int_{-a}^{a} f(x)\,\mathrm{d}x$ is called the **Cauchy principal value** and denoted

$$PV \int_{-\infty}^{\infty} f(x)\,\mathrm{d}x.$$

If $\int_{-\infty}^{\infty} f(x)\,\mathrm{d}x$ exists, then $PV \int_{-\infty}^{\infty} f(x)\,\mathrm{d}x$ necessarily exists. But the converse does not hold as the example $f(x) = x^3$ shows.

Case (ii). We write $\int_{a+0}^{b} f(x)\,\mathrm{d}x$ or simply $\int_a^b f(x)\,\mathrm{d}x$ and call it an **improper integral of the second kind**. This is defined as the limit as $\varepsilon \to +0$ of the integral $\int_{a+\varepsilon}^{b} f(x)\,\mathrm{d}x$:

$$\int_{a+0}^{b} f(x)\,\mathrm{d}x = \lim_{\varepsilon\to+0} \int_{a+\varepsilon}^{b} f(x)\,\mathrm{d}x.$$

This, however, reduces to (i) by the change of variable $x = a + \frac{1}{t}$, $t > 0$. If the function $f(x)$ is unbounded at the right end-point b, then we define

$$\int_a^{b-0} f(x)\,\mathrm{d}x = \lim_{\varepsilon\to+0} \int_a^{b-\varepsilon} f(x)\,\mathrm{d}x,$$

but this also reduces to the case (i) by the change of variable $x = b - \frac{1}{t}$, $t > 0$.

If a primitive function can be easily found, then we may apply Corollary 11.3 and take the limit. E.g.

$$\int_{0+}^{1} \log x\,\mathrm{d}x = \lim_{\varepsilon\to+0} \left[x\log x - x\right]_{\varepsilon}^{1} = -1 \tag{11.106}$$

by (7.7).

Similarly,

$$\int_1^\infty x^{-r}\,\mathrm{d}x = \begin{cases} \lim\limits_{b\to\infty}\left[\frac{x^{1-r}}{1-r}\right]_1^b, & r\neq 1 \\ \lim\limits_{b\to\infty}[\log x]_1^b, & r=1 \end{cases} = \begin{cases} \dfrac{-1}{1-r}, & r>1 \\ \infty, & r\le 1 \end{cases},$$

and

$$\int_{0+}^1 x^{-r}\,\mathrm{d}x = \begin{cases} \dfrac{1}{1-r}, & r<1 \\ \infty, & r\ge 1 \end{cases}.$$

These shift to each other under $x = {}^1\!/_t$.

Instead of series with positive terms, we take infinite integrals of positive-valued functions $f(x)$. Then the **comparison test** reads: for positive-valued functions $f(x)$, $g(x)$, $f(x)\le g(x)$ for $a\le x$,

$$\int_a^\infty g(x)\,\mathrm{d}x < \infty \Rightarrow \int_a^\infty f(x)\,\mathrm{d}x < \infty.$$

Hence by comparing with $\int_1^\infty x^{-r}\,\mathrm{d}x$, which is an analogue of the sum of negative powers of natural numbers $\sum\limits_{n=1}^{\infty} n^{-r}$, we may verify the convergence.

Also if $\lim\limits_{x\to\infty}\frac{f(x)}{g(x)} = 0$ and $\int_a^\infty g(x)\,\mathrm{d}x<\infty$, then we have $\int_a^\infty f(x)\,\mathrm{d}x < \infty$.

For a complex-valued function $f(x)$ we consider its modulus and apply an analogue of the **Weierstrass M-test**:

Theorem 11.8. (Weierstrass M-test) *For the function $f(x,y)$, $x\in[a,\infty)$ of two variables [or on (a,b)] and $y\in Y$, there exists a positive-valued function $M(x)$ such that for $\forall y\in Y$, we have $|f(x,y)|\le M(x)$ and $\int_a^\infty M(x)\,\mathrm{d}x<\infty$ [or $\int_a^b M(x)\,\mathrm{d}x<\infty$], then $\int_a^\infty f(x,y)\,\mathrm{d}x$ [or $\int_a^b f(x,y)\,\mathrm{d}x$] converges absolutely and uniformly on Y and defines an analytic function there.*

Example 11.1. (i) (Euler)

$$\int_0^\infty \frac{\log x}{x^2+1}\,\mathrm{d}x = 0.$$

This is the sum of improper integrals of the first and the second kind. By the change of variable $x={}^1\!/_t$, we have $\int_{0+}^1 \frac{\log x}{x^2+1}\,\mathrm{d}x = -\int_1^\infty \frac{\log t}{t^2+1}\,\mathrm{d}t$. Hence it suffices to show that $\int_1^\infty \frac{\log t}{t^2+1}\,\mathrm{d}t < \infty$. By L'Hospital's rule (Corollary 7.1), we obtain

$$0 \le \lim_{x\to\infty} \frac{\log x}{x^2+1} x^{3/2} \le \lim_{x\to\infty} \frac{\log x}{\sqrt{x}} = \lim_{x\to\infty} \frac{\frac{1}{x}}{\frac{1}{2\sqrt{x}}} = \lim_{x\to\infty} \frac{2}{\sqrt{x}} = 0.$$

Hence by $\int_1^\infty x^{-3/2}\,\mathrm{d}x < \infty$ and the comparison test, the integral is convergent. Hence

$$\int_0^\infty \frac{\log x}{x^2+1}\,\mathrm{d}x = \int_0^1 + \int_1^\infty = -\int_1^\infty \frac{\log x}{x^2+1}\,\mathrm{d}x + \int_1^\infty \frac{\log x}{x^2+1}\,\mathrm{d}x = 0.$$

We note that such a computation is not possible if convergence is not assured.

(ii) The **Mellin transform** of e^{-t}

$$\Gamma(s) = \mathcal{M}[e^{-t}](s) = \int_0^\infty t^s e^{-t} \frac{\mathrm{d}t}{t} \tag{11.107}$$

is absolutely and uniformly convergent for $\sigma = \operatorname{Re} s > 0$ and defines an analytic function, called the **gamma function**. This is also the sum of improper integrals of the first and the second kind. The modulus of the integrand being $t^{\sigma-1}e^{-t}$, we have

$$\Gamma(\sigma) = \int_0^\infty t^{\sigma-1}e^{-t}\,\mathrm{d}t = \int_{0+}^1 t^{\sigma-1}e^{-t}\,\mathrm{d}t + \int_1^\infty t^{\sigma-1}e^{-t}\,\mathrm{d}t.$$

With $Y \subset [c,d]$, we use the Majorants

$$t^{\sigma-1}e^{-t} \le t^{c-1}e^{-t} \quad (0 < t < 1), \quad t^{\sigma-1}e^{-t} \le t^{d-1}e^{-t} \quad (1 \le t)$$

and note that

$$\int_1^\infty t^{d-1}e^{-t}\,\mathrm{d}t = O\left(\int_1^\infty t^{-2}\,\mathrm{d}t\right) = O(1).$$

Along with the Mellin transform we introduce the Laplace transform in anticipation of applications in §13.3.

Definition 11.3. For functions to be Laplace-transformed, we consider only "forgetting the past functions" to the effect that the function $f(t)$ is 0 for $t < 0$, called **causal functions**. Let $H = H(t)$ be the **unit step function** or the Heaviside function $H(t) = 0, t < 0$, $H(t) = 1, t \ge 0$ (cf. Example 15.7). By multiplying this all the functions may be thought of as causal functions. Suppose $y(t) = O(e^{at})$, $t \to \infty$ for an $a \in \mathbb{R}$. The **Laplace transform** $Y(s) = \mathcal{L}[y](s)$ of $y = y(t)$ is defined by

$$Y(s) = \mathcal{L}[y](s) = \int_0^\infty e^{-st}y(t)\,\mathrm{d}t, \quad \operatorname{Re} s > a. \tag{11.108}$$

The integral converges absolutely in $\operatorname{Re} s > a$ and by the Weierstrass M-test, Theorem 11.8, it represents an analytic function there. Thus the domain of definition of a Laplace transform is a half-plane in the first instance, which is sometimes expressed as $\mathcal{RHP}$. The Laplace transform is a (convenient form of the) Fourier transform.

Example 11.2. Let $\alpha \in \mathbb{C}$. Then

$$\mathcal{L}[e^{\alpha t}](s) = \frac{1}{s-\alpha}, \tag{11.109}$$

valid for $\operatorname{Re} s > \operatorname{Re} \alpha$ in the first instance. The right-hand side of (11.109) gives a meromorphic continuation of the left-hand side to the domain $\mathbb{C} - \{\alpha\}$. If $\alpha = u + iv$ is a root of the polynomial $z^2 + bz + c$ which has a conjugate root $\bar{\alpha}$. Then (11.109) reads

$$\mathcal{L}[e^{ut} \sin vt](s) = \frac{v}{s^2 + bs + c}, \quad \mathcal{L}[e^{ut} \cos vt](s) = \frac{s-u}{s^2 + bs + c}. \tag{11.110}$$

Furthermore, (11.109) with $\alpha = i\omega$, $\omega \in \mathbb{R}$ reads

$$\mathcal{L}[\sin \omega t](s) = \frac{\omega}{s^2 + \omega^2} \tag{11.111}$$

and

$$\mathcal{L}[\cos \omega t](s) = \frac{s}{s^2 + \omega^2}. \tag{11.112}$$

For inverting the Laplace transform we also need

$$\mathcal{L}[te^{\omega t}](s) = \frac{1}{(s-\omega)^2}. \tag{11.113}$$

Proof. By definition, (11.109) clearly holds true in the mentioned range. Since the right-hand side is analytic in $\mathbb{C}\backslash\{\omega\}$, the consistency theorem establishes the second assertion. (8.117) establishes (11.111) and (11.112). □

Example 11.3. We evaluate the integral in Example 11.4 by an elementary method. Suppose $b, c \in \mathbb{R}$ satisfy $b^2 - 4c < 0$. Then we have

$$\int_{-\infty}^{\infty} \frac{1}{(x^2 + bx + c)^2} \, dx = \frac{4\pi\sqrt{4c - b^2}}{(4c - b^2)^2}. \tag{11.114}$$

Proof. Recall the partial fraction expansion (13.28):

$$\frac{1}{(x^2+bx+c)^2} = \frac{-\frac{1}{4c-b^2}}{(x-\alpha)^2} + \frac{-\frac{2\sqrt{4c-b^2}}{(4c-b^2)^2}i}{x-\alpha} + \frac{-\frac{1}{4c-b^2}}{(x-\bar{\alpha})^2} + \frac{\frac{2\sqrt{4c-b^2}}{(4c-b^2)^2}i}{x-\bar{\alpha}}$$
$$= \frac{-\frac{1}{4c-b^2}}{(x-\alpha)^2} + \frac{-\frac{1}{4c-b^2}}{(x-\bar{\alpha})^2} - \frac{2\sqrt{4c-b^2}i}{(4c-b^2)^2}\frac{\alpha-\bar{\alpha}}{(x-\alpha)(x-\bar{\alpha})}, \tag{11.115}$$

where $\alpha = \frac{-b+\sqrt{4c-b^2}i}{2}$ and $\alpha - \bar{\alpha} = \sqrt{4c-b^2}i$. Hence

$$\frac{1}{(x^2+bx+c)^2} = -\frac{1}{4c-b^2}\left(\frac{1}{(x-\alpha)^2} + \frac{1}{(x-\bar{\alpha})^2}\right) + \frac{2}{4c-b^2}\frac{1}{x^2+bx+c}. \tag{11.116}$$

Recalling

$$\int \frac{1}{x^2+bx+c}\,\mathrm{d}x = \int \frac{1}{\left(x+\frac{b}{2}\right)^2 + \sqrt{\frac{4c-b^2}{4}}^2}\,\mathrm{d}x$$
$$= \frac{2}{\sqrt{4c-b^2}}\arctan\frac{2x+b}{\sqrt{4c-b^2}} + C, \tag{11.117}$$

we integrate (11.116) to obtain

$$\int \frac{1}{(x^2+bx+c)^2}\,\mathrm{d}x = \frac{1}{4c-b^2}\left(\frac{1}{x-\alpha} + \frac{1}{x-\bar{\alpha}}\right)$$
$$+ \frac{4\sqrt{4c-b^2}}{(4c-b^2)^2}\arctan\frac{2x+b}{\sqrt{4c-b^2}} + C \tag{11.118}$$
$$= \frac{1}{4c-b^2}\frac{b}{x^2+bx+c} + \frac{4\sqrt{4c-b^2}}{(4c-b^2)^2}\arctan\frac{2x+b}{\sqrt{4c-b^2}} + C.$$

Hence

$$\int_{-\infty}^{\infty} \frac{1}{(x^2+bx+c)^2}\,\mathrm{d}x \tag{11.119}$$
$$= \left[\frac{1}{4c-b^2}\frac{b}{x^2+bx+c} + \frac{4\sqrt{4c-b^2}}{(4c-b^2)^2}\arctan\frac{2x+b}{\sqrt{4c-b^2}}\right]_{-\infty}^{\infty}$$
$$= \frac{4\sqrt{4c-b^2}}{(4c-b^2)^2}(\arctan(\infty) - \arctan(-\infty)),$$

which leads to (11.114). □

Example 11.4. Suppose $b, c \in \mathbb{R}$ satisfy $b^2 - 4c < 0$. We find the value of the integral

$$\int_{-\infty}^{\infty} \frac{1}{(x^2+bx+c)^2}\,\mathrm{d}x = \frac{4\pi\sqrt{4c-b^2}}{(4c-b^2)^2}. \tag{11.120}$$

Proof. We integrate the function $f(z) = \frac{1}{(z^2+bz+c)^2}$ along the contour C consisting of the upper semi-circle $C_R : z = Re^{i\theta}$ of radius R going to infinity and the line segment $[-R, R]$. Solving the equation $z^2 + bz + c = 0$, we get the solutions $\alpha, \bar{\alpha}$, where

$$\alpha = \frac{-b + \sqrt{4c - b^2}i}{2}, \tag{11.121}$$

with the bar denoting the complex conjugate. he following partial fraction expansion (13.28) will be found in Example 13.1.

$$\frac{1}{(z^2 + bz + c)^2} = \frac{-\frac{1}{4c-b^2}}{(z - \alpha)^2} + \frac{-\frac{2\sqrt{4c-b^2}}{(4c-b^2)^2}i}{z - \alpha} + h(z), \tag{11.122}$$

where $h(z)$ is the holomorphic part at $z = \alpha$ and is given by $\frac{-\frac{1}{4c-b^2}}{(z-\bar{\alpha})^2} + \frac{\frac{2\sqrt{4c-b^2}}{(4c-b^2)^2}i}{z-\bar{\alpha}}$. Since $f(z)$ has only one second order pole at $z = \alpha$ inside C, we only need the coefficient B of $\frac{1}{z-\alpha}$. By the residue theorem,

$$\int_C f(z)\,\mathrm{d}z = 2\pi i B = \frac{4\pi\sqrt{4c - b^2}}{(4c - b^2)^2}.$$

□

On the other hand,

$$\int_C f(z)\,\mathrm{d}z = \int_{C_R} f(z)\,\mathrm{d}z + \int_{-R}^{R} f(x)\,\mathrm{d}x \to \int_{-\infty}^{\infty} f(x)\,\mathrm{d}x \tag{11.123}$$

as $R \to \infty$ because $\int_{C_R} f(z)\,\mathrm{d}z = O(R^{-3})$.

11.7 Exchange of limit processes

In the case of real functions, convergence is a delicate problem and there are some theorems known assuring termwise differentiation and integration.

The main ingredient is the uniform convergence which is often assured by absolute convergence but not always. The following are often used. The restriction that the functions are real-valued is immaterial and the theorems are applicable to complex-valued functions. Termwise integration is easier and differentiation is more involved.

Theorem 11.9. (Termwise integration) *Suppose that each f_k is a continuous function on $[a, b]$ and that $\sum_{k=1}^{\infty} f_k(x) = f(x)$ uniformly on $[a, b]$. Then term by term integration is true:*

$$\sum_{k=1}^{\infty} \int_a^b f_k(x)\,\mathrm{d}x = \int_a^b f(x)\,\mathrm{d}x = \int_a^b \sum_{k=1}^{\infty} f_k(x)\,\mathrm{d}x.$$

Proof. Let $S_n(x) = \sum_{k=1}^{n} f_k(x)$. By uniformity of convergence, given $\varepsilon > 0$ there exists an $N = N(\varepsilon) \in \mathbb{N}$ such that for $n > N$, $|S_n(x) - f(x)| < \varepsilon$ for all $x \in [a, b]$. Hence

$$\left| \int_a^b f(x)\,\mathrm{d}x - \int_a^b S_n(x)\,\mathrm{d}x \right| \le \int_a^b |f(x) - S_n(x)|\,\mathrm{d}x < (b-a)\varepsilon,$$

whence

$$\int_a^b f(x)\,\mathrm{d}x = \lim_{n\to\infty} \int_a^b S_n(x)\,\mathrm{d}x.$$

Since the right-hand side of this equality is $\lim_{n\to\infty} \int_a^b \sum_{k=1}^{n} f_k(x)\,\mathrm{d}x$, the integral and the finite sum may be exchanged and amounts to $\lim_{n\to\infty} \sum_{k=1}^{n} \int_a^b f_k(x)\,\mathrm{d}x = \sum_{k=1}^{\infty} \int_a^b f_k(x)\,\mathrm{d}x$ and the result follows. □

Theorem 11.10. [Apostol (1957), Theorem 13-14, p. 403] *Assume that each f_k is a real-valued function defined and differentiable at each $x \in (a, b)$. Assume that for at least one point $x_0 \in (a, b)$, the series $\sum_{k=1}^{\infty} f_k(x_0)$ is convergent. Assume further that there exists a function g such that $\sum_{k=1}^{\infty} f_k'(x) = g(x)$ uniformly on (a, b). Then there exists a function f such that $\sum_{k=1}^{\infty} f_k(x) = f(x)$ uniformly on (a, b) and that if $x \in (a, b)$, then the derivative $f'(x)$ exists and equals $\sum_{k=1}^{\infty} f_k'(x) = f'(x)$.*

For differentiation under integral sign, we have

Theorem 11.11. *Suppose that $f(x, t)$ is integrable in x on $[a, b]$ for all t near the point α and that the limit function $\lim_{t\to\alpha} f(x, t) = f(x, \alpha)$ is integrable on $[a, b]$. Further suppose that the limit is uniform in $x \in [a, b]$. Then we may take the limit under the integral sign:*

$$\lim_{t\to\alpha} \int_a^b f(x, t)\,\mathrm{d}x = \int_a^b \lim_{t\to\alpha} f(x, t)\,\mathrm{d}x. \tag{11.124}$$

Proof. By the uniformity of convergence, given $\varepsilon > 0$, there exists a $\delta = \delta(\varepsilon) > 0$ such that for $0 < |t - \alpha| < \delta$,

$$|f(x, t) - f(x, \alpha)| < \varepsilon \tag{11.125}$$

for all $x \in [a, b]$. Hence

$$\left| \int_a^b f(x, t)\,\mathrm{d}x - \int_a^b f(x, \alpha)\,\mathrm{d}x \right| \le \varepsilon \int_a^b \mathrm{d}x = (b-a)\varepsilon,$$

which completes the proof. □

Corollary 11.9. *Suppose that $f(x,t)$ is integrable in x on $[a,b]$ for all t in a certain domain T, that $f(x,t)$ is differentiable in $t \in T$ for every $x \in [a,b]$ and that $f_t = \frac{\partial f}{\partial t}$ is continuous in both the variables x and t. Then we may differentiate under the integral sign:*

$$\frac{\mathrm{d}}{\mathrm{d}t}\int_a^b f(x,t)\,\mathrm{d}x = \int_a^b \frac{\partial}{\partial t} f(x,t)\,\mathrm{d}x. \tag{11.126}$$

Proof. By Theorem 11.11, it suffices to prove that

$$\lim_{h\to 0}\frac{\Delta f}{h} = f_t(x,t)$$

uniformly in $x \in [a,b]$, where $\Delta f = f(x,t+h) - f(x,t)$. By the mean value theorem, there exists a θ, $0 \le \theta \le 1$ such that

$$\Delta f = f(x,t+h) - f(x,t) = f_t(x,t+\theta h).$$

Since $f_t(x,t)$ is continuous in $x \in [a,b]$, it is uniformly continuous on $[a,b]$. Hence if $0 < |h|$ is small enough, then

$$\left|\frac{\Delta f}{h} - f_t(x,t)\right| = |f_t(x,t+\theta h) - f_t(x,t)|$$

is arbitrarily small uniformly in $x \in [a,b]$. Hence the convergence is uniform, and the proof is complete. □

Example 11.5. We find the value of the probability integral, cf. Exercise 105 and Remark 11.3.

$$\int_{-\infty}^{\infty} e^{-t^2}\,\mathrm{d}t = \sqrt{\pi}. \tag{11.127}$$

The integral exists by the comparison test. Consider

$$f(x) = \left(\int_0^x e^{-t^2}\,\mathrm{d}t\right)^2.$$

Then $f'(x) = 2e^{-x^2}\int_0^x e^{-t^2}\,\mathrm{d}t$. Along with f consider

$$g(x) = \int_0^1 \frac{e^{-x^2(t^2+1)}}{t^2+1}\,\mathrm{d}t.$$

Differentiating under the integral sign, we obtain

$$\begin{aligned} g'(x) &= \frac{\mathrm{d}}{\mathrm{d}x}\int_0^1 \frac{1}{t^2+1} e^{-x^2(t^2+1)}\,\mathrm{d}t \\ &= \int_0^1 \frac{1}{t^2+1}\frac{\partial}{\partial x} e^{-x^2(t^2+1)}\,\mathrm{d}t = -2x\int_0^1 e^{-x^2(t^2+1)}\,\mathrm{d}t, \end{aligned}$$

which becomes, by the change of variable $xt = u$, $g'(x) = -2e^{-x^2}\int_0^x e^{-t^2}\,dt$. Hence it follows that $f'(x) + g'(x) = 0$. We conclude from the Newton-Leibniz rule (Corollary 7.2) that $f(x)+g(x) = C$. The value of the constant C may be found by putting $x = 0$: $f(0) = 0$, $g(0) = \int_0^1 \frac{1}{t^2+1}\,dt = \arctan 1 = \frac{\pi}{4}$, i.e. $C = \frac{\pi}{4}$. Hence $f(x) + g(x) = \frac{\pi}{4}$. Letting $x \to \infty$, we conclude that $f(\infty)+g(\infty) = \frac{\pi}{4}$, provided that the limits exist. $f(\infty)$ exists and by taking the limit under the integral sign, we obtain $g(\infty) = 0$. Hence $f(\infty) = \frac{\pi}{4}$ and (11.127) follows.

11.8 Multiple integrals

The computation of single integrals can be reduced to finding a primitive function in view of Corollary 11.3 (which depends on the fundamental theorem of infinitesimal calculus). For n-ple integrals, we repeat the same process n-times.

Such a computation is possible when the domain is a sum of some ordinate sets which are defined in the following

Definition 11.4. A planar domain D is called a **vertical ordinate set** if any vertical line (a line parallel to the y-axis) meets the boundary C of D at most two points. In this case, D can be expressed in the form $a \le x \le b,\ \ \phi_1(x) \le y \le \phi_2(x)$. Hereafter we assume that $\phi_1(x), \phi_2(x)$ are continuous.

Similarly, if any horizontal line (a line parallel to the x-axis) meets the boundary C of D at most two points, then D is called a **horizontal ordinate set**. In this case D can be expressed as $\psi_1(y) \le x \le \psi_2(y)$, $c \le y \le d$ and we assume that $\psi_1(y), \psi_2(y)$ are continuous.

Theorem 11.12. (double integral $\to$ repeated integral) *Suppose that the ordinate set $D : a \le x \le b,\ \ \phi_1(x) \le y \le \phi_2(x)$ has the area and that f is continuous on D. Then the double integral $\iint f(x,y)\,dxdy$ may be computed as a repeated integral*

$$\iint_D f(x,y)\,dxdy = \int_a^b \left[\int_{\phi_1(x)}^{\phi_2(x)} f(x,y)\,dy\right] dx.$$

Remark 11.1. In the above repeated integral, we view $f(x,y)$ as a function in y only and find the primitive function (in y). Then using Corollary 11.3 thereby substituting $\phi_1(x), \phi_2(x)$, we find the function in x only and then find the primitive function and then use Corollary 11.3 again. The process is as "repeating anti-partial differentiation".

In case D is a horizontal ordinate set, we work with

$$\iint_D f(x, y) \mathrm{d}x\mathrm{d}y = \int_c^d \left[\int_{\psi_1(y)}^{\psi_2(y)} f(x, y) \, \mathrm{d}x \right] \mathrm{d}y.$$

Lemma 11.1. *Suppose $\mathbb{I}$ is a closed interval in $\mathbb{R}^2$ and that f is continuous on $\mathbb{I}$ except for an exceptional set E of measure 0. Suppose that for a fixed k no vertical lines meet E at more than k points. Then*

$$\iint_{\mathbb{I}} f(\boldsymbol{x}) \, \mathrm{d}\boldsymbol{x} = \int_a^b \mathrm{d}x \int_c^d f(x, y) \, \mathrm{d}y. \tag{11.128}$$

Proof. Since f is continuous on $[c, d]$ except for at most k points, the integral $F(x) = \int_c^d f(x, y) \, \mathrm{d}y$ exists and defines a function on $[a, b]$. Let $\Delta_y : c = y_0 < \cdots < y_m$ be the division points on $[c, d]$ with $|\Delta_y| < |\Delta_x|$. The lines $x = x_i$, $y = y_j$ determine a grid (or a net) which divides the interval $\mathbb{I}$ into mn subintervals $\mathbb{I}_{ij}$. We divide $\mathbb{I}$ into two D_0 and D_1, where D_0 is the union of those intervals which contain points in E and D_1 is the rest of D.

Then for an arbitrarily fixed $\xi_i \in [a, b]$ $(\Xi = (\xi_i))$, let $S(\Delta_x, F, \Xi)$ be the Riemann sum in (11.1). We have

$$F(\xi_i) = \int_c^d f(\xi_i, y) \mathrm{d}x\mathrm{d}y = \int_{y_0}^{y_m} f(\xi_i, y) \mathrm{d}x\mathrm{d}y = \sum_{j \le m-1} \int_{y_j}^{y_{j+1}} f(\xi_i, y) \, \mathrm{d}x\mathrm{d}y. \tag{11.129}$$

Since $f(\xi_i, y)$ is a continuous function in y at all but k points, we may apply the first mean value theorem for integrals, Theorem 11.4 and obtain terms of the form $f(\xi_{ij})$, where $\xi_{ij} = (\xi_i, Y_j(\xi_i))$. For the remaining k terms, we have bounds $M\Delta y_{ij}$, where M is the upper bound of $|f|$ on $\mathbb{I}$.

Hence

$$F(\xi_i) = \sum_{\mathbb{I}_{ij} \in \subset \mathbb{I}_1} f(\xi_{ij}) \mu(\mathbb{I}_{ij}) + O(\sum_{\mathbb{I}_{ij} \in \subset D_0} \mu(D_{ij})). \tag{11.130}$$

Since $\int \int_D f \, \mathrm{d}\boldsymbol{x}$ exists and D_0 is the circumference set of E with measure 0, it follows that

$$\lim_{|\Delta| \to 0} S(F, \Xi, \Delta_x) = \iint_D f \, \mathrm{d}\boldsymbol{x}. \tag{11.131}$$

□

If the domain D of f is a vertical ordinate set, we extend f to a closed interval that contains D by **0-extension**, i.e. $f(x) = 0$ for $x \notin D$. Theorem 11.12 follows from Lemma 11.1.

Exercise 102. Prove Corollary 5.2.

Solution. We apply Lemma 11.1. Let $P_1 = (a, c)$, $P_2 = (b, c)$, $P_3 = (a, d)$, $P_4 = (b, d)$ be the vertices of $\mathbb{I}$. Then

$$\iint_{\mathbb{I}} f_{xy}\, \mathrm{d}x\mathrm{d}y = \int_a^b \mathrm{d}x \int_c^d \frac{\partial}{\partial y}\left(\frac{\partial f}{\partial x}\right) \mathrm{d}y = \int_a^b \mathrm{d}x \left[\frac{\partial f}{\partial x}\right]_c^d$$
$$= \int_a^b \mathrm{d}x\,(f_x(x,d) - f_x(x,c)) = f(P_4) - f(P_3) - (f(P_2) - f(P_1))$$

by Corollary 11.3. We have the same result for f_{yx}, so that

$$\iint_{\mathbb{I}} (f_{xy} - f_{yx})\, \mathrm{d}x\mathrm{d}y = 0,$$

and so $f_{xy} = f_{yx}$ throughout $\mathbb{I}$. For any choice of the rectangle, we have this equality, so that it holds throughout D.

Theorem 11.12 involves not only the decomposition into repeated single integrals but also the change of order of integration which is known as Fubini's theorem. Since we do not dwell on measure theory save for a brief description of Borel measure in §15.1, we take it for granted and refer to it as **Fubini's theorem**. Cf. e.g. [Hewitt (1965), Chapter 3], [Kestelman (1960)].

Theorem 11.13. (Fubini's theorem) *For two measurable spaces (X_i, μ_i, E_i), their Cartesian product $(X_1 \times X_2, \mu_1 \times \mu_2, E)$ is also measurable and we have*

$$\mu(f) = \mu_1(\mu_2(f)) = \mu_2(\mu_1(f)) \tag{11.132}$$

for any integrable function $f \in L^1$.

Example 11.6. Let D be the triangle with vertices at $(0,0)$, $(1,1)$, $(0,1)$. Then we evaluate the double integral $\iint_D e^{-y^2}\, \mathrm{d}x\mathrm{d}y$. We may express D as a horizontal ordinate set $0 \le x \le y$, $0 \le y \le 1$ (if we use the vertical ordinate set $0 \le x \le 1$, $x \le y \le 1$, we cannot go on):

$$\iint_D e^{-y^2}\, \mathrm{d}x\mathrm{d}y = \int_0^1 e^{-y^2} \mathrm{d}y \int_0^y \mathrm{d}x = \int_0^1 y e^{-y^2} \mathrm{d}y = \left[-\frac{1}{2}e^{-y^2}\right]_0^1 = \frac{1}{2}\left(1 - \frac{1}{e}\right).$$

Example 11.7. Let D be the square with vertexes at $(1,0)$, $(0,1)$, $(-1,0)$, $(0,-1)$. Then we evaluate the double integral $I = \iint_D (x^2 + y^2)\, \mathrm{d}x\mathrm{d}y$.

Since by symmetry, the value is 4 times $\iint_{D_1}$, where $D_1 : 0 \le x \le 1,\ 0 \le y \le -x+1$, we obtain

$$\begin{aligned} I &= 4\iint_{D_1} (x^2+y^2)\,\mathrm{d}x\mathrm{d}y = 4\int_0^1 \mathrm{d}x \int_0^{-x+1} (x^2+y^2)\,\mathrm{d}y \\ &= 4\int_0^1 \left[x^2 y + \frac{1}{3}y^3\right]_0^{-x+1} \mathrm{d}x = -4\int_0^1 \left(x^2(x-1) + \frac{1}{3}(x-1)^3\right)\mathrm{d}x. \end{aligned}$$

From here it may be simpler as in Exercise 90 to use the Taylor expansion of the integrand $\frac{4}{3}(x-1)^3 + 2(x-1)^2 + x - 1$ to arrive at the value $\frac{2}{3}$.

Theorem 11.14. (change of variable for multiple integrals) *Suppose that $f : D \to \mathbb{R}$ is continuous and that the domain $D \subset \mathbb{R}^n$ is the image of the topological mapping (homeomorphism) $\boldsymbol{\phi}$ of class C^1, i.e. the vector-valued function $\boldsymbol{\phi} : D' = \boldsymbol{\phi}^{-1}(D) \approx D \subset \mathbb{R}^n$ $\boldsymbol{\phi}(\boldsymbol{t}) = \begin{pmatrix}\phi_1(\boldsymbol{t})\\ \vdots \\ \phi_n(\boldsymbol{t})\end{pmatrix}$ is such that the Jacobian $J_{\boldsymbol{\phi}} = \frac{\partial(x_1,\cdots,x_n)}{\partial(t_1,\cdots,t_n)} \ne 0$ on D'. Then by the* **change of variable**

$$\boldsymbol{x} = \boldsymbol{\phi}(\boldsymbol{t}),\ \boldsymbol{t} = \begin{pmatrix} t_1 \\ \vdots \\ t_n \end{pmatrix} \in \boldsymbol{\phi}^{-1}(D), \tag{11.133}$$

we have

$$\int_D f(\boldsymbol{x})\,\mathrm{d}\boldsymbol{x} = \int_{\boldsymbol{\phi}^{-1}(D)} f(\boldsymbol{\phi}(\boldsymbol{t}))\,|J_{\boldsymbol{\phi}}|\,\mathrm{d}\boldsymbol{t} \tag{11.134}$$

$$= \int_{\boldsymbol{\phi}^{-1}(D)} f(\phi_1(t_1,\cdots,t_n),\cdots,\phi_n(t_1,\cdots,t_n)) \left|\frac{\partial(x_1,\cdots,x_n)}{\partial(t_1,\cdots,t_n)}\right| \mathrm{d}t_1\cdots\mathrm{d}t_n,$$

where $\boldsymbol{\phi}^{-1}(D) = D'$ is a domain in the $\boldsymbol{t}$-domain corresponding to the domain D in $\mathbb{R}^n$. The formula also holds true if $J_{\boldsymbol{\phi}} = 0$ on subsets with measure 0 *of $\boldsymbol{\phi}^{-1}(D)$.*

Proof. 1°. For any point $\boldsymbol{t}_0 \in D'$, we write

$$\boldsymbol{\phi}(\boldsymbol{t}) = \boldsymbol{\phi}(\boldsymbol{t}_0) + \boldsymbol{\psi}(\boldsymbol{t}),$$

where $\boldsymbol{\psi}(t)$ satisfies the same condition as $\boldsymbol{\phi}$ and $\boldsymbol{\psi}(\boldsymbol{t}_0) = \boldsymbol{o}$. Hence we may suppose that $\boldsymbol{\psi} \in C^1(D')$ and $\boldsymbol{\psi}(\boldsymbol{t}_0) = \boldsymbol{o}$.

We call a mapping $\boldsymbol{p} : D \to \mathbb{R}^n$ primitive if there exists a $j, 1 \le j \le n$ such that

$$(\boldsymbol{p}(\boldsymbol{x}), \boldsymbol{e}_i) = (\boldsymbol{x}, \boldsymbol{e}_i)$$

for all $i \neq j$, i.e. $\boldsymbol{p}$ changes only the jth component. Also let $\boldsymbol{n}$ denote a linear transformation changing at most two components. We shall prove in 3° that there exists a neighborhood $V - \{\boldsymbol{\phi}(\boldsymbol{t}_0)\} \subset D$ such that for all $\boldsymbol{x} \in V$ we have

$$\boldsymbol{\phi}(\boldsymbol{x}) = (\boldsymbol{p}_n \circ \boldsymbol{n}_n \circ \cdots \boldsymbol{p}_1 \circ \boldsymbol{n}_1)(\boldsymbol{x}), \tag{11.135}$$

where $\boldsymbol{p}_j$, $\boldsymbol{n}_i$ are defined above and $\boldsymbol{p}_j(\boldsymbol{t}_0) = \boldsymbol{n}_i(\boldsymbol{t}_0) = \boldsymbol{o}$.

2°. If the change of variable is made successively as $\boldsymbol{\psi} \circ \boldsymbol{\phi}$ (i.e. $\boldsymbol{x} = (\boldsymbol{\psi} \circ \boldsymbol{\phi})(\boldsymbol{u})$), then substituting this in (11.134),

$$\int_D f(\boldsymbol{x})\, \mathrm{d}\boldsymbol{x} = \int_{(\boldsymbol{\psi}\circ\boldsymbol{\phi})^{-1}(D)} f(\boldsymbol{\psi} \circ \boldsymbol{\phi}(\boldsymbol{u}))\, |J_{\boldsymbol{\phi}}(\boldsymbol{t})|\, |J_{\boldsymbol{\psi}}(\boldsymbol{u})|\, \mathrm{d}\boldsymbol{u}$$

where the product of Jacobians becomes $J_{\boldsymbol{\psi}}(\boldsymbol{u}) J_{\boldsymbol{\phi}}(\boldsymbol{t}) = J_{\boldsymbol{\psi}\circ\boldsymbol{\phi}}(\boldsymbol{t})$ by Corollary 6.1. Hence (11.134) is valid with $\boldsymbol{\phi}$ replaced by $\boldsymbol{\psi} \circ \boldsymbol{\phi}$ and so if $\boldsymbol{\phi}$ is of the form (11.135), it suffices to check with $\boldsymbol{p}$ and $\boldsymbol{n}$. For $\boldsymbol{p}_m = \boldsymbol{p}_m(x_1, \cdots, x_n) = (x_1, \cdots, p(x_m), \cdots, x_n)$, we decompose the multiple integral into repeated single integrals by Theorem11.13. Since $J_{\boldsymbol{p}_m} = g'$, we may apply Corollary 11.4 the change of variable formula for a single integral to conclude the assertion.

If $\boldsymbol{n}$ changes only two components, then again Fubini's theorem, Theorem 11.13 established the assertion.

3°. Proof of (11.135) by induction. Suppose

$$(\boldsymbol{\phi}_m(\boldsymbol{x}), \boldsymbol{e}_i) = (\boldsymbol{x}, \boldsymbol{e}_i) \tag{11.136}$$

for $1 \leq i \leq m-1$, i.e. $\boldsymbol{\phi}_m$ does not change the first $m-1$ components.

Let

$$(a_{ij}) := A_m := \nabla \boldsymbol{\phi}_m(\boldsymbol{o}) = \frac{\partial(x_1, \cdots, x_{m-1}, \cdots, x_n)}{\partial(x_1, \cdots, x_{m-1}, t_m, \cdots, t_n)}. \tag{11.137}$$

Then

$$A_m = \begin{pmatrix} E_{m-1} & * \\ O & * \end{pmatrix},$$

so that

$$(A_m \boldsymbol{x}, \boldsymbol{e}_i) = (\boldsymbol{x}, \boldsymbol{e}_i) = x_i, \quad 1 \leq i \leq m-1. \tag{11.138}$$

By Exercise 144 and (11.138),

$$a_{ij} = (A_m \boldsymbol{e}_j, \boldsymbol{e}_i) = (\boldsymbol{e}_j, \boldsymbol{e}_i) = 0, \quad 1 \leq i \leq m-1,\ i \geq m.$$

If for all $i \geq m$, we had $a_{im} = 0$, then we would have

$$A_m \boldsymbol{e}_i = \boldsymbol{a}_i = \sum_{k=1}^{n} a_{ki} \boldsymbol{e}_k = \sum_{k=m+1}^{n} a_{ki} \boldsymbol{e}_k.$$

By assumption, A_m is regular and so $A_m e_i$, $m \leq i \leq n$ are linearly independent, which contradicts the above linear relation among $n - m + 1$ vectors. Hence we may assume that for some $k \geq m$, we have $a_{km} \neq 0$.

Define the linear maps by

$$\boldsymbol{q}_m(\boldsymbol{e}_i) = \begin{cases} \boldsymbol{e}_i & i \neq m \\ \boldsymbol{o} & i = m \end{cases}, \tag{11.139}$$

$$\boldsymbol{n}_m(\boldsymbol{e}_i) = \begin{cases} \boldsymbol{e}_i & i \neq m, i \neq k \\ \boldsymbol{e}_k & i = m \\ \boldsymbol{e}_m & i = k \end{cases}$$

and

$$\boldsymbol{p}_m(\boldsymbol{x}) = \boldsymbol{q}_m(\boldsymbol{x}) + (\boldsymbol{e}_m, (\boldsymbol{\phi}_m \circ \boldsymbol{n}_m)(\boldsymbol{x}))\boldsymbol{e}_m. \tag{11.140}$$

Then for $i \neq m$, we have

$$(\boldsymbol{p}_m(\boldsymbol{x}), \boldsymbol{e}_i) = (\boldsymbol{q}_m, \boldsymbol{e}_i) = (\boldsymbol{x}, \boldsymbol{e}_i),$$

so that $\boldsymbol{p}_m$ is primitive changing only the mth component.

Next we show that in some neighborhood U_m of $\boldsymbol{o}$, the Jacobi matrix $\nabla \boldsymbol{p}_m$ is regular, which in turn implies that $J_{\boldsymbol{\phi}_m} \neq \boldsymbol{o}$. Then Theorem 8.3 entails that $\boldsymbol{\phi}_m$ has its inverse of class C^1 in U_m. Consider the system of linear equations $\nabla \boldsymbol{p}_m \boldsymbol{h} = \boldsymbol{o}$. Then by the chain rule, Theorem 6.1, $\nabla(\boldsymbol{p}_m \circ \boldsymbol{n}_m)(\boldsymbol{o}) = A_m N_m$, where $N_m = \nabla \boldsymbol{n}_m(\boldsymbol{o})$ and so

$$\boldsymbol{o} = \nabla \boldsymbol{\phi}_m \boldsymbol{h} = Q_m \boldsymbol{h} + (\boldsymbol{e}_m, A_m N_m \boldsymbol{h}))\boldsymbol{e}_m,$$

where $Q_m = \nabla \boldsymbol{q}_m(\boldsymbol{o})$. Since $Q_m \boldsymbol{h}$ has no component in $\boldsymbol{e}_m$, it follows that

$$Q_m \boldsymbol{h} = \boldsymbol{o}, \quad (\boldsymbol{e}_m, A_m N_m \boldsymbol{h}) = 0. \tag{11.141}$$

From the first equality, we have $\boldsymbol{h} = c(\boldsymbol{e}_m)$ with $c \in \mathbb{R}$ a constant. Substituting this in the second of we obtain

$$0 = (\boldsymbol{e}_m, A_m N_m c \boldsymbol{e}_m) = c(\boldsymbol{e}_m, A_m \boldsymbol{e}_k)$$

by (11.139). The inner product is $(A_m \boldsymbol{e}_k, \boldsymbol{e}_m) = a_{km}$, so that we have

$$c a_{km} = 0$$

and so $c = 0$. Hence $\boldsymbol{h} = \boldsymbol{o}$ by Theorem A.4. This implies non-vanishing of the Jacobi matrix in $U_m = U_m(\boldsymbol{o})$. We set

$$V_m = \boldsymbol{p}_m(U_m) \subset \mathbb{R}^n \tag{11.142}$$

and define for any $\boldsymbol{y} = \boldsymbol{p}_m(\boldsymbol{x}) \in V_m$

$$\boldsymbol{\phi}_{m+1}(\boldsymbol{y}) = \boldsymbol{\phi}_m(\boldsymbol{z}), \quad \boldsymbol{z} = \boldsymbol{n}_m((\boldsymbol{p}_m^{-1})(\boldsymbol{y})) = \boldsymbol{n}_m((\boldsymbol{x})). \tag{11.143}$$

Then

$$(\boldsymbol{e}_m, \boldsymbol{\phi}_{m+1}(\boldsymbol{y})) = (\boldsymbol{e}_m, \boldsymbol{\phi}_m(\boldsymbol{n}_m(\boldsymbol{x}))) = (\boldsymbol{e}_m, \boldsymbol{p}_m(\boldsymbol{x})) = (\boldsymbol{e}_m, \boldsymbol{y})$$

by (11.140).

For $i < m$

$$(\boldsymbol{e}_i, \boldsymbol{\phi}_{m+1}(\boldsymbol{y})) = (\boldsymbol{e}_i, \boldsymbol{\phi}_m(\boldsymbol{n}_m(\boldsymbol{x}))) = (\boldsymbol{e}_i, \boldsymbol{n}_m(\boldsymbol{x}))$$

since $\boldsymbol{\phi}_m$ does not change first $(m-1)$ components. But the last term is

$$(\boldsymbol{e}_i, \boldsymbol{x}) = (\boldsymbol{e}_i, \boldsymbol{q}_m(\boldsymbol{x})) = (\boldsymbol{e}_i, \boldsymbol{p}_m(\boldsymbol{x})) = (\boldsymbol{e}_i, \boldsymbol{y}).$$

Hence $\boldsymbol{\phi}_{m+1}$ satisfies the same condition as $\boldsymbol{\phi}_m$ that it does not change first $(m-1)$ components.

We may express (11.143) as

$$\boldsymbol{\phi}_m = \boldsymbol{\phi}_{m+1} \circ \boldsymbol{p}_m \circ \boldsymbol{n}_m \tag{11.144}$$

since $\boldsymbol{n}_m^{-1} = \boldsymbol{n}_m$. Hence we may apply (11.144) inductively starting $\boldsymbol{\phi}_1 = \boldsymbol{\phi}$ as follows.

$$\boldsymbol{\phi}_1 = \boldsymbol{\phi}_2 \circ \boldsymbol{p}_1 \circ \boldsymbol{n}_1 = \cdots = \boldsymbol{\phi}_{n+1} \circ \boldsymbol{p}_n \circ \boldsymbol{n}_n \circ \cdots \boldsymbol{p}_1 \circ \boldsymbol{n}_1. \tag{11.145}$$

Since $\boldsymbol{\phi}_{n+1}$ is the identity, (11.145) leads to (11.135).

4°. We consider a continuous function f with compact support $\operatorname{supp} f \subset D$. □

Cf. (14.11) for Theorem 11.14. The following lemma helps to interpret Theorem 11.14 from the point of view of differential forms and works as a precursor of §14.1.

Lemma 11.2. *Suppose the functions $x = x(u,v)$, $y = y(u,v)$ are of class C^1 on a domain D. Then we have*

$$\mathrm{d}x\mathrm{d}y = \frac{\partial(x,y)}{\partial(u,v)}\mathrm{d}u\mathrm{d}v, \tag{11.146}$$

where the computation of differential forms is to follow the alternating rules in §14.1:

$$\mathrm{d}v\mathrm{d}u = -\mathrm{d}u\mathrm{d}v \tag{11.147}$$

and

$$\mathrm{d}u\mathrm{d}u = \mathrm{d}v\mathrm{d}v = 0, \tag{11.148}$$

where (11.147) *is a special case of* (14.21) *and shows a relationship between areas with sign of two infinitesimal parallelograms with* dx, dy *as adjacent sides, while* (11.147) *means the areas of degenerated parallelograms and follows from* (11.147).

Proof. By (5.21),

$$\mathrm{d}x = \frac{\partial x}{\partial u}\mathrm{d}u + \frac{\partial x}{\partial v}\mathrm{d}v, \ \mathrm{d}y = \frac{\partial y}{\partial u}\mathrm{d}u + \frac{\partial y}{\partial v}\mathrm{d}v.$$

Hence

$$\mathrm{d}x\mathrm{d}y = \frac{\partial x}{\partial u}\frac{\partial y}{\partial v}\mathrm{d}u\mathrm{d}v + \frac{\partial y}{\partial u}\frac{\partial x}{\partial v}\mathrm{d}v\mathrm{d}u = \left(\frac{\partial x}{\partial u}\frac{\partial y}{\partial v} - \frac{\partial y}{\partial u}\frac{\partial x}{\partial v}\right)\mathrm{d}u\mathrm{d}v = \frac{\partial\,(x,y)}{\partial\,(u,v)}\mathrm{d}u\mathrm{d}v.$$

□

Remark 11.2. Lemma 11.2 has an apparent generalization to the case of higher dimension. In applying Lemma 11.2, we need to pay attention to the sign of the integral since the change of order $\mathrm{d}x$ and $\mathrm{d}y$ gives rise to the sign changes. Once we make a convention that $\int_D f\mathrm{d}x\mathrm{d}y$ gives the area $\mu(D)$ for $f = 1$, then to make the sign consistent, we are to take the absolute value of the Jacobian. Then Theorem 11.14 follows from Lemma 11.2 in view of the local-global principle. Cf. (14.11).

For more details on (11.147) and (11.148), cf. Proposition 14.1, §14.1. (11.148) follows from (11.147) since $\operatorname{char}\mathbb{R} = 0$.

Hereafter we often denote the correspondence between the domains under the mapping ϕ by

$$D \leftrightarrow \phi^{-1}(D). \tag{11.149}$$

Example 11.8. Let D denote the unit disc (with radius 1 and center at the origin). Then we evaluate the integral

$$\iint_D e^{-(x^2+y^2)}\,\mathrm{d}x\mathrm{d}y.$$

By the polar coordinate $\begin{pmatrix} x \\ y \end{pmatrix} = \boldsymbol{z} = \phi(r,\theta) = \begin{pmatrix} r\cos\theta \\ r\sin\theta \end{pmatrix}$

$$D \leftrightarrow \phi^{-1}(D) = \left\{ \begin{pmatrix} r \\ \theta \end{pmatrix} \middle| \, 0 \le r \le 1, \ 0 \le \theta \le 2\pi \right\}.$$

Since the Jacobian is r by (5.39) which vanishes only at the origin. Substituting $\mathrm{d}x\mathrm{d}y = r\mathrm{d}r\mathrm{d}\theta$,

$$\begin{aligned} \iint_D e^{-(x^2+y^2)}\,\mathrm{d}x\mathrm{d}y &= \iint_{\phi^{-1}(D)} e^{-r^2}\, r\mathrm{d}r\mathrm{d}\theta = \int_0^{2\pi}\mathrm{d}\theta\int_0^1 re^{-r^2}\mathrm{d}r \\ &= \int_0^{2\pi}\left[-\frac{1}{2}e^{-r^2}\right]_0^1\mathrm{d}\theta = \pi\left(1-\frac{1}{e}\right). \end{aligned}$$

Example 11.9. Let D denote the $\frac{3}{8}$ sector of the unit disc, i.e. in the polar coordinates, $\frac{\pi}{4} \leq \theta \leq \pi$. Then we evaluate the integral

$$I = \iint_D xy^2 \, \mathrm{d}x\mathrm{d}y.$$

As in Example 11.8,

$$I = \int_0^1 r^4 \, \mathrm{d}r \int_{\frac{\pi}{4}}^{\pi} \cos\theta \sin^2\theta \, \mathrm{d}\theta = \left[\frac{1}{5}r^5\right]_0^1 \left[\frac{1}{3}\sin -3\theta\right]_0^1 = -\frac{\sqrt{2}}{60}.$$

Viewing D as a vertical ordinate set, we have

$$\begin{aligned} I &= \int_{-1}^{0} x \, \mathrm{d}x \int_0^{\sqrt{1-x^2}} y^2 \, \mathrm{d}y + \int_0^{\frac{1}{\sqrt{2}}} x \, \mathrm{d}x \int_x^{\sqrt{1-x^2}} y^2 \, \mathrm{d}y \qquad (11.150) \\ &= \frac{1}{3}\int_{-1}^{0} x(1-x^2)^{\frac{3}{2}} \, \mathrm{d}x + \frac{1}{3}\int_0^{\frac{1}{\sqrt{2}}} x\left((1-x^2)^{\frac{3}{2}} - x^3\right) \mathrm{d}x. \end{aligned}$$

By the change of variable $x = \sin\theta$,

$$\int x(1-x^2)^{\frac{3}{2}} \, \mathrm{d}x = -\frac{1}{5}\cos^5\theta + C.$$

The corresponding ranges of integrals in θ are $\left[-\frac{\pi}{2}, 0\right]$ and $\left[0, \frac{\pi}{4}\right]$, and after simple calculation we find that the last integral in (11.150) is $-\frac{\sqrt{2}}{60}$.

Exercise 103. Let D be a domain on the xy-plane with the boundaries $x = 2, \ y = x, \ y = -x^2$.
(i) Then find the area of D. Also find the domain $\boldsymbol{\phi}^{-1}(D)$ on the uv-plane which maps onto D under the transformation $\begin{pmatrix} x \\ y \end{pmatrix} = \boldsymbol{z} = \boldsymbol{\phi}(u, v) = \begin{pmatrix} u + v \\ -u^2 + v \end{pmatrix}$ whence find the area of D again.

(ii) Evaluate $\iint \frac{\mathrm{d}x \, \mathrm{d}y}{(x-y+1)^2}$.

Exercise 104. For the domain D with boundaries $x = 0, \ y = 0, \ x + y = 1$, evaluate

$$\iint \exp\left(\frac{x-y}{x+y}\right) \mathrm{d}x\mathrm{d}y$$

by the change of variable $\begin{pmatrix} x \\ y \end{pmatrix} = \boldsymbol{z} = \boldsymbol{\phi}(u, v) = \begin{pmatrix} \frac{1}{2}(u+v) \\ \frac{1}{2}(u-v) \end{pmatrix}$.

Exercise 105. Prove the formula

$$\Gamma(\alpha)\,\Gamma(\beta) = \Gamma(\alpha+\beta)\,\mathrm{B}(\alpha,\beta), \tag{11.151}$$

whence in particular

$$\Gamma\left(\frac{1}{2}\right) = \sqrt{\pi}, \tag{11.152}$$

or the value of the probability integral $\int_0^\infty e^{-x^2}\,\mathrm{d}x = \frac{\sqrt{\pi}}{2}$ in (11.158) below.

Solution. The gamma function, as the Mellin transform of e^{-x} is expounded in Example 11.1, (ii). The improper integral is absolutely and uniformly convergent in the wide sense in $\sigma > 0$, whence it follows that $\Gamma(s)$ is analytic in the right half-plane $\sigma > 0$.

This is also referred to as the **Eulerian integral of the second kind** and we state the definition again for comparison with the integral of the first kind:

$$\Gamma(s) = \int_0^\infty e^{-x}x^{s-1}\,\mathrm{d}x \tag{11.153}$$

for $\sigma > 0$.

Let the **beta function** $\mathrm{B}(\alpha,\beta)$ be defined by the **Eulerian integral of the first kind**

$$\mathrm{B}(\alpha,\beta) = \int_0^1 t^{\alpha-1}(1-t)^{\beta-1}\,\mathrm{d}t, \quad \mathrm{Re}\,\alpha > 0, \mathrm{Re}\,\beta > 0. \tag{11.154}$$

First, in (11.154), put $t = \sin^2\theta$ to obtain

$$\mathrm{B}(\alpha,\beta) = 2\int_0^{\frac{\pi}{2}} \sin^{2\alpha-1}\theta\,\cos^{2\beta-1}\theta\,\mathrm{d}\theta. \tag{11.155}$$

If in (11.153), we put $t = x^2$, then

$$\Gamma(\alpha) = 2\int_0^\infty x^{2\alpha-1}e^{-x^2}\,\mathrm{d}x,$$

whence for $\mathrm{Re}\,\alpha > 0$, $\mathrm{Re}\,\beta > 0$,

$$\begin{aligned}
\Gamma(\alpha)\,\Gamma(\beta) &= 4\int_0^\infty x^{2\alpha-1}e^{-x^2}\,\mathrm{d}x\int_0^\infty y^{2\beta-1}e^{-y^2}\,\mathrm{d}y \\
&= 4\lim_{X\to\infty}\left(\int_0^X x^{2\alpha-1}e^{-x^2}\,\mathrm{d}x\int_0^X y^{2\beta-1}e^{-y^2}\,\mathrm{d}y\right) \\
&= 4\lim_{X\to\infty}\int_0^X\int_0^X x^{2\alpha-1}y^{2\beta-1}\,e^{-(x^2+y^2)}\,\mathrm{d}x\,\mathrm{d}y \\
&= 4\lim_{X\to\infty}\iint_D x^{2\alpha-1}y^{2\beta-1}\,e^{-(x^2+y^2)}\,\mathrm{d}x\,\mathrm{d}y, \qquad (11.156)
\end{aligned}$$

where $D = \left\{ \binom{x}{y} \middle| 0 \le x, y, \sqrt{x^2+y^2} \le X \right\}$. By the change of variable $x = r\cos\theta,\ y = r\sin\theta$, we have the correspondence

$$D \leftrightarrow \phi^{-1}(D) = \left\{ \begin{pmatrix} r \\ \theta \end{pmatrix} \middle| 0 \le r \le X,\ 0 \le \theta \le \frac{\pi}{2} \right\}, \tag{11.157}$$

where the absolute value of the Jacobian of this transformation is r by (5.39), which is $\neq 0$ except at the origin. Hence

$$\begin{aligned} \Gamma(\alpha)\,\Gamma(\beta) &= 4 \lim_{X\to\infty} \iint_{\tilde{D}} r^{2\alpha+2\beta-2}\, e^{-r^2} \sin^{2\alpha-1}\theta \cos^{2\beta-1}\theta\, r\, drd\theta \\ &= 2\int_0^\infty r^{2\alpha+2\beta-1} e^{-r^2}\, dr \cdot 2\int_0^{\frac{\pi}{2}} \sin^{2\alpha-1}\theta \cos^{2\beta-1}\theta\, d\theta \\ &= \Gamma(\alpha+\beta)\, \mathrm{B}(\alpha,\beta) \end{aligned}$$

by (11.155) above. This completes the proof of (11.151).

Putting now $\alpha = \beta = \frac{1}{2}$, we obtain

$$\Gamma\left(\frac{1}{2}\right)^2 = \Gamma(1)\, 2\int_0^{\frac{\pi}{2}} d\theta = \pi,$$

whence (11.152) follows. It follows that

$$\int_{-\infty}^{\infty} e^{-\frac{x^2}{2}}\, dx = \sqrt{2\pi}, \tag{11.158}$$

the **probability integral** which is used to normalize the distribution function of the Gaussian (or normal) distribution.

Remark 11.3. An ordinary procedure for proving (11.152) is to use (11.156) for $\alpha = \beta = \frac{1}{2}$:

$$\left(\int_0^\infty e^{-x^2}\, dx \right)^2 = \iint_D e^{-(x^2+y^2)}\, dxdy$$

by Fubini's theorem, where D is the limit of the quadrant of a disc with center at origin and radius $R \to \infty$. Hence

$$= \lim_{R\to\infty} \left(\int_0^R r\, e^{-r^2}\, dr \int_0^{\frac{\pi}{2}} d\theta \right) = \left[-\frac{1}{2} e^{-r^2} \right]_0^\infty \frac{\pi}{2} = \frac{1}{4}\pi$$

as in Example 11.8.

Thus we see that if we generalize the problem by introducing the parameters α and β, we achieve a wider perspective.

Exercise 106. Prove (11.152) by the expression

$$\mathrm{B}(\alpha,\beta) = \int_0^\infty \frac{y^{\alpha-1}}{(1+y)^{\alpha+\beta}}\,\mathrm{d}y. \tag{11.159}$$

Solution. Making the change of variable $t = \frac{y}{1+y}$ in (11.154), we obtain (11.159). By the change of variable $x \leftrightarrow x\xi$ with $\operatorname{Re}\xi > 0$ in (11.153) gives a useful formula

$$\Gamma(s) = \xi^s \int_0^\infty e^{-x\xi} x^{s-1}\,\mathrm{d}x \tag{11.160}$$

which reads with $\xi = 1+y, s = \alpha+\beta$

$$\Gamma(\alpha+\beta) = (1+y)^{\alpha+\beta} \int_0^\infty e^{-x(1+y)} x^{\alpha+\beta-1}\,\mathrm{d}x,$$

whence

$$\frac{y^{\alpha-1}}{(1+y)^{\alpha+\beta}} = \frac{y^{\alpha-1}}{\Gamma(\alpha+\beta)} \int_0^\infty e^{-x(1+y)} x^{\alpha+\beta-1}\,\mathrm{d}x. \tag{11.161}$$

Integrating this over $(0,\infty)$ and changing the order of integration, we deduce that the right-hand side of (11.159) is

$$\frac{1}{\Gamma(\alpha+\beta)} \int_0^\infty x^{\alpha+\beta-1} e^{-x}\,\mathrm{d}x \int_0^\infty y^{\alpha-1} e^{-xy}\,\mathrm{d}y.$$

Since the inner integral is $x^{-\alpha}\Gamma(\alpha)$ by (11.160), it can be factored out and we are left with $\frac{\Gamma(\alpha)}{\Gamma(\alpha+\beta)} \int_0^\infty x^{\beta-1} e^{-x}\,\mathrm{d}x$, which amounts to $\frac{\Gamma(\alpha)\Gamma(\beta)}{\Gamma(\alpha+\beta)}$, proving (11.151).

Example 11.10. Using the method in Remark B.1, we prove a generalization of the probability integral. Let A be a positive definite real symmetric matrix.

$$I = \int_{\mathbb{R}^n} \exp\left(-\frac{1}{2}(A\boldsymbol{x},\boldsymbol{x})\right)\,\mathrm{d}\boldsymbol{x} = (2\pi)^{\frac{n}{2}} |A|^{-\frac{1}{2}}. \tag{11.162}$$

Suppose

$${}^tPAP = \begin{pmatrix} \lambda_1 & & O \\ & \ddots & \\ O & & \lambda_n \end{pmatrix} = D \tag{11.163}$$

with $P \in O_3$ (orthogonal matrix of degree 3). Let

$$\boldsymbol{x} = \varphi(\boldsymbol{y}) = P\boldsymbol{y}. \tag{11.164}$$

Then the Jacobian is $J_\varphi = |P| = 1$ and (writing $\boldsymbol{y} = {}^t(x_1, \cdots, x_k)$)

$$\exp\left(-\frac{1}{2}(A\boldsymbol{x}, \boldsymbol{x})\right) = \exp\left(-\frac{1}{2}(D\boldsymbol{y}, \boldsymbol{y})\right) = \exp\left(-\frac{1}{2}\sum_{k=1}^{n} \lambda_k x_k^2\right) = \prod_{k=1}^{n} \exp\left(-\frac{1}{2}\lambda_k x_k^2\right).$$

Hence the integral I reduces to the product of n integrals of the form

$$I_k := \int_{-\infty}^{\infty} \exp(-\frac{1}{2}\lambda_k x_k^2)\, \mathrm{d}x_k = 2\int_0^{\infty}. \tag{11.165}$$

The last integral amounts to the probability integral (11.158), so that

$$I_k = \sqrt{2\pi}\sqrt{\lambda_k}^{-1}. \tag{11.166}$$

Appealing to the fact that $|A| = \prod_{k=1}^{n} \lambda_k$, we conclude the assertion.

Exercise 107. Let $D : 0 \le x \le \sqrt{\pi},\ x \le y \le \sqrt{\pi}$. Then prove that

$$\iint_D \frac{1}{y^2} \sin y^2\, \mathrm{d}x\mathrm{d}y = \frac{1}{2} G_1. \tag{11.167}$$

Solution. Since D is an ordinate set, we obtain

$$\iint_D \mathrm{d}x\mathrm{d}y = \int_0^{\sqrt{\pi}} \mathrm{d}x \int_x^{\sqrt{\pi}} \frac{1}{y^2} \sin y^2\, \mathrm{d}y. \tag{11.168}$$

Hence substituting the Maclaurin expansion and integrating termwise, we see that it is $\frac{1}{2}G_1$.

Exercise 108. Let $D : x^2 + y^2 \le 1,\ \ x \ge 0$. Then prove the identity

$$\iint_D \sqrt{x}\mathrm{d}x\mathrm{d}y = \frac{8}{5}\int_0^1 \frac{t^2}{\sqrt{1-t^4}}\mathrm{d}t. \tag{11.169}$$

Solution. By the change of variable $x = r\cos\theta$ $y = r\sin\theta$, we have the correspondence similar to (11.157)

$$D \leftrightarrow \phi^{-1}D = \left\{\begin{pmatrix} r \\ \theta \end{pmatrix} \middle| 0 \le r \le 1, -\frac{\pi}{2} \le \theta \le \frac{\pi}{2}\right\}$$

where the absolute value of the Jacobian of this transformation is $r \ne 0$ save for the origin. Hence

$$\iint_D \sqrt{x}\, \mathrm{d}x\mathrm{d}y = \int_0^1 r^{\frac{3}{2}}\, \mathrm{d}r \int_{-\frac{\pi}{2}}^{\frac{\pi}{2}} \sqrt{\cos\theta}\, \mathrm{d}\theta = \frac{2}{5}\int_{-\frac{\pi}{2}}^{\frac{\pi}{2}} \sqrt{\cos\theta}\, \mathrm{d}\theta. \tag{11.170}$$

By the change of variable $\sqrt{\cos\theta} = t$, we arrive at (11.169).

If we use the known value $\int_0^1 \frac{t^2}{\sqrt{1-t^4}}dt = 2.62205756\cdots$, then the approximate value of the integral is $= 10.48823\cdots$.

Exercise 109. Evaluate the following double integrals in two ways viewing the domain as a vertical and a horizontal ordinate set. Plot the graphic surfaces.

(1) $\iint_D xy\,dxdy = \frac{1}{12}$, where $D: 0 \le x \le 1, x^2 \le y \le \sqrt{x}$.
(2) $\iint_D (x^2+y^2)\,dxdy = \frac{1}{9}e^3 + e - \frac{19}{9}$, where $D: 0 \le x \le 1, 0 \le y \le e^x$.
(3) $\iint_D xy\,dxdy = \frac{1}{8}\pi^2$, where $D: 0 \le x \le \pi, 0 \le y \le \sin x$.
(4) $\iint_D y^2 \log x\,dxdy = \frac{8}{3}\log 2 - \frac{5}{8}$, where $D: 1 \le x \le 2, -x \le y \le x$.
(5) $\iint_D x\cos y\,dxdy = \frac{1}{2} + \sin 1 - \frac{3}{2}\cos 1$, where $D: 0 \le x \le 1, -x \le y \le x^2$.
(6) $\iint_D xy\,dxdy = \frac{1}{24}$, where $D: 0 \le x, 0 \le y, \quad x+y \le 1$.
(7) $\iint_D xy^2\,dxdy = \frac{1}{6480}$, where $D: 0 \le x, 0 \le y, \quad 2x+3y \le 1$.
(8) $\iint_D (x^2+3y^2)\,dxdy = \frac{1}{3}$, where $D: 0 \le x, 0 \le y, \quad x+y \le 1$.
(9) $\iint_D xy\,dxdy = \frac{1}{24}$, where $D: 0 \le x, 0 \le y, \quad x+y \le 1$.
(10) $\iint_D x^2\,dxdy = \frac{1}{3}$, where $D: -1 \le x+y \le 1, -1 \le x-y \le 1$.
(11) $\iint_D e^{x+y}\,dxdy = e - \frac{1}{e}$, where $D: -1 \le x+y \le 1, -1 \le x-y \le 1$.
(12) $\iint_D e^{x-y}(x+y)\,dxdy = \frac{1}{4}(e-1)$, where $D: 0 \le x+y \le 1, 0 \le x-y \le 1$.
(13) $\iint_D (x+y)\,dxdy = 6$, where $D: 2 \le x+y \le 4, \quad 0 \le x-y \le 2$.
(14) $\iint_D xy\,dxdy = \frac{3}{2}\log 2$, where $D: 1 \le xy \le 2, 0 < \frac{1}{2}x \le y \le 2x$.
(15) $\iint_D y^2\,dxdy = \frac{9}{8}$, where $D: 1 \le xy \le 2, 0 < \frac{1}{2}x \le y \le 2x$.
(16) $\iint_D (x^2+y^2)\,dxdy = \frac{9}{8}$, where $D: 1 \le xy \le 2, 0 < x \le y \le 2x$.
(17) $\iint_D x^2y^2\,dxdy = \frac{1}{24}\pi$, where $D: x^2+y^2 \le 1$.
(18) $\iint_D x^2\,dxdy = 4\pi$, where $D: x^2+y^2 \le 4$.
(19) $\iint_D y^2\,dxdy = \frac{15}{4}\pi$, where $D: 1 \le x^2+y^2 \le 4$.
(20) $\iint_D xy\,dxdy = \frac{1}{8}$, where $D: 0 \le x, 0 \le y, x^2+y^2 \le 1$.
(21) $\iint_D xy^2\,dxdy = \frac{2}{15}$, where $D: 0 \le x, x^2+y^2 \le 1$.
(22) $\iint_D (x-y)\,dxdy = -\frac{2\sqrt{2}}{3}$, where $D: x \le y, x^2+y^2 \le 1$.
(23) $\iint_D \frac{xy}{x^2+y^2+1}\,dxdy = 0$, where $D: 0 \le y, x^2+y^2 \le 1$.
(24) $\iint_D \sqrt{x^2+y^2}\,dxdy = \frac{1}{6}\pi$, where $D: 0 \le x, 0 \le y, x^2+y^2 \le 1$.
(25) $\iint_D xy^2\,dxdy = \frac{1}{3}$, where $D: |y| \le x \le 2-|y|$.
(26) $\iint_D 2y\log x\,dxdy = \frac{64}{15}\log 2 - \frac{421}{225}$, where $D: 1 \le x \le 2, x^2 \le y \le 2x$.
(27) $\iint_D y\,dxdy = -\frac{9}{4}$, where $D: y \le x-2, y^2 \le 4-x$.
(28) $\iint_D \frac{x}{y^2}\,dxdy = \frac{3}{2} - \log 2$, where $D: 1 \le x \le 2, 1 \le y \le x^2$.

11.9 Triple integrals

Exercise 110. Evaluate the following triple integrals in two ways viewing the domain as a vertical and a horizontal ordinate set. Plot the graphic surface of the domain.
(1) $\iiint_D z^3 \, dxdydz = \frac{31}{60}$, where $D: 0 \le x \le 1, 0 \le y \le 1, 0 \le z \le x+y$.
(2) $\iiint_D (x^2+y^2) \, dxdydz = \frac{112}{45}$, where $D: -1 \le x \le 1, -1 \le y \le 1, 0 \le z \le x^2+y^2$.
(3) $\iiint_D (x+y+z) \, dxdydz = \frac{33}{2}$, where $D: 0 \le x, 0 \le y, 0 \le z, x+2y+3z \le 6$.
(4) $\iiint_D (x^2+y^2+z^2) \, dxdydz = \frac{1}{20}$, where $D: 0 \le x \le 1, 0 \le y \le 1, 0 \le z, x+y+z \le 1$.
(5) $\iiint_D x^2+y^2 \, dxdydz = \frac{2}{3}$, where $D: -1 \le x+y \le 1, -1 \le x-y \le 1, 0 \le z \le 1$.
(6) $\iiint_D \cos(x+y+z) \, dxdydz = \frac{\pi^2}{8} - 1$, where $D: 0 \le x, 0 \le y, 0 \le z, x+y+z \le \frac{\pi}{2}$.
(7) $\iiint_D (x^2+y^2+z^2) \, dxdydz = \frac{4\pi}{5}$, where $D: x^2+y^2+z^2 \le 1$.
(8) $\iiint_D \frac{z}{x^2+y^2} \, dxdydz = \frac{1}{2}\log 2 - \frac{5}{16}$, where $D: 0 \le x, 0 \le y, 0 \le z, x+y+z \le 1$.
(9) $\iiint_D (x^2+y^2) \, dxdydz = \frac{8\pi}{15}$, where $D: x^2+y^2+z^2 \le 1$.
(10) $\iiint_D z \, dxdydz = \frac{\pi}{4}$, where $D: x^2+y^2+z^2 \le 1, 0 \le z$.
(11) $\iiint_D x^2 z \, dxdydz = \frac{\pi}{24}$, where $D: x^2+y^2+z^2 \le 1, 0 \le z$.
(12) $\iiint_D x^2 \, dxdydz = \frac{2\pi}{15}$, where $D: x^2+y^2+z^2 \le 1, 0 \le z$.
(13) $\iiint_D \frac{1}{(x^2+y^2+z^2+1)^2} \, dxdydz = 2\pi\left(\arctan a - \frac{a}{1+a^2}\right)$, where $D: x^2+y^2+z^2 \le a^2, a > 0$.
(14) $\iiint_D \frac{1}{\sqrt{1-x^2-y^2-z^2}} \, dxdydz = \pi^2$, where $D: x^2+y^2+z^2 < 1$.
(15) $\iiint_D z^2 \, dxdydz = \frac{1}{3}\pi$, where $D: x^2+y^2 \le 1, 0 \le z \le 1$.
(16) $\iiint_D \frac{z}{x^2+y^2} \, dxdydz = \pi \log 2$, where $D: 1 \le x^2+y^2 \le 4, 0 \le z \le 1$.
(17) $\iiint_D x^2 \, dxdydz = \frac{1}{12}\pi$, where $D: x^2+y^2 \le z \le 1$.
(18) $\iiint_D z \, dxdydz = \pi$, where $D: x^2+y^2 \le z \le 2-x^2-y^2$.
(19) $\iiint_D (x^2+y^2-z^2) \, dxdydz = \frac{\pi}{3}a^5$, where $D: x^2+y^2 \le a^2, |z| \le a (a > 0)$.
(20) $\iiint_D (2x^2-y^2)z \, dxdydz = \frac{\pi}{48}a^8$, where $D: 0 \le z \le a^2-x^2-y^2 (a > 0)$.
(21) $\iiint_D z \, dxdydz = \frac{1}{4}\pi$, where $D: 0 \le z \le \sqrt{x^2+y^2} \le 1$.
(22) $\iiint_D \frac{1}{x^2+y^2+z^2+1} \, dxdydz = 4\pi - \pi^2$, where $D: x^2+y^2+z^2 \le 1$.
(23) $\iiint_D \frac{1}{x^2+y^2+z^2+1} \, dxdydz = 2\pi - \frac{1}{2}\pi^2$, where $D: x^2+y^2+z^2 \le 1, z \ge 0$.
(24) $\iiint_D (x^2+y^2) \, dxdydz = \frac{8-5\sqrt{2}}{30}\pi$, where $D: x^2+y^2+z^2 \le 1, z \ge \sqrt{x^2+y^2}$.

(25) $\iiint_D z^2\,\mathrm{d}x\mathrm{d}y\mathrm{d}z = \frac{4-\sqrt{2}}{30}\pi$, where $D: x^2+y^2+z^2 \le 1, z \ge \sqrt{x^2+y^2}$.
(26) $\iiint_D \frac{1}{(x^2+y^2+z^2)^{3/2}}\,\mathrm{d}x\mathrm{d}y\mathrm{d}z = 4\pi\log 2$, where $D: 1 \le x^2+y^2+z^2 \le 4$.
(27) $\iiint_D \frac{1}{(x^2+y^2+z^2+1)^{3/2}}\,\mathrm{d}x\mathrm{d}y\mathrm{d}z = 4\pi\left(\log\frac{2+\sqrt{5}}{1+\sqrt{2}} - \frac{2}{\sqrt{5}} + \frac{1}{\sqrt{2}}\right)$, where D : $1 \le x^2+y^2+z^2 \le 4$.
(28) $\iiint_D (x^2+y^2+z^2)\,\mathrm{d}x\mathrm{d}y\mathrm{d}z = \frac{13}{192}$, where $D: 0 \le x+y+z \le 1, 0 \le -x+y+z \le 1, 0 \le x+2z \le 1$.
(29) $\iiint_D (x^2+y^2+z^2)\,\mathrm{d}x\mathrm{d}y\mathrm{d}z = \frac{7}{24}$, where $D: 0 \le x-y+z \le 1, 0 \le x+z \le 1, 0 \le y+2z \le 1$.
(30) $\iiint_D x\,\mathrm{d}x\mathrm{d}y\mathrm{d}z = 0$, where $D: -1 \le x+y+z \le 1, -1 \le -2y+z \le 1, 0 \le z, -1 \le -x+y+z \le 1$.

Chapter 12

Limit values and summability

Recall (11.56).

$$\arctan x = \sum_{n=0}^{\infty} \frac{(-1)^n}{2n+1} x^{2n+1}. \tag{12.1}$$

The series

$$\sum_{n=1}^{\infty} \frac{(-1)^{n-1}}{n}$$

is convergent in view of Corollary 2.3. To find its value, we appeal to the Abel continuity theorem, Theorem 12.2.

Since it is the value of the limit function $\log(1+z)$ at $z=1$, it follows from Corollary 12.2 that

$$\sum_{n=1}^{\infty} \frac{(-1)^{n-1}}{n} = \lim_{z \to 1-0} \log(1+z) = \log 2. \tag{12.2}$$

However, if we appeal to Corollary 12.2, we obtain a much stronger result

$$\sum_{n=1}^{\infty} \frac{e^{2\pi n x}}{n} = -\lim_{z \to e^{2\pi x}} \log(1-z) = -\log\left(1-e^{2\pi x}\right) \tag{12.3}$$

for $0 < x < 1$, where z approaches $e^{2\pi x}$ along a Stoltz path within the unit circle. For further development, cf. §12.2 below.

Various methods of summation have been invented so as to assign a value to a divergent series including Fourier series.

The process T of assigning a meaningful value α to a divergent series $\sum\limits_{n=1}^{\infty} a_n$ is called a **summability method** or a **method of summation** and denoted $\sum\limits_{n=1}^{\infty} a_n = \alpha \quad (T)$. It is often said that the T-**mean** or **average** is A.

Definition 12.1. A summability method (T) is called **regular** if every convergent series is summable.

A typical one is $(C,1)$-mean (Césaro mean of the first order) which we apply to the series

$$s_1 = \sum_{n=0}^{\infty} (-1)^n. \tag{12.4}$$

To assign this expected value to s_1, we consider the sequence of its partial sums $\{S_n\}$, which consists of $S_{2n} = 0, S_{2n+1} = 1$. The arithmetic mean of this sequence is $\sigma_{2n} := \frac{1}{2n}\sum_{k=1}^{2n} S_k = \frac{n}{2n} = \frac{1}{2}$ and $\sigma_{2n+1} := \frac{1}{2n+1}\sum_{k=1}^{2n+1} S_k = \frac{n+1}{2n+1} \to \frac{1}{2}$ as $n \to \infty$. Hence $\lim_{n\to\infty} \sigma_n = \frac{1}{2}$ and we say that the series s_1 is **Cesàro summable** or $(C,1)$-summable to $\frac{1}{2}$ and denoted by

$$s_1 = \frac{1}{2} \quad (C,1). \tag{12.5}$$

On the other hand, the power series corresponding to (12.4) is

$$f(z) = \sum_{n=1}^{\infty} (-1)^{n-1} z^n = \frac{z}{1+z},$$

absolutely convergent for $|z| < 1$.

Hence we may assign the **Abel mean** to the divergent series s_1

$$\sum_{n=1}^{\infty} (-1)^{n-1} = f(1) = \frac{1}{2} \qquad (A). \tag{12.6}$$

Here we state a *rather unique treatment of the Abel continuity theorem based on the convergence of generalized Dirichlet series and conformality of an analytic function.* This shows the power of generalization and complex analysis. Following tradition, we use $s = \sigma + it$ for the complex variable. Let $\{\lambda_n\} \subset \mathbb{R}$ be an increasing sequence for which we may suppose $\lambda_1 > 0$. For a sequence $\{a_n\} \subset \mathbb{C}$, the series

$$f(s) = \sum_{n=1}^{\infty} a_n e^{-\lambda_n s} \tag{12.7}$$

convergent in some half-plane, is called a **generalized Dirichlet series**.

1. If $\lambda_n = \log n$ with log denoting the principal value, $f(s) = \sum_{n=1}^{\infty} \frac{a_n}{n^s}$ is (an ordinary) Dirichlet series.
2. If $\lambda_n = n$ and $e^{-s} = w$, $f(z) = f(z) = \sum_{n=1}^{\infty} a_n w^n$ is the power series.

Theorem 12.1. *If the series (12.7) is convergent for $s = s_0 = \sigma_0 + it_0$, then $f(s)$ is uniformly convergent in the right half-plane $\sigma > \sigma_0$ in the wide sense and represents an analytic function there. More precisely, let D be an angular domain*

$$\sigma - \sigma_0 \geq 0, \quad \arg(s - s_0) \leq \delta \tag{12.8}$$

with $0 < \delta < \frac{\pi}{2}$. Then $f(s)$ is uniformly convergent on D in the wide sense.

Corollary 12.1. (Counterpart of Abel continuity theorem) *$f(s)$ approaches to $f(s_0)$ as $s \to s_0$ in the angular domain (12.8).*

Corollary 12.2. (Abel continuity theorem—complex case) *Suppose a power series $f(z) = \sum_{n=1}^{\infty} a_n z^n$ converges at the point z_0 on its circle of convergence. Draw two chords (inside the circle) that start from z_0 and form an angle δ with the tangent at z_0 of the circle ($0 < \delta < \frac{\pi}{2}$). Let Δ be the (closure of) intersection of this angular subdomain and the disc of convergence. Then as $f(z)$ approaches $f(z_0)$ as $z \to z_0$ in the angular domain inside Δ. This is called the limit along a* **Stoltz path**.

Proof. For the special case 2 above, i.e. $\lambda_n = n$ and $e^{-s} = z$, Corollary 12.1 amounts to continuity of power series. The corresponding domain under this analytic map is the unit disc since $|e^{-s}| = e^{-\sigma}$. To make this in conformity with the radius of convergence, we need to magnify it. The angle is preserved since this is a conformal mapping. □

Theorem 12.2. (Abel continuity theorem—real case) *Suppose the power series $f(s) = \sum_{n=0}^{\infty} a_n x^n$ is convergent with radius of convergence $r > 0$ and that it is convergent at $x = r$ $[x = -r]$, then it is uniformly convergent for $0 \leq x \leq r$ $[-r \leq x \leq 0]$ and we have continuity*

$$\lim_{x \to r} f(z) = f(r) \tag{12.9}$$

and similarly for $x \to -r$.

12.1 Averaged Toeplitz transformation

Theorem 12.3. *Suppose* $\lim_{n\to\infty} a_n = a$, *i.e.*

$$a_n = a + o(1) \quad \text{for} \quad n > N = N(\varepsilon) \tag{12.10}$$

for every $\varepsilon > 0$ and that the weight sequence $\{w_n\}$ satisfies the conditions ($W_n > 0$, w constant)

$$\sum_{k=1}^{M} |w_k| = o(W_n), \ \sum_{k=M+1}^{n} |w_k| = O(W_n), \sum_{k=1}^{n} w_k = wW_n(1 + o(1)) \tag{12.11}$$

for a suitably chosen $M = M(n, N) \in \mathbb{N}$ such that

$$N < M < n. \tag{12.12}$$

Then

$$\lim_{n\to\infty} \frac{1}{W_n} \sum_{k=1}^{n} w_k a_k = wa. \tag{12.13}$$

Proof. Note that $\{a_n\}$ is bounded: $a_n = O(1)$ for all n. Hence the sum of first M terms $A_M = \sum_{k=1}^{M} w_k a_k$ is $O(\sum_{k=1}^{M} |w_k|) = o(W_n)$. For the remaining sum $B_M = \sum_{k=M+1}^{n} w_k a_k$ we substitute (12.10) to obtain $B_M = a\sum_{k=M+1}^{n} w_k + o\left(\sum_{k=M+1}^{n} |w_k|\right) = a\sum_{k=M+1}^{n} w_k + o(1)$ by condition (12.11). Adding A_M, we have

$$A_M + B_M = a\sum_{k=1}^{n} w_k + o(W_n) = awW_n(1 + o(1)), \tag{12.14}$$

which is (12.13). □

Example 12.1. The case $w_k = 1$ and $W_n = n$. For condition (12.11) to be satisfied, we must have $M = o(n)$. We may choose $M = n^{1-\delta}$ for a $\delta > 0$. For (12.12) to hold, we must have $n > N^{\frac{1}{1-\delta}}$. Then

$$\lim_{n\to\infty} \frac{1}{n} \sum_{k=1}^{n} a_k = a. \tag{12.15}$$

This is regularity of the $(C, 1)$-method of summation, called Cauchy's formula. Cauchy's formula amounts to

$$\frac{1}{n} \sum_{k=M+1}^{n} a_k = a\frac{n-M}{n} + o\left(\frac{n-M}{n}\right) = a\frac{n-M}{n} + o(1) \tag{12.16}$$

$$\frac{1}{n} \sum_{k=1}^{M} a_k = a\frac{M}{n} + o(1) \tag{12.17}$$

for any $N < M < n$. This example is a special case of the following more general result.

Example 12.2. The case $w_k = k^l$ and $W_n = n^{l+1}$. For condition (12.11) to be satisfied, we must have $M = o(n)$. We may choose $M = n^{1-\delta}$ for a

$\delta > 0$. For (12.12) to hold, we must have $n > N^{\frac{1}{1-\delta}}$. Recall the well-known formula

$$\sum_{k=1}^{L} k^{l+1} = \frac{1}{l+1}\left(B_{l+1}(L) - B_{l+1}\right) \tag{12.18}$$

where $B_k(x) = \sum_{r=0}^{k} \binom{n}{r} B_{k-r} x^r$ is the kth Bernoulli polynomial and B_r is the rth Bernoulli number. Hence with $W_n = n^{l+1}$ and $w = \frac{1}{l+1}$ Condition (12.11) reads

$$\sum_{k=1}^{M} w_k = M^{l+1} = o(n^{l+1}), \sum_{k=M+1}^{n} w_k = O(n^{l+1}) = O(W_n), \sum_{k=1}^{n} w_k = wW_n(1+o(1)).$$

Hence choosing $M = n^{1-\delta} = o(n)$, say we conclude that

$$\lim_{n\to\infty} \frac{1}{n^{l+1}} \sum_{k=1}^{n} k^l a_k = \frac{1}{l+1} a. \tag{12.19}$$

The case $l = 0$ is Example 12.1. The case $l = 1$ reads

$$\lim_{n\to\infty} \frac{1}{n^2} \sum_{k=1}^{n} k a_k = \frac{1}{2} a.$$

Example 12.3. (Cesàro) Assume that $\lim_{n\to\infty} a_n = a$ and $\lim_{n\to\infty} b_n = b$. Then we have

$$\lim_{n\to\infty} \frac{\sum_{k=1}^{n} a_k b_{n-k+1}}{n} = ab. \tag{12.20}$$

By (12.17), $A_M = \frac{1}{n}\sum_{k=1}^{M} b_{n-k+1} a_k$ is $ab\frac{M}{n} + o(1)$ while

$$B_M = \frac{1}{n} \sum_{k=M+1}^{n} b_{n-k+1} a_k = \frac{a}{n} \sum_{k=M+1}^{n} b_{n-k+1} + o\left(\frac{n-M}{n}\right)$$
$$= \frac{a}{n} \sum_{k=1}^{n-M} b_k + o(1) = ab\frac{n-M}{n} + o(1).$$

Adding these, we conclude the assertion.

Direct solution. We fix an $N \in \mathbb{N}$ for which $a_n = a + o(1)$ and $\frac{\sum_{k=1}^{n} a_k}{n} = a + o(1)$ and the same for $\{b_n\}$. Note that $\frac{\sum_{k=M+1}^{n} a_k}{n}$: $b_n = b + o(1)$ and $\frac{\sum_{k=1}^{n} b_k}{n} = b + o(1)$ for $cn - Mna + o(1)$. Divide the sum $\frac{\sum_{k=1}^{n} a_k b_{n-k+1}}{n}$ as $B_M + A_M$, where $B_M = \frac{1}{n}\sum_{k=1}^{M} a_k b_{n-k+1}$ and $A_M = \frac{1}{n}\sum_{k=M+1}^{n} a_k b_{n-k+1}$. We must have $n - M + 1 > N$ in B_M and $M + 1 > N$ in A_M, which implies $n > N + M + 1 > 2N$. In order for M to lie in the interval $N - 1 < M < n - N - 1$ we choose the average $M = \left[\frac{n+1}{2}\right]$. Then

$$B_M = b\frac{1}{n}\sum_{k=1}^{M} a_k = b\frac{1}{n}\sum_{k=1}^{n} a_k - b\frac{1}{n}\sum_{k=M+1}^{n} a_k = \frac{M}{n}ab = \frac{1}{2}ab,$$

where we omitted the errors $o(1)$. Similarly for the A_M, whence (12.20).

Example 12.4. Assume that $\lim_{n\to\infty} a_n = a$. Then we have

$$\lim_{n\to\infty} \frac{\sum_{k=0}^{n}\binom{n}{k}a_k}{2^n} = a. \tag{12.21}$$

For recall that

$$2^n = e^{n\log 2} \geq \frac{1}{(M+1)!}n^{M+1}. \tag{12.22}$$

Note that for $k, M \in \mathbb{N}$, $k \leq M$, we have

$$\frac{1}{n^M}\binom{n}{k} = \frac{1}{n^{M-k}}\frac{1}{k!}\left(1-\frac{1}{n}\right)\cdots\left(1-\frac{k-1}{n}\right) \leq \frac{1}{n^{M-k}}\frac{1}{k!}, \tag{12.23}$$

and

$$c_k(n) := \frac{1}{n^k}\binom{n}{k} = \frac{1}{k!}\left(1-\frac{1}{n}\right)\cdots\left(1-\frac{k-1}{n}\right). \tag{12.24}$$

Hence by (12.23)

$$\sum_{k=0}^{M}\frac{1}{n^{M+1}}\binom{n}{k} = \frac{1}{n^{M+1}} + \sum_{k=1}^{M}\frac{1}{n^{M+1}}\binom{n}{k} = \frac{1}{n}o\left(1+\sum_{k=1}^{M}\frac{1}{k!}\right) = O\left(\frac{1}{n}\right) = o(1). \tag{12.25}$$

Hence by (12.22) we have *a fortiori*

$$\frac{1}{2^n}\sum_{k=0}^{M}\binom{n}{k} = o(1).$$

Hence Theorem 12.3 applies.

Remark 12.1. (12.24) may be used to prove the existence of the limit

$$e = \lim_{n\to\infty}\left(1+\frac{1}{n}\right)^n, \tag{12.26}$$

the **base to the natural logarithm** denoted e. Cf. (1.2). For we write

$$a_n = \left(1+\frac{1}{n}\right)^n = 2 + \sum_{k=2}^{n} c_k(n). \tag{12.27}$$

Then note that $c_k(n)$ in (12.24) is for $2 \le k \le n$

$$c_k(n) = \frac{1}{k!}\left(1-\frac{1}{n}\right)\cdots\left(1-\frac{k-1}{n}\right) \le \begin{cases} c_k(n+1) \\ \frac{1}{k!}. \end{cases} \tag{12.28}$$

Hence (12.27) is estimated as

$$a_n = 2 + \sum_{k=2}^{n} c_k(n) \le \begin{cases} 2 + \sum_{k=2}^{n} c_k(n+1) + c_{n+1}(n+1) = a_{n+1} \\ 2 + \sum_{k=2}^{n} \frac{1}{k!} \le 2 + \sum_{k=2}^{\infty} \frac{1}{2^{k-1}} = 3. \end{cases} \tag{12.29}$$

Hence $\{a_n\}$ is a bounded monotone increasing sequence and so by R_3 it is convergent.

Exercise 111. Prove that the sequence $\{b_n\}$ where b_n is defined by

$$b_n := \sum_{k=0}^{n} \frac{1}{k!} < 3 \tag{12.30}$$

converges to e.

Solution. (12.29) incidentally proves that it is bounded from above. Since it is an increasing sequence, it converges to b, say. We have $e \le b$. On the other hand, we have for

$$a_n \ge 2 + \sum_{k=1}^{M} \frac{1}{k!}\left(1-\frac{1}{n}\right)\cdots\left(1-\frac{k-1}{n}\right) = \sum_{k=1}^{M} \frac{1}{k!} + o(1) = b + o(1) \tag{12.31}$$

for $n \ge M > N$. Hence $e = b$ and

$$e = \sum_{k=0}^{\infty} \frac{1}{k!}. \tag{12.32}$$

This is the value at $x = 1$ of the exponential function e^x. Cf. §11.3.

Exercise 112. Prove that the sequences $\{a_n\}$, $\{b_n\}$ where a_n, b_n are defined by

$$a_n = n^{\frac{1}{n}}, \quad b_n = a^{\frac{1}{n}} \quad (a > 0) \tag{12.33}$$

converge to 1.

Solution.

$$\frac{a_{n+1}}{a_n} = \left(\frac{(n+1)^n}{n^{n+1}}\right)^{\frac{1}{n(n+1)}} = \left(\left(1+\frac{1}{n}\right)^n \frac{1}{n}\right)^{\frac{1}{n(n+1)}} < \left(\frac{3}{n}\right)^{\frac{1}{n(n+1)}}. \tag{12.34}$$

Hence for $n > 3$, $\{a_n\}$ is monotone decreasing. Further, since

$$0 \le n-1 = \left(n^{\frac{1}{n}} - 1\right)\left(n^{\frac{n-1}{n}} + \cdots + 1\right), \tag{12.35}$$

we have $a_n \ge 1$. By R_3 the sequence converges to $1 \le \inf\{a_n\} = \ell$, say. If $\ell = 1+h$, $h > 0$, then $a_n \ge 1+h$ and

$$n \ge (1+h)^n > \frac{h^2}{2} n(n-1)$$

which is impossible for $n > \frac{2}{h^2} + 1$. Hence $\ell \le 1$, so that $\ell = 1$.

Similarly, for $a > 1$

$$\frac{b_{n+1}}{b_n} = a^{-\frac{1}{n(n+1)}} < 1,$$

so that $\{b_n\}$ is monotone decreasing and bounded by 1 from below. Hence the sequence converges to $1 \le \inf\{b_n\} = \ell$, say. If $\ell = 1+h$, $h \ge 0$, then $b_n \ge 1+h$ and

$$a \ge (1+h)^n > nh,$$

whence

$$h < \frac{a}{n} \to 0 \quad (n \to \infty).$$

Hence $h \le 0$. It follows that $h = 0$ and $\ell = 1$. The case $0 < a < 1$ follows from $\left(\frac{1}{a}\right)^{\frac{1}{n}}$.

Consider an infinite degree matrix

$$T = \begin{pmatrix} t_{00} & t_{01} & \cdots & t_{0l} & \cdots \\ t_{10} & t_{11} & \cdots & t_{1l} & \cdots \\ \cdots \\ t_{n0} & t_{n1} & \cdots & t_{nl} & \cdots \\ \cdots \end{pmatrix}. \tag{12.36}$$

Given a sequence $\{a_n\}$ we form the sequence with nth row of T

$$\sigma_n = \sum_{l=0}^{\infty} t_{nl} a_l \tag{12.37}$$

called an inner transformation of $\{a_n\}$ by T. If $\lim_{n\to\infty} \sigma_n = a$, then we say that $\{a_n\}$ is T summable to a and write $a_n \to a \quad (T)$. If the matrix T satisfies the following Toeplitz condition, it is called regular:

(i) $\sum_{n=0}^{M} |t_{nl}| = o(1)$ for each fixed l and M.

(ii) $\sum_{l=0}^{\infty} |t_{nl}| = O(1)$.

(iii) $\sum_{l=0}^{\infty} t_{nl} = w$.

Theorem 12.4. *If* $\lim_{n\to\infty} a_n = a$, *then* $\lim_{n\to\infty} \sigma_n = wa$.

Proof. Choose $N \in \mathbb{N}$ such that $a_l = a + o(1)$ for $l > N$. Divide the sum in (12.37) into two: $l \le N$, $N+1 < l$. Substituting $a_l = a + o(1)$ in the second sum, we have

$$\sigma_n = \sum_{l=0}^{N} + \sum_{l=N+1}^{\infty} t_{nl}a_l = a \sum_{l=N+1}^{\infty} t_{nl} + o(\sum_{l=N+1}^{\infty} |t_n|) + \sum_{l=0}^{N} t_{nl}a_l \tag{12.38}$$

$$= wa - a\sum_{l=0}^{N} t_{nl} + O(\sum_{l=0}^{N} |t_{nl}||a_l|)$$

$$= wa + O(\sum_{l=0}^{N} |t_{nl}|) = wa + o(1)$$

say, by (i)–(iii). □

Toeplitz condition (i) is usually stated as $\lim_{n\to\infty} t_{nl} = 0$ and so the above theorem is an averaged version.

12.2 Uniformly convergent function series

Lemma 12.1. (Generalization of Dirichlet's test) *Suppose that the partial sums of* $a_n(x)$ *are bounded uniformly in* x, $\lim_{n\to\infty} b_n(s) = 0$ *uniformly in* $\Re s = \sigma > 0$ *and* $|b_n(s) - b_{n+1}(s)| \le c_n$ *and* $\sum_{n=1}^{\infty} c_n < \infty$. *Then the series* $\sum_{n=1}^{\infty} a_n(x)b_n(s)$ *is uniformly convergent in* $\Re s = \sigma > 0$ *and* $x \in \mathrm{I} = [a,b]$.

This is a generalization of Theorem 2.2 and proof can be given using Corollary 10.1.

Exercise 113. Prove that the **polylogarithm function of order** s or the **Lerch zeta-function**

$$l_s(x) = \sum_{n=1}^{\infty} \frac{e^{2\pi inx}}{n^s} \tag{12.39}$$

is absolutely convergent for $\Re s = \sigma > 1$, $x \in \mathbb{R}$ or $\mathrm{Im}\, x > 0$, $s \in \mathbb{C}$. Prove that for $s = 1$, the series on the right of (12.39) is uniformly convergent in an interval not containing an integer and defines the polylogarithm function of order 1, which is indeed, the monologarithm function:

$$l_1(x) = \sum_{n=1}^{\infty} \frac{e^{2\pi i n x}}{n} = -\log\left(1 - e^{2\pi i x}\right). \tag{12.40}$$

Theorem 12.5. (Dirichlet-Abel) *Let q be a fixed modulus > 1. Let $R_n(x)$ denote a complex-valued function defined on $\mathrm{I} = [0,1]$ such that $R_n(x) = R_k(x)$ for $n \equiv k \pmod q$, and a fortiori, there are q different functions. Assume that each $R_k(x)$ is of Lipschitz α, $\alpha \geq 1$, [in symbol $R_k(x) \in \mathrm{Lip}\,\alpha$] and the vanishingness condition*

$$\sum_{k=1}^{q} R_k(x) = 0 \tag{12.41}$$

for each $x \in \mathrm{I}$. Then the Dirichlet series

$$F(s) = F(s,x) = \sum_{n=1}^{\infty} \frac{R_n(x^n)}{n^s} \tag{12.42}$$

is uniformly convergent in $\mathrm{Re}\, s = \sigma > 0$ and $x \in \mathrm{I}$.

If, further, R_k are all continuous on I, then $F(s)$ is also continuous on I and

$$F(1,1) = \sum_{n=1}^{\infty} \frac{R_n(1)}{n} = -\frac{1}{q}\sum_{k=1}^{q} R_k(1)\psi\left(\frac{k}{q}\right) = \sum_{k=1}^{q} \hat{R}_k(1) l_1\left(\frac{k}{q}\right), \tag{12.43}$$

where $l_1(x)$ is (12.40) and ψ is the Euler digamma function

$$\psi(x) = \frac{\Gamma'}{\Gamma}(x). \tag{12.44}$$

For a Dirichlet series $F(s) = \sum_{n=1}^{\infty} \frac{a_n}{n^s}$ with periodic coefficients $\{a_n\}$ of period q, (12.43) reads

$$\lim_{s\to 1}\left(F(s) - \frac{\frac{\hat{a}_q}{\sqrt{q}}}{s-1}\right) = -\frac{1}{q}\sum_{k=1}^{q} a_k \psi\left(\frac{k}{q}\right) = \sum_{k=1}^{q-1} \hat{a}_k \ell_1\left(\frac{k}{q}\right) + \frac{\hat{a}_q}{\sqrt{q}}\gamma, \tag{12.45}$$

where

$$\hat{a}_n = \frac{1}{\sqrt{q}}\sum_{k=1}^{q} e^{-2\pi i k \frac{n}{q}} a_k, \quad \hat{a}_1 = \sqrt{q}. \tag{12.46}$$

Since for any fixed positive integer $m > 1$, $\sum_{j=1}^{m} e^{2\pi i \frac{j}{m}} = \frac{e^{2\pi i \frac{1}{m}}(1-e^{2\pi i})}{1-e^{2\pi i \frac{1}{m}}} = 0$, therefore by applying Lemma 12.1, we conclude that $l_s(z)$ is uniformly convergent for $\mathrm{Re} s > 0$ if $z = \frac{k}{m} \in \mathbb{Q}, 0 < k < m$.

Proof. We apply Lemma 12.1. Since with $b_n(s) = n^{-s}$ for $\sigma > 0$, $\lim_{n\to\infty} b_n(s) = 0$ and $|b_n(s) - b_{n+1}(s)| = \left|\frac{1}{s}\int_n^{n+1} t^{-\sigma-1}\mathrm{d}t\right| = O(n^{-\sigma-1})$, we have $\sum_{n=1}^{\infty} c_n = \zeta(\sigma+1) < \infty$ for $\sigma > 0$.

The boundedness of the partial sum of $a_n(x) = R_n(x)$ follows from the following remark.

Writing $n = jm + k$, $1 \le k \le m$, $0 \le j \le \left[\frac{n}{m}\right]$ we may express $F(s,x)$ as

$$F(s,x) = \sum_{k=1}^{m} \sum_{\substack{n=1 \\ n\equiv k \pmod m}}^{\infty} \frac{R_n(x^n)}{n^s} = m^{-s} \sum_{k=1}^{m}\sum_{j=0}^{\infty} \frac{R_k(x^{mj+k})}{(j+\frac{k}{m})^s}. \tag{12.47}$$

Hence it suffices to consider the partial zeta-function

$$F_k(s,x) = \sum_{\substack{n=1 \\ n\equiv k \pmod m}}^{\infty} \frac{R_k(x^n)}{n^s}.$$

By assumption

$$\left|R_k(x^{jm+k}) - R_k(x^{jm})\right| = O(x^{jm\alpha}\left|x^k - 1\right|^{\alpha}).$$

Hence if $m \mid N$

$$\sum_{n=1}^{N} R_n(x^n) = \sum_{j=0}^{[N/m]}\sum_{k=1}^{m} R_k(x^{jm+k}) = \sum_{j=0}^{[N/m]}\sum_{k=1}^{m} R_k(x^{jm})$$
$$+ O\left(\sum_{j=0}^{[N/m]} x^{\alpha jm} \sum_{k=1}^{m}(1-x^k)^{\alpha}\right).$$

Now

$$\sum_{j=0}^{[N/m]} (x^{\alpha m})^j \sum_{k=1}^{m}(1-x^k)^{\alpha} \tag{12.48}$$

$$\le \begin{cases} 0, & x = 1 \\ \frac{1}{1-x^{\alpha m}}(1-x)^{\alpha} \sum_{k=1}^{m}(1+x+\cdots+x^{k-1})^{\alpha}, & 0 \le x < 1 \end{cases}$$

$$= \begin{cases} 0, & x = 1 \\ O\left(\left(\frac{1-x}{(1-x^{\alpha m})^{\frac{1}{\alpha}}}\right)^{\alpha}\right), & 0 \le x < 1 \end{cases}$$

(the last line being valid because $\alpha > 0$).

Now note that

$$\lim_{x\to 1^-} \frac{1-x}{(1-x^{\alpha m})^{\frac{1}{\alpha}}} = \lim_{x\to 1^-} \frac{(1-x^{\alpha m})^{1-\frac{1}{\alpha}}}{m x^{\alpha m-1}} = 0$$

for $1 - \frac{1}{\alpha} > 0$, i.e. $\alpha > 1$, and that for $\alpha = 1$, the above sum is $O(1)$. Hence altogether it follows that

$$\sum_{n=1}^{N} R_n(x^n) = \sum_{k=1}^{m} \sum_{j=0}^{[N/m]} R_k(x^{jm}) + O(1) = 0 + O(1) = O(1)$$

by assumption. Hence the series for $F(s,x)$ is uniformly convergent for $\sigma > 0$, $x \in \mathrm{I}$.

If $R_k(x)$ are continuous, then the uniformly convergent series of continuous functions being continuous, we conclude that $F(s,x)$ is continuous in $x \in \mathrm{I}$ for each s in $\{s|\ \Re s = \sigma > 0\}$. Hence in particular,

$$F(1,x) = \sum_{n=1}^{\infty} \frac{R_n(x^n)}{n}$$

being continuous in $x \in \mathrm{I}$, we conclude (12.43). □

Chapter 13

ODE and FDE

In this chapter, we shall treat FDE (Finite Difference Equations, also called linear recurrence sequences) and ODE (Ordinary Differential Equations) on equal footing. This is possible because of the Pincherle duality to be described in §13.4. We shall state the use of Laplace transforms and some applications of the difference operator as a counterpart of the differential operator.

13.1 Fibonacci sequence and Binet's formula

The sequence $\{F_n\}$ defined by the recurrence

$$F_{n+2} = F_{n+1} + F_n \tag{13.1}$$

with the initial terms $F_1 = 1$, $F_2 = 1$ is called the **Fibonacci sequence** for which we have

Theorem 13.1. (Binet's formula) *The general term is given by*

$$F_n = \frac{1}{\sqrt{5}}\left(\tau^n - (-\tau^{-1})^n\right), \tag{13.2}$$

where $\tau = \frac{1+\sqrt{5}}{2} = 1.618\cdots$ *is the golden ratio* (8.125).

Proof. We prove the above formula by the diagonalization of the matrix

$$A = \begin{pmatrix} 0 & 1 \\ 1 & 1 \end{pmatrix}. \tag{13.3}$$

Putting $\boldsymbol{x}_n = \begin{pmatrix} F_n \\ F_{n+1} \end{pmatrix}$, we obtain

$$\boldsymbol{x}_{n+1} = A\boldsymbol{x}_n \tag{13.4}$$

and so the situation is the same as the case of a **geometric sequence** $\{a_n\}$ with $a_{n+1} = ra_n$ whose general term is $a_n = r^{n-1}a_1$. Hence **analogously**,

$$\boldsymbol{x}_{n+1} = A^n \boldsymbol{x}_1 \tag{13.5}$$

and it suffices to find a general form of A^n. To this end, we use the method of diagonalization.

Step I. First we find the eigenvalues of A. The eigen equation (characteristic equation) is

$$0 = |A - \lambda E| = \begin{vmatrix} -\lambda & 1 \\ 1 & 1-\lambda \end{vmatrix} = \lambda^2 - \lambda - 1,$$

and so the eigenvalues are $\lambda = \tau$ and $\lambda = \rho = -\tau^{-1}$. We use the following data.

$$\tau + \rho = 1, \tau\rho = -1, \tau - \rho = \sqrt{5}, 1 - \tau^{-1} = \tau{-}2, \tau - 1 = \tau^{-1}.$$

Step II. Secondly, we find the eigenspace belonging to each root of the eigen equation:

$$E_A(\lambda) = \left\{ \boldsymbol{x} = \begin{pmatrix} x_1 \\ x_2 \end{pmatrix} \in \mathbb{C}^2 \middle| (A - \lambda E)\boldsymbol{x} = \mathbf{o} \right\}. \tag{13.6}$$

By simple calculation using fundamental operations, we find

$$E_A(\tau) = \mathbb{R}\begin{pmatrix} \tau - 1 \\ 1 \end{pmatrix} = \mathbb{R}\begin{pmatrix} \tau^{-1} \\ 1 \end{pmatrix}, \; E_A(\rho) = \mathbb{R}\begin{pmatrix} \rho - 1 \\ 1 \end{pmatrix} = \mathbb{R}\begin{pmatrix} -\tau \\ 1 \end{pmatrix}. \tag{13.7}$$

Step III. Putting

$$P = \begin{pmatrix} \tau^{-1} & -\tau \\ 1 & 1 \end{pmatrix}, \tag{13.8}$$

we conclude that it is a regular matrix by the theory of eigenspaces and we should have the diagonalization

$$P^{-1}AP = \begin{pmatrix} \tau & 0 \\ 0 & -\tau^{-1} \end{pmatrix}, \tag{13.9}$$

whence we have

$$A^n = P \begin{pmatrix} \tau^n & 0 \\ 0 & (-\tau^{-1})^n \end{pmatrix} P^{-1}. \tag{13.10}$$

Up here there is no need to know the value of the determinant $|P|$, which can be found to be $|P| = \tau^{-1} + \tau = \sqrt{5} \neq 0$ and

$$P^{-1} = \frac{1}{\tau + \tau^{-1}} \begin{pmatrix} 1 & \tau \\ -1 & \tau^{-1} \end{pmatrix}. \tag{13.11}$$

By a direct computation we obtain

$$A^n = \frac{1}{\sqrt{5}} \begin{pmatrix} \tau^{n-1} - (-\tau^{-1})^{n-1} & \tau^n - (-\tau^{-1})^n \\ \tau^n - (-\tau^{-1})^n & \tau^{n+1} - (-\tau^{-1})^{n+1} \end{pmatrix}, \tag{13.12}$$

Substituting (13.12) and (13.12), we obtain

$$\begin{aligned} F_{n+1} &= \frac{1}{\sqrt{5}}(\tau^{n-1}(\tau+1) - (-\tau^{-1})^{n-1}(-\tau^{-1}+1)) \\ &= \frac{1}{\sqrt{5}}\{\tau^{n+1} - (-\tau^{-1})^{n+1}\} \end{aligned} \tag{13.13}$$

which leads to (13.2) immediately in view of the above data, completing the proof. □

Remark 13.1. The Fibonacci sequence is often given with initial values $F_0 = 1, F_1 = 1$, in which case (13.2) should read

$$F_{n-1} = \frac{1}{\sqrt{5}} \left(\tau^n - (-\tau^{-1})^n\right). \tag{13.14}$$

13.2 The operator method

In this section we shall give another method for solving the FDE by the operator. The **difference operator** Δ is important in many disciplines as a substitute for the differential operator D:

$$\Delta y_n = y_{n+1} - y_n. \tag{13.15}$$

Let E denote the **shift operator**: $Ey_n = y_{n+1}$. Then, $\Delta = E - I$, where I indicates the identity operator, which is sometimes omitted, or $E = \Delta + I$. Hence any recurrence equation may be expressed as an FDE, and so we adopt the terminology FDE in what follows.

With the shift operator, (13.1) may be expressed as

$$(E^2 - E - I)y_n = 0, \tag{13.16}$$

It is important to recall the **Laplace transform** method in ODE (ordinary differential equations), in which there are two fundamental ingredients, viz., the existence of the inverse Laplace operator and the **partial fraction expansion.** In our case, the partial fraction expansion of the inverse operator is

$$\begin{aligned}\frac{1}{E^2 - E - I} &= \frac{1}{\sqrt{5}}\left(\frac{1}{E-\tau} - \frac{1}{E-\rho}\right) \\ &= \frac{1}{\sqrt{5}}\left((E-\tau)^{-1} - (E-\rho)^{-1}\right),\end{aligned} \tag{13.17}$$

with $(E-\alpha)^{-1}$ indicating the inverse operator, which we assume to exist. Now we may argue as follows. We think of the solutions of (13.16) as the **result with the inverse operator operated on** 0, i.e. we view the formal operation

$$y_n = (E-\alpha)^{-1}0 \tag{13.18}$$

as finding $\{y_n\}$ such that

$$0 = (E-\alpha)y_n = y_{n+1} - \alpha y_n. \tag{13.19}$$

This gives rise to $y_n = a_1\alpha^n$, a **geometric sequence**, where a_1 is the initial term. Substituting this in (13.17) with $\alpha = \tau, -\tau^{-1}$, we conclude

$$y_n = \frac{1}{\sqrt{5}}\left(\tau^n - (-\tau^{-1})^n\right), \tag{13.20}$$

which is (13.2), completing the proof.

Remark 13.2.

(1) It is also useful to remember the fact that the Laplace transform is invertible because it is a sort of **Fourier transform**. Save for the differential and integral operators, only the Fourier transform is **invertible**, which explains itself its importance in science.

(2) This method is rather instructive and leads to the Pincherle duality.

(3) The partial fraction expansion (13.17) should not be viewed as a mere petty computation. It is indeed the **residue calculus**.

13.3 The Laplace transform method

To solve linear DE with constant coefficients, we may use the Laplace transform method. Recall Definition 11.3. A general procedure of solving a DE by the Laplace transform method consists in finding the inverse Laplace transform such as (11.109) or (11.113). The reader can see many examples in literature where the Laplace transform is a rational function. Since the signals are usually exponential functions, this case will cover many practical applications.

Corollary 13.1. (Inverse Laplace transform) *Suppose $F(s)$ is analytic except for finitely many poles in $\mathbb{C}$ and that $F(s) = o(1)$, $|s| \to \infty$. Let $c > 0$. Suppose on* $\mathrm{Re}\, s = c$ *there is no pole of F and let $s_1, \cdots, s_n$ be the poles of F on the left of* $\mathrm{Re}\, s = c$. *Then for the Laplace transform $F(s) = \mathcal{L}[f](s)$ any $t > 0$*

$$f(t) = \mathcal{L}^{-1}[F](t) = \frac{1}{2\pi i}\,\mathrm{PV}\int_{Br} e^{st}F(s)\,\mathrm{d}s = \sum_{k=1}^{n} \operatorname*{Res}_{s=s_k} e^{st}F(s), \qquad (13.21)$$

where more precisely, the left-hand side is $\frac{1}{2}(f(t+0)+f(t-0))$ and PV *resp. Br means the* **Cauchy principal value** *and the vertical path $c - i\infty \to c + i\infty$, called* **Bromwich path** *compiled in*

$$\mathrm{PV}\int_{Br} e^{st}F(s)\,\mathrm{d}s = \lim_{R\to\infty}\int_{c-iR}^{c+iR} e^{st}F(s)\,\mathrm{d}s. \qquad (13.22)$$

Example 13.1. Suppose $b > 0, c \in \mathbb{R}$ satisfy $b^2 - 4c < 0$. Then we find the current $y = y(t)$ described by the DE

$$y'' + by' + cy = e^{-\frac{b}{2}t} \sin \frac{\sqrt{4c - b^2}}{2} t \qquad (13.23)$$

where the initial values are assumed to be 0: $y(0) = 0$, $y'(0) = 0$ is

$$y(t) = e^{-\frac{b}{2}t}\left(-\frac{1}{\sqrt{4c-b^2}}t\cos\sqrt{4c-b^2}t + \frac{2}{4c-b^2}\sin\sqrt{4c-b^2}t\right). \qquad (13.24)$$

Proof. Let $Y(s) = L[y](s)$ be the Laplace transform of $y(t)$. Then we have

$$\frac{2}{\sqrt{4c-b^2}}Y(s) = \frac{1}{(s^2+bs+c)^2}.$$

Solving the equation $s^2 + bs + c = 0$, we get the solutions $\alpha, \bar{\alpha}$, where $\alpha = \frac{-b+\sqrt{4c-b^2}i}{2}$, with the bar denoting the complex conjugate. And we are to find the partial fraction expansion

$$\frac{1}{(s^2+bs+c)^2} = \frac{1}{(s-\alpha)^2(s-\bar{\alpha})^2} = \frac{A}{(s-\alpha)^2} + \frac{B}{s-\alpha} + \frac{\bar{A}}{(s-\bar{\alpha})^2} + \frac{\bar{B}}{s-\bar{\alpha}}. \qquad (13.25)$$

The coefficients can be found by residue calculus as follows, thereby we take into account the following properties: $\alpha^2 + b\alpha + c = 0$, $\alpha - \bar{\alpha} = 2i \operatorname{Im} \alpha = \sqrt{4c - b^2}i$. To find A, B we clear the denominators by multiplying both sides of (13.25) by $(\alpha - \bar{\alpha})^2$ to find

$$\frac{1}{(s-\bar{\alpha})^2} = \frac{(s-\bar{\alpha})^2}{(s^2+bs+c)^2} \tag{13.26}$$
$$= A + B(s-\alpha) + (s-\alpha)^2 \left(\frac{\bar{A}}{(s-\bar{\alpha})^2} + \frac{\bar{B}}{(s-\bar{\alpha})} \right),$$

the right-hand side being the Taylor expansion of the left-hand side. Letting $s \to \alpha$, we find the y-intercept A: $A = \lim_{s\to\alpha} \frac{1}{(s-\bar{\alpha})^2} = \frac{1}{(\alpha-\bar{\alpha})^2} = -\frac{1}{4c-b^2}$. To determine B we differentiate both sides of (13.26) to find that

$$\frac{-2}{(s-\bar{\alpha})^3} = \frac{\mathrm{d}}{\mathrm{d}t} \frac{1}{(s-\bar{\alpha})^2} \tag{13.27}$$
$$= B + (s-\alpha)(\cdots),$$

whence $B = \lim_{s\to\alpha} \frac{-2}{(s-\bar{\alpha})^3} = \frac{-2}{(\alpha-\bar{\alpha})^3} = -\frac{2\sqrt{4c-b^2}}{(4c-b^2)^2}i$. Hence we conclude that

$$\frac{1}{(s^2+bs+c)^2} = \frac{-\frac{1}{4c-b^2}}{(s-\alpha)^2} + \frac{-\frac{2\sqrt{4c-b^2}}{(4c-b^2)^2}i}{s-\alpha} + \frac{-\frac{1}{4c-b^2}}{(s-\bar{\alpha})^2} + \frac{\frac{2\sqrt{4c-b^2}}{(4c-b^2)^2}i}{s-\bar{\alpha}}. \tag{13.28}$$

We have therefore

$$\frac{2}{\sqrt{4c-b^2}} y(t) = \frac{2}{\sqrt{4c-b^2}} L^{-1}[L[y]](t) \tag{13.29}$$
$$= \frac{2}{4c-b^2} t e^{-\frac{b}{2}t} \cos\sqrt{4c-b^2}t + \frac{4}{(4c-b^2)^{3/2}} e^{-\frac{b}{2}t} \sin\sqrt{4c-b^2}t, \tag{13.30}$$

or (13.24), completing the proof. □

Elucidation of (13.26).

We shall show that this is simply the Taylor expansion of $\frac{1}{(s-\bar{\alpha})^2}$ divided by $\frac{1}{(s-\alpha)^2}$.

Since $f(s) = \frac{1}{(s-\bar{\alpha})^2}$ is regular at $s = \alpha$, it can be expanded into the Taylor series. Recalling (13.33), we have

$$f^{(k)}(s) = k! \binom{-2}{k} (s-\bar{\alpha})^{-2-k}, \quad f^{(k)}(\alpha) = (k+1)!(-1)^k (\alpha-\bar{\alpha})^{-2-k}. \tag{13.31}$$

Hence the Taylor expansion of $\frac{1}{(s-\bar{\alpha})^2}$ at $s = \alpha$ is

$$f(s) = (\alpha - \bar{\alpha})^{-2} - 2(\alpha - \bar{\alpha})^{-3}(s-\alpha) + \sum_{k=2}^{\infty}(k+1)(-1)^k(\alpha - \bar{\alpha})^{-2-k}(s-\alpha)^k \tag{13.32}$$

whence we obtain the Laurent expansion at $s = \alpha$

$$\frac{1}{(s^2 + bs + c)^2} = P(s) + h(s), \tag{13.33}$$

where

$$P(s) = P_\alpha(s) = \frac{-\frac{1}{4c-b^2}}{(s-\alpha)^2} + \frac{-\frac{2\sqrt{4c-b^2}}{(4c-b^2)^2}i}{s-\alpha}$$

and

$$h(s) = h_\alpha(s) = \sum_{k=2}^{\infty}(k+1)(-1)^k(\alpha - \bar{\alpha})^{-2-k}(s-\alpha)^{k-2}$$

are the principal part and the holomorphic at $s = \alpha$, respectively.

In the same way, we have the Laurent expansion at $s = \bar{\alpha}$:

$$\frac{1}{(s^2 + bs + c)^2} = P_{\bar{\alpha}}(s) + h_{\bar{\alpha}}(s) \tag{13.34}$$

with $P_{\bar{\alpha}}(s) = \frac{-\frac{1}{4c-b^2}}{(s-\bar{\alpha})^2} + \frac{\frac{2\sqrt{4c-b^2}}{(4c-b^2)^2}i}{s-\bar{\alpha}}$. Hence we conclude (13.28) up to a holomorphic part. But

$$\frac{1}{(s^2 + bs + c)^2} - P_\alpha(s) - P_{\bar{\alpha}}(s)$$

is an integral function (i.e. analytic all over the complex plan) which is bounded (since it approaches 0 as $|s| \to \infty$. Hence Liouville's theorem implies that it is a constant. It must be 0 since it approaches 0.

Remark 13.3. Although, the DE (13.23) may be considered for any b. The reason why we restricted b to be positive can be seen from (13.24) in which the current goes to 0 only when $b > 0$. This case is known as a **transient state** which is the state of a circuit without current in the beginning $y(0) = 0$ and exerted current in one instant $e(t) = e^{-\frac{b}{2}t} \sin \frac{\sqrt{4c-b^2}}{2}t$ and immediate shut-down of switch. Then the exponential decrease of the current in the circuit obeys the law (13.24).

13.4 The Pincherle duality

In this section we shall state the Pincherle duality between ODE and FDE and elucidate the transition from ODE to FDE illustrated above. Pincherle [Pincherle (1888)] considered the correspondence between the following ODE and FDE to analyze two different methods of Pochhammer and Goursat toward the introduction of the generalized hypergeometric functions and established the duality.

$$\sum_{h=0}^{m}\sum_{k=0}^{p} a_{hk}e^{-kt}\psi^{(k)}(t)=0, \quad \sum_{k=0}^{p}\sum_{\ell=0}^{p} a_{\ell k}(x+k)^{\ell} f(x+k)=0, \tag{13.35}$$

where the functions ψ and f are required to be related under the Laplace type transforms.

$$f(x)=\int_{\ell} e^{-xt}\psi(t)\,\mathrm{d}t, \quad \psi(t)=\int_{L} e^{-st} f(s)\,\mathrm{d}s, \tag{13.36}$$

where ℓ and L are suitable integration paths in the complex domain.

What is the most interesting is Pincherle's approach to Goursat function which translates the FDE of the first order ($k \le 1$ in (13.35)) to the ODE of the first order. We follow [Mainardi and Pagnini (2003)] to allow the different exponents p and q. The FDE reads

$$\left(\sum_{\ell=0}^{p} a_{\ell 0}x^{\ell}\right) f(x)+\left(\sum_{\ell=0}^{q} a_{\ell 1}(x+1)^{\ell}\right) f(x+1)=0, \tag{13.37}$$

which corresponds to

$$\sum_{\ell=0}^{p} a_{\ell 0}\psi^{(\ell)}(t)+e^{-t}\left(\sum_{\ell=0}^{q} a_{\ell 1}\psi^{(\ell)}(t)\right)=0. \tag{13.38}$$

Using Mellin's results, the solutions of (13.37) are expressed as

$$f(x)=c^{x}\frac{\prod_{j=1}^{p}(x-\rho_j)}{\prod_{k=1}^{q}(x-\sigma_k)}, \tag{13.39}$$

where $c=-\frac{a_{p0}}{a_{q1}}$ is the complex constant. This serves as the integrand in (13.36):

$$\psi(t)=\int_{L} e^{st} f(s)\,\mathrm{d}s, \tag{13.40}$$

where

$$f(s)=c^{s}\frac{\prod_{j=1}^{m}(s-\rho_j)\prod_{k=1}^{n}(1+\sigma_k-s)}{\prod_{k=n+1}^{p}(s-\sigma_k)\prod_{j=m+1}^{q}(1+\rho_j-s)}. \tag{13.41}$$

By the change of variable

$$-\rho_j = b_j\,(1 \le j \le m), 1+\rho_j = b_j\,(m+1 \le j \le q),$$
$$1+\sigma_k = a_k\,(1 \le k \le n), -\sigma_k = a_k\,(n+1 \le k \le p),$$

(13.41) amounts to the integral for the Meijer G-function:

$$f(s) = c^s \frac{\prod_{j=1}^{m}(b_j+s)\prod_{k=1}^{n}(a_k-s)}{\prod_{k=n+1}^{p}(s+a_k)\prod_{j=m+1}^{q}(b_j-s)}. \tag{13.42}$$

Definition 13.1. We use the abbreviation. FDE = Finite Difference Equation, ODE = ordinary differential equation both with constant coefficients, gen. power ser. = generating power series, op. = operator, transf. = transform, lin.alg. = linear algebra. The equations are given in the form

$$f(D)y = 0, \quad f(E)y_1 = 0, \tag{13.43}$$

where $f(X) = c_n X^n + \cdots + c_1 X + c_0$ and $D^k f = f^{(k)}$, $E^k y_1 = y_{k+1}$. We consider only the case where the associated matrix A is diagonizable as $P^{-1}AP = \Lambda$ with Λ the diagonal matrix with eigenvalues entries.

Eq	op.	bases	method	inv. op.	lin.alg.
FDE	E	λ_j^n	gen. power ser.	Δ^{-1}	$P\Lambda^n P^{-1}\boldsymbol{x}(0)$
ODE	D	$e^{\lambda_j t}$	Laplace transf.	D^{-1}	$Pe^{\Lambda t}P^{-1}\boldsymbol{x}(0)$

Table 3. The Pincherle duality

13.5 Miscellany on the difference operator

In addition to operator method in §13.2, we shall give a symbolic method in the form of (13.44). It is used to find particular solutions (cf. e.g. [Levy and Lessman (1992), pp. 106–107]) based on the difference operator Δ. We shall state the applications of Δ in diverse settings.

$$(E-\alpha)^{-1} = (\Delta + 1 - \alpha)^{-1} = -\frac{1}{\alpha-1}\left(1 - \frac{\Delta}{\alpha-1}\right)^{-1} \tag{13.44}$$
$$= -\frac{1}{\alpha-1}\sum_{k=0}^{\infty}\frac{1}{(\alpha-1)^k}\Delta^k,$$

where the operation of Δ^k follows the rule [Comtet (1974), (6d), p. 14]

$$\Delta^k y_n = \sum_{j=0}^{k}(-1)^{k-j}\binom{k}{j}y_{n+j}. \tag{13.45}$$

If we apply to a polynomial in n, then it works normally. E.g. if $y_n = n^2$, then $\Delta^3 n^2 = 0$ and $\Delta^2 n^2 = 2$, $\Delta n^2 = 2n + 1$.

Example 13.2. (i) For $n \in \mathbb{N}$ we have

$$\Delta^k n^{-s} = \sum_{j=0}^{k} (-1)^{k-j} \binom{k}{j} (n+j)^{-s}, \tag{13.46}$$

and in particular,

$$\Delta^k 1^{-s} = \sum_{j=0}^{k} (-1)^{k-j} \binom{k}{j} (j+1)^{-s}, \tag{13.47}$$

which is due to Hasse [Hasse (1930), p. 460]. In particular, if $s = -n$, $n \in \mathbb{N} \cup \{0\}$, then $\sum_{k=0}^{n} \frac{1}{k+1} \Delta^k 1^{-s}$ is essentially the Bernoulli number.
(ii) Let $\langle x \rangle$ denote a principal unit in $\mathbb{Z}_p$ (cf. Remark 16.2), i.e. a unit such that $\langle x \rangle \equiv 1 \pmod{q}$, where $q = p$ for odd prime p and $q = 4$ for $p = 2$. Then

$$\Delta^k \langle x \rangle^{n-1} \equiv 1 \pmod{q^k}. \tag{13.48}$$

(13.48) follows from

$$\Delta^k \langle x \rangle^{n-1} = \sum_{j=0}^{k} (-1)^{k-j} \binom{k}{j} \langle x \rangle^{n-1+j} = \langle x \rangle^{n-1} (\langle x \rangle - 1)^k.$$

By applying the difference operator to the input function, we find the solutions to the FDE.

Subsequently, we shall give some applications of the difference operator to other sequences.

If we apply the difference operator to $y_n = \alpha^n$, we obtain

$$\Delta^k \alpha^n = \alpha^n \sum_{j=0}^{k} (-1)^{k-j} \binom{k}{j} \alpha^j = \alpha^n (\alpha - 1)^k. \tag{13.49}$$

Substituting (13.49) in (13.44) leads to an absurd result

$$(E - \alpha)^{-1} \alpha^n = \alpha^n \frac{-1}{\alpha - 1} \sum_{k=0}^{\infty} 1.$$

Most probably we cannot apply the infinite series of operators.

We recall the Frobenius theorem ([Comtet (1974), Theorem E, p. 244])

$$A_n(\alpha) = \alpha \sum_{k=0}^{n} k! S(n,k) (\alpha - 1)^{n-k}, \tag{13.50}$$

where $A(\alpha)$ is the **Eulerian polynomial** (introduced by generatingfunctionology).

Also we recall another formula, which was the motivation for introducing Eulerian polynomials ([Comtet (1974), Theorem F, p. 245])

$$\frac{1}{(1-\alpha)^{n+1}} A_n(\alpha) = \sum_{\ell=0}^{\infty} \ell^n \alpha^\ell. \tag{13.51}$$

Theorem 13.2. *Under the above notation, we have*

$$(E-\alpha)^{-1} 0^n = (-1)^n \sum_{\ell=0}^{\infty} \ell^n \alpha^{\ell-1}. \tag{13.52}$$

Proof. We appeal to the formula ([Comtet (1974), (1b), p. 204])

$$\Delta^k 0^n = \sum_{j=0}^{k} (-1)^{k-j} \binom{k}{j} j^n = k! S(n,k), \tag{13.53}$$

where $S(n,k)$ is the **Stirling number of the second kind**. Substituting (13.53) in (13.44) and noting that $S(n,k) = 0$ for $k > n$, we deduce that

$$(E-\alpha)^{-1} 0^n = -\frac{1}{(\alpha-1)^{n+1}} \sum_{k=0}^{n} k! S(n,k) (\alpha-1)^{n-k}. \tag{13.54}$$

By (13.50), we may express (13.54) as

$$(E-\alpha)^{-1} 0^n = -\frac{1}{\alpha(\alpha-1)^{n+1}} A_n(\alpha). \tag{13.55}$$

□

This is not of the desired form $((E-\alpha)^{-1})$ though it may be of some interest on its own.

Theorem 13.3. (Hasse) *For all $s \in \mathbb{C}$ and $n \in \mathbb{N}$, the formula*

$$\sum_{k=0}^{\infty} \frac{1}{k+1} \Delta^k n^{-s} = s \left(\zeta(s+1) - \sum_{\nu=1}^{n-1} \nu^{s+1} \right), \tag{13.56}$$

hold true, where $\zeta(s)$ indicates the Riemann zeta-function. (13.56) *is to read for $n = 1$*

$$\sum_{k=0}^{\infty} \frac{1}{k+1} \Delta^k 1^{-s} = s\zeta(s+1). \tag{13.57}$$

(13.56) *and* (13.57) *give the analytic continuation of the Riemann zeta-function over the whole plane.*

Proof. We apply the Hecke **gamma transform** which reads for $\lambda > 0$ and $\sigma > 0$

$$\Gamma(s)\lambda^{-s} = \int_0^\infty t^s e^{-\lambda t}\,\frac{dt}{t}. \tag{13.58}$$

Recalling (13.46), we deduce that

$$\Gamma(s)\sum_{k=0}^\infty \frac{1}{k+1}\Delta^k n^{-s} = \int_0^\infty \frac{1}{1-e^{-t}} e^{-nt} t^{s+1}\,\frac{dt}{t} \tag{13.59}$$

where we use the absolutely convergent series

$$\frac{1}{r}\sum_{k=0}^\infty \frac{1}{k+1} r^{k+1} = -\frac{1}{r}\log(1-r)$$

with $r = 1 - e^{-t}$ and $\frac{1}{1-e^{-t}} = \sum_{m=0}^\infty e^{-mt}$. Applying (13.58) implies (13.56). □

13.6 Generating power series

There are many generating function of a given sequence. In this section we restrict to generating power series of some interest and use.

1°. We consider a generating power series $f(z)$ associated with a sequence $\{a_n\}$:

$$f(z) = \sum_{n=0}^\infty a_n z^n = \frac{1}{z^2 + bz + c} \tag{13.60}$$

$b, c \in \mathbb{R}$, $b^2 - 4c > 0$. Let $\alpha = \frac{-b+\sqrt{b^2-4c}}{2}$ and $\beta = \frac{-b-\sqrt{b^2-4c}}{2}$ be the roots of the denominator $z^2 + bz + c$. Then the partial fraction expansion of $f(z)$ is

$$\begin{aligned}\frac{1}{z^2+bz+c} &= \frac{1}{\alpha-\beta}\left(\frac{1}{z-\alpha} - \frac{1}{z-\beta}\right) \\ &= \frac{1}{\alpha-\beta}\left(\frac{-\alpha^{-1}}{1-\frac{z}{\alpha}} - \frac{-\beta^{-1}}{1-\frac{z}{\beta}}\right).\end{aligned} \tag{13.61}$$

Hence for $|z| < \min|\alpha|, |\beta|$ we have the geometric series expansion

$$\begin{aligned}f(z) &= \frac{1}{\alpha-\beta}\left(\frac{1}{z-\alpha} - \frac{1}{z-\beta}\right) \\ &= -\frac{1}{\alpha-\beta}\left(\alpha^{-1}\sum_{n=0}^\infty \left(\frac{z}{\alpha}\right)^n - \beta^{-1}\sum_{n=0}^\infty \left(\frac{z}{\beta}\right)^n\right) \\ &= \frac{1}{\alpha-\beta}\sum_{n=0}^\infty \frac{\alpha^{n+1}-\beta^{n+1}}{(\alpha\beta)^{n+1}} z^n.\end{aligned} \tag{13.62}$$

Comparing the coefficients, we conclude that

$$a_n = \frac{1}{\alpha - \beta} \frac{\alpha^{n+1} - \beta^{n+1}}{c^{n+1}}. \tag{13.63}$$

In the case of Fibonacci sequence, we have $\alpha = \tau$, $\beta = -\tau^{-1}$ and $\alpha - \beta = \sqrt{5}$, so that (13.63) leads to Binet's formula (13.2).

Example 13.3. The generating power series for the Chebyshëv polynomials of the second kind is given by

$$f(z) = \sum_{n=0}^{\infty} U_n(x) z^n = \frac{1}{z^2 + 2xz + 1}. \tag{13.64}$$

Hence the universal expression valid for all complex $x \neq \pm 1$ follows:

$$U_n(x) = \frac{1}{2\sqrt{x^2-1}} \left(\left(x + \sqrt{x^2-1} \right)^{n+1} - \left(x - \sqrt{x^2-1} \right)^{n+1} \right), \tag{13.65}$$

where $\sqrt{x^2-1}$ means a branch which is positive for real $|x| > 1$.

2°. Let $f(n)$ denote the number of parentheses operated in the product of n elements in a non-associative algebraic structure where there is defined a product. Cf. §1.4. Defining $f(0) = 0$ and $f(1) = 1$, we find that

$$f(n) = \sum_{m=1}^{n-1} f(m) f(n-m) = \sum_{l+m=n} f(l) f(m). \tag{13.66}$$

For $|z| \ll 1$ (small enough) define

$$F(z) = \sum_{n=1}^{\infty} f(n) z^n. \tag{13.67}$$

Then

$$F(z)^2 = \sum_{n=2}^{\infty} \sum_{m=1}^{n-1} f(m) f(n-m) z^n = F(z) - z, \tag{13.68}$$

whence

$$F(z) = \frac{1}{2} \left(1 - \sqrt{1-4z} \right),$$

say, whence we see that we may suppose $|z| < \frac{1}{4}$. Invoking the binomial expansion (11.62), we conclude that

$$F(z) = \frac{1}{2} \sum_{n=1}^{\infty} \frac{(2n-3)!!}{(2n)!!} 4^n z^n, \tag{13.69}$$

whence

$$f(n) = \frac{(2n-3)!!}{2(2n)!!} 4^n. \tag{13.70}$$

13.7 Bernoulli and Euler polynomials

Definition 13.2. The product $\frac{z}{e^z-1}e^{xz}$ is analytic in $|z| < 2\pi$ and has the power series expansion of the form:

$$\frac{ze^{xz}}{e^z-1} = \sum_{n=0}^{\infty} \frac{B_n(x)}{n!} z^n, \quad |z| < 2\pi. \tag{13.71}$$

The nth coefficient $B_n(x)$ is called the nth **Bernoulli polynomial**.

Apparently $\frac{z}{e^z-1}e^{xz}$ is the Abel convolution in §8.3 and since (13.73) gives the power series coefficients, it follows that $\frac{B_n(x)}{n!} = \sum_{k+l=n} \frac{B_k}{k!}\frac{x^l}{(l)!}$, whence that

$$B_n(x) = \sum_{k=0}^{n} \binom{n}{k} B_k x^{n-k}. \tag{13.72}$$

$B_n = B_n(0)$ is called the nth **Bernoulli number**. They are therefore generated by the power series

$$\frac{z}{e^z-1} = \sum_{n=0}^{\infty} \frac{B_n}{n!} z^n, \quad |z| < 2\pi. \tag{13.73}$$

To check that the odd-indexed Bernoulli numbers are 0 we notice that

$$\sum_{n=0}^{\infty} \frac{B_n}{n!}(-z)^n = \frac{-z}{e^{-z}-1} = e^z \frac{z}{e^z-1} = z + \frac{z}{e^z-1} = \sum_{n=0}^{\infty} \frac{\tilde{B}_n}{n!} z^n, \tag{13.74}$$

where $\tilde{B}_n = B_n$ for $n \neq 1$ and $\tilde{B}_1 = B_1 + 1$. For odd n, we have $(-1)^n B_n = B_n$, so that $B_{2k+1} = 0$. (13.74) may be thought of as a special case of (13.74) with $x = 1$, so that we have in conjunction with (13.72),

$$\sum_{k=0}^{n} \binom{n}{k} B_k = B_n(1) = \tilde{B}_n \tag{13.75}$$

which is a special case of (10.50). Or

$$\sum_{k=0}^{n-1} \binom{n}{k} B_k = 0 \tag{13.76}$$

save for $n = 1$, in which case $B_1(1) = B_1 + 1 = \frac{1}{2}$. We may calculate Bernoulli numbers from the recursion (13.76)

$$B_0 = 1,\ B_1 = -\frac{1}{2},\ B_2 = \frac{1}{6},\ B_4 = -\frac{1}{30},\ B_{2k+1} = 0\ (k = 1, 2, \cdots). \tag{13.77}$$

The nth Bernoulli polynomial can be expressed as $(B+x)^n$: (13.72) may be viewed as an **umbral calculus** formula (Lucas 1891):

$$B_n(x) = (B+x)^n = \sum_{k=0}^{n} \binom{n}{k} B_{n-k}\, x^k, \tag{13.78}$$

where, by umbral calculus, we mean that after expanding the binomial, the exponent of B is to be degraded to subscript.

The first a few Bernoulli polynomials are:

$$\begin{aligned}
B_0(x) &= 1,\ B_1(x) = B_0 x + B_1 = x - \frac{1}{2},\\
B_2(x) &= B_0 x^2 + 2B_1 x + B_2 = x^2 - x + \frac{1}{6},\\
B_3(x) &= B_0 x^3 + 3B_1 x^2 + 3B_2 x + B_3 = x^3 - \frac{3}{2}x^2 + \frac{1}{2}x,\\
B_4(x) &= B_0 x^4 + 4B_1 x^3 + 6B_2 x^2 + 4B_3 x + B_4\\
&= x^4 - 2x^3 + x^2 - \frac{1}{30},\ B_4 = -\frac{1}{30},\\
B_5(x) &= x^5 - \frac{5}{2}x^4 + \frac{5}{3}x^3 - \frac{1}{6}x\\
B_6(x) &= x^6 - 3x^5 + \frac{5}{2}x^4 - \frac{1}{2}x^2 + \frac{1}{42},\ B_6 = \frac{1}{42}.
\end{aligned}$$

For its importance and ubiquity, we make special mention of the **first periodic Bernoulli polynomial** once and for all and use it freely without further mentioning with $[x]$ denoting the greatest integer function in Corollary 4.3:

$$\bar{B}_1(x) = x - [x] - \frac{1}{2} = \frac{1}{2\pi i} \sum_{\substack{k=-\infty \\ k\neq 0}}^{\infty} \frac{e^{2\pi i k x}}{k} = -\frac{1}{\pi} \sum_{n=1}^{\infty} \frac{\sin 2\pi n x}{n}, \tag{13.79}$$

where the equality holds for $x \notin \mathbb{Z}$ and the saw-tooth Fourier series on the far-right side is boundedly convergent. For $x \in \mathbb{Z}$, $\bar{B}_1(x) = -\frac{1}{2}$ while the Fourier series converges to 0 and care must be taken in this case. The saw-tooth Fourier series is often denoted by $\psi(x)$ or $((x))$ in literature.

$$\bar{B}_1(x) = \psi(x) = ((x)), \qquad x \notin \mathbb{Z}.$$

A proof of (13.79) is given in Exercise 129 below. Recently there is a discovery [Wang *et al.* (2019)] to the effect that the saw-tooth Fourier series is an intrinsic property of the polylogarithm function of order 1 or the Lerch zeta-function at $s = 1$, cf. (12.40).

More generally we also define the nth periodic Bernoulli polynomial once and for all by

$$\bar{B}_n(x) = B_n(x - [x]) = \sum_{k=0}^{n} \binom{n}{k} B_{n-k}\{x\}^k \tag{13.80}$$

which has an absolutely convergent Fourier series

$$\bar{B}_n(x) = -\frac{n!}{(2\pi i)^n} \sum_{\substack{k=-\infty \\ k\neq 0}}^{\infty} \frac{e^{2\pi i k x}}{k^n}. \tag{13.81}$$

This follows by termwise integration (assured by Theorem 11.9) of (13.79) on appealing to the uniformity of convergence as proved in Exercise 129.

In the same way, we may define the nth **Euler polynomials** $E_n(x)$ by

$$\frac{2e^{xz}}{e^z+1} = \sum_{n=0}^{\infty} \frac{E_n(x)}{n!} z^n, \quad |z| < \pi \tag{13.82}$$

the series being absolutely and uniformly convergent in $|z| < \pi$ (the nearest singularity from the origin being $z = \pm i\pi$). See below.

As stated on [Serre (1973), p. 90, Footnote (2)], there are several definitions of Bernoulli numbers. The most commonly used ones are those in the b-notation in [Serre (1973)] while Leopoldt's definition differs only at one value B_1 which is defined to be $\frac{1}{2}$ rather than $-\frac{1}{2}$. Here we followed Washington's notation [Washington (1982)] and introduced the B_n by (13.73). For a systematic account of Bernoulli polynomials, we refer to [Kanemitsu and Tsukada (2007), Chapter 1] and we freely use the results from it. There are many generalizations of Bernoulli numbers and polynomials. Most of them have been introduced so as to express the special values of the relevant zeta- and L-functions at negative integral argument, whence at certain positive integral arguments, while the Bernoulli numbers themselves were used to express the sum of powers of natural numbers up to n, say, which were used by Euler to solve the Basler problem $\zeta(2) = \frac{\pi^2}{6}$.

Along with the Bernoulli numbers, we introduce **Euler numbers** $\{E_{2n}\}$ either by

$$\frac{1}{\cosh z} = \frac{2}{e^z + e^{-z}} = \sum_{n=0}^{\infty} \frac{E_{2n}}{(2n)!} z^{2n}, \quad |z| < \frac{\pi}{2} \tag{13.83}$$

or by

$$\sec z = \frac{2}{e^{iz} + e^{-iz}} = \sum_{n=0}^{\infty} \frac{(-1)^n E_{2n}}{(2n)!} z^{2n}, \quad |z| < \frac{\pi}{2}. \tag{13.84}$$

These two are equivalent and shift to each other under the rotation of the complex plane by $\frac{\pi}{2}$. One can easily evaluate the following:

$$E_0 = 1, E_2 = -1, E_4 = 5, E_6 = -61, E_8 = 1385, E_{10} = -50521 \cdots . \tag{13.85}$$

Note that unlike Bernoulli numbers, Euler numbers are all integers and

$$E_n = 2^n E_n \left(\frac{1}{2}\right), \tag{13.86}$$

where the nth Euler polynomial defined by (13.82). (13.86) clearly follows from (13.83) and (13.82) on noting that $\frac{2}{e^z+e^{-z}} = \frac{2e^z}{e^{2z}+1}$.

On writing

$$(\sec z)^2 = \sum_{n=0}^{\infty} \frac{(-1)^n E_{2n}^{(2)}}{(2n)!} z^{2n}, \quad |z| < \frac{\pi}{2}, \tag{13.87}$$

$E_{2n}^{(2)}$ being called the $2n$th Euler number of order 2 it follows from (13.84) that

$$E_{2n}^{(2)} = (2n)! \sum_{v_1+v_2=n} \frac{E_{2v_1} E_{2v_2}}{(2v_1)!(2v_2)!}. \tag{13.88}$$

Comparing (5.34) to the tangent numbers in (13.87) (Exercise 47), we find that

$$E_{2n}^{(2)} = (-1)^n A_{2n+1}. \tag{13.89}$$

Using (13.88), (13.89), we see immediately

$$A_5 = 4!\left(2 \cdot \frac{E_0 E_4}{4!} + \frac{E_2 E_2}{2!2!}\right) = 16,$$

$$A_7 = -2 \cdot 6!\left(\frac{E_0 E_6}{6!} + \frac{E_2 E_4}{2!4!}\right) = 272,$$

so that

$$\frac{A_5}{5!} = \frac{2}{15}, \quad \frac{A_7}{7!} = \frac{17}{315}, \cdots .$$

Fig. 13.1: Jacobi

Chapter 14

Vector analysis

14.1 Differential forms

In this section we partially follow [Maurin (1973)] and [Buck (1965)] and give rudiments of the theory of differential forms in an elementary and concrete way. This enables us to elucidate higher differentials in Definition 8.1, the alternating property in Lemma 11.2, etc. (5.21) is a 1-form which is the differentiation of the 0-form (function) $y = y(x_1, \cdots, x_n)$.

Setting aside analytic aspects, we may also state some results up to graded algebra in Definition 14.2 from rather algebraic point of view, cf. Appendix E.

We write $\mathrm{d}\boldsymbol{x} = {}^t(\mathrm{d}x_1, \cdots, \mathrm{d}x_n)$. Then the differential of order k in (8.2) is the (differential) k-form arising from the 0-form f.

$$\mathrm{d}^k f(\boldsymbol{x}; \boldsymbol{h}) = (\boldsymbol{h} \cdot \nabla)^k f(\boldsymbol{x}) = \sum_{i_1, \cdots, i_k = 1}^{n} D_{i_1, \cdots, i_k} f(\boldsymbol{x}) \mathrm{d}x_{i_1} \cdots \mathrm{d}x_{i_k}, \tag{14.1}$$

where

$$D_{i_1, \cdots, i_k} f = \frac{\partial^k}{\partial x_{i_1} \cdots \partial x_{i_k}} f. \tag{14.2}$$

In general, let I denote the k-ple index

$$I = I_k = \{i_1, \cdots, i_k\}, \tag{14.3}$$

where $1 \le k \le n$ and define

$$\mathrm{d}\boldsymbol{x}_I = \mathrm{d}x_{i_1} \wedge \cdots \wedge \mathrm{d}x_{i_k} \leftrightarrow \mathrm{d}x_{i_1} \cdots \mathrm{d}x_{i_k}. \tag{14.4}$$

Let $\Omega \subset \mathbb{R}^n$ be an open domain. Then with coefficients $a_I = a_I(\boldsymbol{x})$ of class C^r, we define the (differential) k**-form** $\omega = \omega^k$ on Ω by

$$\omega = \omega^k(\boldsymbol{x}) = \sum_I a_I(\boldsymbol{x}) \mathrm{d}\boldsymbol{x}_I \tag{14.5}$$

$$= \sum_{i_1=1}^{n} \cdots \sum_{i_k=1}^{n} a_{i_1, \cdots, i_k}(\boldsymbol{x}) \mathrm{d}x_{i_1} \wedge \cdots \wedge \mathrm{d}x_{i_k}.$$

To give a meaning as a domain-functional to this expression, we recall the fixed k-dimensional cube in (11.2).

$$\mathbb{I} = \mathbb{I}^k = [a_1, b_1] \times \cdots \times [a_k, b_k]. \tag{14.6}$$

Let

$$\phi : \mathbb{I}^k \to \Omega \tag{14.7}$$

be a vector-valued function of C^r, $r \geq 0$ (on some open neighborhood of $\mathbb{I}^k$).

We call the pair

$$s_k = (\mathbb{I}^k, \phi), \tag{14.8}$$

or sometimes the function (14.7) the **singular cube** in Ω of dimension k. s_0 means a point in Ω. In view of Proposition 14.1, (v), we may assume $k \leq n$ (which is predicted by (14.1)). Let

$$|s_k| = \operatorname{Im} \phi = \phi(\mathbb{I}^k) \tag{14.9}$$

and call it the **carrier** (or support) of s_k. Hence this subdomain of Ω is parametrized by

$${}^t(x_1, \cdots, x_n) = \boldsymbol{x} = \phi(\boldsymbol{t}), \quad \boldsymbol{t} = {}^t(t_1, \cdots, t_k) \in \mathbb{I}^k \tag{14.10}$$

which we assume throughout. Also we shall stick to a special notation $\langle s_k, \omega^k \rangle$ for the inner product to emphasize that it means the integral (14.11) in the following theorem.

Theorem 14.1. *The k-form ω on Ω is a domain-functional which associates to s_k the value (which is indeed an inner product)*

$$\begin{aligned} \langle s_k, \omega^k \rangle &= \underbrace{\int \cdots \int_\Omega}_{k} \omega = \sum_I \int_{\mathbb{I}^k} a_I(\phi(\boldsymbol{t})) J_{\phi_I} \mathrm{d}\boldsymbol{t} \\ &= \sum_{i_1=1}^n \cdots \sum_{i_k=1}^n \int_{\mathbb{I}^k} a_{i_1, \cdots, i_k}(\phi(\boldsymbol{t})) \det \left(\frac{\partial x_{i_l}}{\partial t_m} \right) \mathrm{d}t_1 \cdots \mathrm{d}t_k, \end{aligned} \tag{14.11}$$

where J_{ϕ_I} indicates the Jacobian ((5.10))

$$J_{\phi_I} = \det \left(\frac{\partial x_{i_l}}{\partial t_m} \right)_{1 \leq l, m \leq k} = \frac{\partial(x_{i_1}, \cdots x_{i_k})}{\partial(t_1, \cdots, t_k)} \tag{14.12}$$

and $\mathrm{d}\boldsymbol{t} = \mathrm{d}t_1 \cdots \mathrm{d}t_k$ is the (Lebesgue) measure of $\mathbb{R}^k$.

Cf. Theorem 11.14 above for $k = n$.

Definition 14.1. Let $\boldsymbol{p}$ denote the $\binom{n}{k}$-dimensional vector with (14.12) as its Ith entry

$$p_I = J_{\phi_I} = \frac{\partial(x_{i_1}, \cdots x_{i_k})}{\partial(t_1, \cdots, t_k)}. \tag{14.13}$$

If

$$\rho(\boldsymbol{t})^2 := |\boldsymbol{p}|^2 = \sum_I \left(\frac{\partial(x_{i_1}, \cdots x_{i_k})}{\partial(t_1, \cdots, t_k)}\right)^2 > 0, \tag{14.14}$$

then the cube s_k is called regular. For a regular cube and continuous function f on it, we define the integral

$$\int f \,\mathrm{d}S = \int_{\mathbb{I}^k} f \,\mathrm{d}S := \int_{\mathbb{I}^k} f(\boldsymbol{\phi}(\boldsymbol{t}))\rho(\boldsymbol{t}) \,\mathrm{d}\boldsymbol{t}, \tag{14.15}$$

with 'area' element $\mathrm{d}S = \rho(\boldsymbol{t}) \,\mathrm{d}\boldsymbol{t}$.

Lemma 14.1. *If ω is a k-form in* (14.11), *then*

$$\int_{s_k} \omega = \int_{\mathbb{I}^k} (\boldsymbol{a}, \frac{1}{|\boldsymbol{p}|}\boldsymbol{p}) \,\mathrm{d}S. \tag{14.16}$$

By Theorem 14.1, the left-hand side is $\int_{\mathbb{I}^k} \sum_I a_I(\boldsymbol{\phi}(\boldsymbol{t}))J_{\phi_I} \mathrm{d}\boldsymbol{t}$ whose integrand is $\sum_I a_I p_I = (\boldsymbol{a}, \frac{1}{|\boldsymbol{p}|}\boldsymbol{p})\rho(\boldsymbol{t})$ which is the right-hand side.

In §14.3 we study the case $k = n - 1$ whose special case $n = 3$, $k = 2$ is the integral over a surface to be expounded in Definition 14.13.

Definition 14.2. Let $V = V^k$ denote the set of all k-forms, which turns into a vector space over $\mathbb{R}$ by introducing the addition and scalar multiplication by

$$\begin{aligned} \langle s_k, \omega_1^k + \omega_2^k \rangle &= \langle s_k, \omega_1^k \rangle + \langle s_k, \omega_2^k \rangle \\ \langle s_k, \alpha\omega^k \rangle &= \alpha \langle s_k, \omega^k \rangle. \end{aligned} \tag{14.17}$$

It becomes an alternating algebra with wedge product in Definition 14.3 below.

Let $\Lambda^k(\Omega) \subset V^k$ be the subspace consisting of all infinitely differentiable k-forms on Ω. Then Λ^k being 0 for $k > n$,

$$\Lambda(\Omega) = \sum_{k=0}^{n} \Lambda^k(\Omega) \tag{14.18}$$

is a graded algebra with wedge product.

We denote the $I = \{i_1, \cdots, i_k\}$-**summand** of ω in (14.5) by

$$\omega_I = \omega_I^k = a_I(\boldsymbol{x})\mathrm{d}\boldsymbol{x}_I = a_{i_1,\cdots,i_k}(\boldsymbol{x})\mathrm{d}x_{i_1} \wedge \cdots \wedge \mathrm{d}x_{i_k}. \tag{14.19}$$

Proposition 14.1. (i) *We have*

$$\omega^k = \sum_I \omega_I^k. \tag{14.20}$$

(ii) *If the index* $J = \{i_1, \cdots, i_{p-1}, i_{p+1}, i_p, \cdots, i_k\}$ *is obtained from* $I = \{i_1, \cdots, i_k\}$ *by changing two indices* i_p, i_{p+1}, *then*

$$a_J \mathrm{d}x_J = -a_I \mathrm{d}x_I. \tag{14.21}$$

I.e. if the order of $\mathrm{d}x_p$ *and* $\mathrm{d}x_{p+1}$ *are changed, then* $\mathrm{d}x_{i_1} \wedge \cdots \wedge \mathrm{d}x_{i_k}$ *changes its sign in conformity with* (11.147). *It follows that if there is a repetition of the index, say* i_p, *then the form* $\mathrm{d}x_{i_1} \wedge \cdots \wedge \mathrm{d}x_{i_p} \wedge \cdots \wedge \mathrm{d}x_{i_p} \cdots \wedge \mathrm{d}x_{i_k}$ *is* 0.
(iii) *If*

$$\omega_1^k = \sum_I a_I \mathrm{d}x_I, \quad \omega_2^k = \sum_I b_I \mathrm{d}x_I,$$

then

$$\omega_1^k + \omega_2^k = \sum_I (a_I + b_I)\mathrm{d}x_I, \quad \alpha\omega_1^k = \sum_I \alpha a_I \mathrm{d}x_I \quad (\alpha \in \mathbb{R}).$$

(iv) *Every form can be reduced to its canonical form*

$$\omega^k = \sum_I \omega_I^k \tag{14.22}$$

where $I = \{i_1, \cdots, i_k\}$ *runs through only those which satisfy* $i_1 < i_2 < \cdots < i_k$.
(v) *For* $k > n$, *all the* k*-forms on* $\Omega \subset \mathbb{R}^n$ *vanish.*

Proof. (ii) follows from the alternating property of the determinant. (v) follows from the remark that in any index $I = \{i_1, \cdots, i_k\}$, there must be a repetition, so that the corresponding form is 0. □

Definition 14.3. For

$$\omega^k = \sum_I \omega_I^k = \sum_I a_I(\boldsymbol{x})\mathrm{d}\boldsymbol{x}_I = \sum_{i_1=1}^n \cdots \sum_{i_k=1}^n a_{i_1,\cdots,i_k}(\boldsymbol{x})\mathrm{d}x_{i_1} \wedge \cdots \wedge \mathrm{d}x_{i_k},$$

$$\omega^\ell = \sum_J \omega_J^\ell = \sum_J b_J(\boldsymbol{x})\mathrm{d}\boldsymbol{x}_J = \sum_{j_1=1}^n \cdots \sum_{j_\ell=1}^n b_{j_1,\cdots,j_\ell}(\boldsymbol{x})\mathrm{d}x_{j_1} \wedge \cdots \wedge \mathrm{d}x_{j_\ell}$$

we define the (inner) or **wedge product** by

$$\omega^{k+\ell} = \sum_{I,J} \omega_I^k \wedge \omega_J^\ell = \omega^k \wedge \omega^\ell = \sum_{I,J} a_I(\boldsymbol{x}) b_J(\boldsymbol{x}) \mathrm{d}\boldsymbol{x}_I \wedge \mathrm{d}\boldsymbol{x}_J \tag{14.23}$$

$$= \sum_{i_1=1}^n \cdots \sum_{i_k=1}^n \sum_{j_1=1}^n \cdots \sum_{j_\ell=1}^n a_{i_1,\cdots,i_k}(\boldsymbol{x}) b_{j_1,\cdots,j_\ell}(\boldsymbol{x}) \mathrm{d}x_{i_1} \wedge \cdots \wedge \mathrm{d}x_{i_k} \wedge \mathrm{d}x_{j_1} \wedge \cdots \wedge \mathrm{d}x_{j_\ell}.$$

Theorem 14.2. *The wedge product is associative and distributive, i.e.*

$$(\omega_1 \wedge \omega_2) \wedge \omega_3 = \omega_1 \wedge (\omega_2 \wedge \omega_3), \quad (\omega_1 + \omega_2) \wedge \omega_3 = \omega_1 \wedge \omega_2 + \omega_2 \wedge \omega_3. \tag{14.24}$$

Also the formula for change of order holds true:

$$\omega^k \wedge \omega^\ell = (-1)^{kl} \omega^\ell \wedge \omega^k. \tag{14.25}$$

Proof. Only (14.25) needs a proof. In (14.23) for $\mathrm{d}x_{j_1}$ to be at the far-left, we need to change k times, amounting to the factor $(-1)^k$. Then for $\mathrm{d}x_{j_2}$ to be immediately after $\mathrm{d}x_{j_1}$, we need to change k times. Since there are l terms, we will have the sign change of $(-1)^{kl}$ to arrive at $\omega^\ell \wedge \omega^k$. □

Exercise 114. We write u_i for $\mathrm{d}x_i$. Or consider the situation as in Theorem E.1. For a matrix $A = (a_{ij})$ of degree n consider the linear transformation $f = f_A$ defined by

$$f(u_i) = \sum_{j=1}^n a_{ij} u_j, \quad 1 \le y \le n, \tag{14.26}$$

where $\{u_1, \cdots, u_n\}$ is a basis and let

$$f(u_1) \wedge \cdots \wedge f(u_n) = \sum_{j_1, j_n = 1}^n a_{1j_1} \cdots a_{nj_n} u_{j_1} \wedge \cdots \wedge u_{j_n} = D(f) u_1 \wedge \cdots \wedge u_n, \tag{14.27}$$

say. Then check that

$$D(f) = D(f_A) = \sum_{\sigma \in S_n} \mathrm{sgn}(\sigma) a_{1\sigma(1)} \cdots a_{n\sigma(n)} \tag{14.28}$$

is the **determinant** of A, where S_n indicates the nth symmetric group consisting of all permutations of order n. For the 3rd symmetric group, see Exercise 28.

Solution. It suffices to notice that rearranging $u_{j_1} \wedge \cdots \wedge u_{j_n}$ into $u_1 \wedge \cdots \wedge u_n$ there occur $\mathrm{sgn}(\sigma)$ sign changes, where $\sigma = \begin{pmatrix} j_1 & \cdots & j_n \\ 1 & \cdots & n \end{pmatrix}$ and sgn is the number of transitions of two symbols. In the case of $M = Ku_1 \oplus \cdots \oplus Ku_n$, the space $\Lambda_n(M)$ is spanned by one element $u_1 \wedge \cdots \wedge u_n$ and $D(A)$ is uniquely determined and it is often denoted by $|A|$ or $\det A$.

As is mentioned in the proof of Proposition 14.1, the determinant has linearity w.r.t. the column, alternating property. With the condition $|E| = 1$, it is uniquely determined.

The 0-form $\omega^0 = a(\boldsymbol{x})$ is a function of class C^r as mentioned above and its inner derivative is given in (5.21):

$$\mathrm{d}\omega^0 = \mathrm{d}a(\boldsymbol{x}) = \sum_{i=1}^{n} \frac{\partial a(\boldsymbol{x})}{\partial x_i} \mathrm{d}x_i. \tag{14.29}$$

Definition 14.4. The (inner) **derivation** (or a differential operator)

$$\mathrm{d} : \Lambda^k(\Omega) \to \Lambda^{k+1}(\Omega), \tag{14.30}$$

where $\Lambda^k(\Omega)$ is the space of infinitely differentiable k-forms in Definition 14.2, is defined for $\omega^k = \sum_I \omega_I^k = \sum_I a_I(\boldsymbol{x})\mathrm{d}\boldsymbol{x}_I$ by

$$\mathrm{d}\omega^k = \sum_I \mathrm{d}\omega_I^k = \sum_I \mathrm{d}a_I \wedge \mathrm{d}\boldsymbol{x}_I \tag{14.31}$$

and

$$\mathrm{d}\omega_I^k = \mathrm{d}a_I \wedge \mathrm{d}\boldsymbol{x}_I = \sum_{i=1}^{n} \frac{\partial a_I(x)}{\partial x_i} \mathrm{d}x_i \wedge \mathrm{d}\boldsymbol{x}_I \tag{14.32}$$

by (14.29).

Theorem 14.3. (i)

$$\mathrm{d}(\omega^k \wedge \omega^l) = \mathrm{d}\omega^k \wedge \omega^l + (-1)^k \omega^k \wedge \mathrm{d}\omega^l. \tag{14.33}$$

(ii) *For every* $\omega \in \Lambda(\Omega)$

$$\mathrm{d}^2\omega = \mathrm{d}(\mathrm{d}\omega) = 0. \tag{14.34}$$

Proof. Let

$$\omega^k = \sum \omega_I^k, \quad \omega_I^k = a_I \mathrm{d}x_I, \quad \omega^l = \sum \omega_J^l, \quad \omega_J^l = b_J \mathrm{d}x_J. \tag{14.35}$$

By (14.31), it suffices to check (14.33) for each I-summand. By definition,

$$\mathrm{d}(\omega_I^k \wedge \omega_J^l) = \mathrm{d}(a_I b_J) \wedge \mathrm{d}x_I \wedge \mathrm{d}x_J = \mathrm{d}(a_I b_J)\mathrm{d}x_I \wedge \mathrm{d}x_J. \tag{14.36}$$

By (14.29) and the differentiation formula for the product,

$$d(a_I b_J) = \sum_{j=1}^{n} \frac{\partial}{\partial x_j}(a_I b_J)dx_j = \left(a_I \sum_{j=1}^{n} \frac{b_J}{\partial x_j}dx_j + b_J \sum_{j=1}^{n} \frac{a_I}{\partial x_j}dx_j \right)$$

$$= b_J da_I + a_I db_J. \tag{14.37}$$

Substituting this in (14.36), we find that

$$d(\omega_I^k \wedge \omega_J^l) = (da_I \wedge dx_I) \wedge b_J dx_J + a_I db_J \wedge (dx_I \wedge dx_J). \tag{14.38}$$

Applying (14.25) to the second summand, we see that it amounts to $(-1)^k a_I dx_I \wedge (db_J \wedge dx_J)$, which is $(-1)^k(\omega_I^k \wedge d\omega_J^l)$. Since the first summand is $d\omega_I^k \wedge \omega_J^l$, we deduce that

$$d(\omega_I^k \wedge \omega_J^l) = d\omega_I^k \wedge \omega_J^l + (-1)^k(\omega_I^k \wedge d\omega_J^l). \tag{14.39}$$

Substituting (14.39) in (14.31), we obtain

$$d(\omega^k \wedge \omega^l) = \sum_{I,J} d(\omega_I^k \wedge \omega_J^l) = \sum_{I,J} d\omega_I^k \wedge \omega_J^l + (-1)^k \sum_{I,J} \omega_I^k \wedge d\omega_J^l. \tag{14.40}$$

By (14.23) the right-hand side amounts to the right-hand side of (14.33). This completes the proof of (i).

For (ii), by (14.31), it suffices to differentiate (14.32).

$$d^2\omega_I = d^2\omega_I^k = \sum_{i=1}^{n} d\left(\frac{\partial a_I}{\partial x_i}dx_i\right) \wedge d\boldsymbol{x}_I = \sum_{j=1}^{n}\sum_{i=1}^{n} \frac{\partial^2 a_I}{\partial x_j \partial x_i} dx_j \wedge dx_i \wedge d\boldsymbol{x}_I. \tag{14.41}$$

Noting that $\frac{\partial^2 a_I}{\partial x_j \partial x_i} = \frac{\partial^2 a_I}{\partial x_i \partial x_j}$, changing the order of summation on writing i for j and j for i and finally applying the sign change property, we see that $d^2\omega_I = -d^2\omega_I$, whence (ii) follows. □

The algebra $\Lambda(\Omega)$ with its wedge product and derivation is called the Cartan algebra.

Example 14.1. Recall the spherical coordinates (or polar coordinates in the space) cf. Example 5.2, (ii)

$$\boldsymbol{x} = \boldsymbol{\phi}(\boldsymbol{t}), {}^t(x_1, x_2, x_3) = {}^t(t_1 \cos t_2 \sin t_3, t_1 \sin t_2 \sin t_3, t_1 \cos t_3), \tag{14.42}$$

where

$$\mathbb{I} = \mathbb{I}^3 = [0, r] \times [0, 2\pi] \times [0, \pi].$$

This parametrizes the ball of radius r with center at the origin. The sphere (the surface of the ball) has been treated in Example 5.2. In the same way as in Lemma 11.2 above we have the rule

$$dx_1 \wedge dx_2 \wedge dx_3 = \frac{\partial(x_1, x_2, x_3)}{\partial(t_1, t_2, t_3)} dt_1 \wedge dt_2 \wedge dt_3. \tag{14.43}$$

Proof. We use the alternating properties of the forms as in Lemma 11.2. Then when forming the wedge product, only those terms remain which have no repetitions of the variable. I.e., if

$$\mathrm{d}x_i = \sum_{j=1}^{3} a_{ij}\mathrm{d}t_j,$$

then

$$\begin{aligned} &\mathrm{d}x_1 \wedge \mathrm{d}x_2 \wedge \mathrm{d}x_3 \\ &= (a_{11}a_{22}a_{33} + a_{12}a_{23}a_{31} + a_{13}a_{32}a_{21} - a_{11}a_{23}a_{32} - a_{13}a_{22}a_{31} - a_{12}a_{21}a_{33}) \\ &\times \mathrm{d}t_1 \wedge \mathrm{d}t_2 \wedge \mathrm{d}t_3, \end{aligned} \tag{14.44}$$

which is (14.43). □

Definition 14.5. A k-form $\omega^k \in \Lambda^k(\Omega)$ is said to be **closed** if $\mathrm{d}\omega^k = 0$. We denote the set of all closed k forms by

$$Z^k(\Omega) = \{\omega^k | \mathrm{d}\omega^k = 0\} \tag{14.45}$$

and call it the k-**cocycle**.

A k-form $\omega^k \in \Lambda^k(\Omega)$ is said to be **exact** (or complete) if there exists a $\omega^{k-1} \in \Lambda^{k-1}(\Omega)$ such that $\omega^k = \mathrm{d}\omega^{k-1}$. We denote the set of all exact k forms by

$$B^k(\Omega) = \{\omega^k | \omega^k = \mathrm{d}\omega^{k-1} \,\text{for}\, \omega^{k-1} \in \Lambda^{k-1}(\Omega)\} \tag{14.46}$$

and call it the k-**coboundary**.

Remark 14.1. If we denote the derivation d from $\Lambda^{k-1}(\Omega) \to \Lambda^k(\Omega)$ by δ^{k-1}, i.e. we consider the dual complex $C = \{\Lambda^k(\Omega), \delta^k\}$,

$$C : \cdots \to \Lambda^{k-1}(\Omega) \xrightarrow{\delta^{k-1}} \Lambda^k(\Omega) \xrightarrow{\delta^k} \Lambda^{k+1}(\Omega) \to \cdots, \tag{14.47}$$

with

$$\delta^k \circ \delta^{k-1} = 0, \tag{14.48}$$

then

$$B^k(\Omega) = \mathrm{Im}\,\delta^{k-1}, \quad Z^k(\Omega) = \mathrm{Ker}\,\delta^k, \tag{14.49}$$

where as in (1.74), Im means the image and $\mathrm{Ker}\,\delta^k$ is, by definition, the inverse image $(\delta^k)^{-1}(\{0\})$ of the identity element 0.

Lemma 14.2. *$Z^k(\Omega)$ and $B^k(\Omega)$ form vector spaces and $B^k(\Omega) \subset Z^k(\Omega)$.*

Proof. The second assertion follows from (14.48) which is (14.34). □

Definition 14.6. The factor group

$$H^k(\Omega) = Z^k(\Omega)/B^k(\Omega) \tag{14.50}$$

in the sense of Exercise 22 is called the k**th cohomology group** of Ω.

When we speak of a cohomology group, we usually mean the graded module $H^*(C) = \sum_{k\in I} \Lambda^k(\Omega)$ in Appendix E.

Definition 14.7. Let T be a map of class C^1 from a domain $\Omega \subset \mathbb{R}^n$ to another domain $\tilde{\Omega} \subset \mathbb{R}^m$. Then the composite mapping $T \circ \phi$ is defined by

$$Ts_k = (\mathbb{I}^k, T \circ \phi) \quad (s_k = (\mathbb{I}^k, \phi)). \tag{14.51}$$

We may define the mapping

$$T^* : \Lambda(\tilde{\Omega}) \to \Lambda(\Omega) \tag{14.52}$$

by

$$\langle s_k, T^*\tilde{\omega}^k \rangle = \langle Ts_k, \tilde{\omega}^k \rangle, \quad \tilde{\omega}^k \in \Lambda(\tilde{\Omega}), \ k = 1, 2, \cdots, \tag{14.53}$$

the **adjoint** (or a dual) of T.

Theorem 14.4. *If*

$$\omega^k(\boldsymbol{y}) = \sum_I a_I(\boldsymbol{y}) \mathrm{d}\boldsymbol{y}_I, \quad \boldsymbol{y} \in \tilde{\Omega} \tag{14.54}$$

and

$$T(\boldsymbol{x}) = (t_1(\boldsymbol{x}), \cdots, t_m(\boldsymbol{x})), \quad \boldsymbol{x} \in \Omega, \tag{14.55}$$

then

$$T^*\omega^k(\boldsymbol{x}) = \sum_{I,J} a_I(T(\boldsymbol{x})) J_T \mathrm{d}\boldsymbol{x}_J, \tag{14.56}$$

where J_T indicates the Jacobian

$$\det\left(\frac{\partial t_{i_l}}{\partial x_{j_p}}\right)_{1\le l,p\le k} = \frac{\partial(t_{i_1}, \cdots t_{i_k})}{\partial(x_{j_1}, \cdots, x_{j_k})}. \tag{14.57}$$

Proof. Since by Theorem 14.1

$$\langle Ts_k, \omega^k \rangle = \sum_I \int_{\mathbb{I}^k} a_I((T \circ \phi(\boldsymbol{u}))) J_{T\circ\phi}(\boldsymbol{u}) \, \mathrm{d}\boldsymbol{u}_I, \tag{14.58}$$

it follows from the multiplication formula for the Jacobian, Corollary 6.1, that

$$J_{T\circ\phi}(\boldsymbol{u}) \, \mathrm{d}\boldsymbol{u}_I = J_T(\boldsymbol{x}) J_\phi(\boldsymbol{u}) \, \mathrm{d}\boldsymbol{u}_I = J_T(\boldsymbol{x}) \, \mathrm{d}\boldsymbol{x}_J, \tag{14.59}$$

which proves (14.56). □

The adjoint map (14.56) is an algebra homomorphism between Cartan algebras:

Theorem 14.5. *The adjoint has the following properties.*
(i) *(Linearity)* $T^*(\lambda_1\omega_1 + \lambda_2\omega_2) = \lambda_1 T^*\omega_1 + \lambda_2 T^*\omega_2$.
(ii) $T^*(\omega_1 \wedge \omega_2) = T^*\omega_1 \wedge T^*\omega_2$.
(iii) *(Invariance under change of variable)*

$$\mathrm{d}T^*(\omega) = T^*(\mathrm{d}\omega). \tag{14.60}$$

(iv) *If* $T_1 : \Omega_1 \to \Omega_2$ *and* $T_2 : \Omega_2 \to \Omega_3$, *then* $(T_1 \circ T_2)^* = T_2^* \circ T_1^*$.

Proof. Only (iii) needs a proof, the others being immediate consequences of Theorem 14.4. For a 0-form $\omega^0(\boldsymbol{y}) = a(\boldsymbol{y})$, we have

$$(T^*a)(\boldsymbol{x}) = a(T(\boldsymbol{x})) = (a \circ T)(\boldsymbol{x}).$$

Hence by (14.29)

$$\mathrm{d}(T^*a)(\boldsymbol{x}) = \mathrm{d}(a \circ T)(\boldsymbol{x}) = \sum_{i=1}^{n} \frac{\partial a \circ T}{\partial x_i}\mathrm{d}x_i.$$

By the chain rule, $\frac{\partial a \circ T}{\partial x_i} = \sum_{j=1}^{m} \frac{\partial a}{\partial t_j}\frac{\partial t_j}{\partial x_i}$. Since $\mathrm{d}a = \sum_{j=1}^{m} \frac{\partial a}{\partial t_j}\mathrm{d}t_j$, (14.56) leads us to

$$\mathrm{d}(T^*a)(\boldsymbol{x}) = T^*(\mathrm{d}a)(\boldsymbol{x}). \tag{14.61}$$

Now, for ω^k, applying (ii), (14.31), (14.61) and (14.31) successively, we conclude (14.60). □

Definition 14.8. A topological space X is called **simply connected** if any continuous closed curve lying in X can be contracted to one point without passing outside of X. For Poincaré's lemma to hold, we need to impose a more stringent condition to the effect that the open set $\Omega \subset \mathbb{R}^n$ can be **differentiably contracted to one point** $\boldsymbol{x}_0$, say, i.e. there exists a continuous C^1 class mapping $h = h(\boldsymbol{x}, t) : \Omega \times [0, 1] \to \Omega$ such that for every $\boldsymbol{x} \in \Omega$,

$$h(\boldsymbol{x}, 0) = \boldsymbol{x}_0, \quad h(\boldsymbol{x}, 1) = \boldsymbol{x}. \tag{14.62}$$

We refer to such a mapping h a homotopy of Ω to $\boldsymbol{x}_0$ and we refer to Ω as **smoothly contractible** in what follows. A domain $\Omega \subset \mathbb{R}^n$ is called **starlike** with respect to a point $\boldsymbol{x}_0$ if for every point $\boldsymbol{x} \in \Omega$ the line segment joining them (8.5) is contained in Ω. If $\boldsymbol{x}_0$ can run all the points, i.e. for any two points, the line segment joining them belong to Ω, then it is called **convex**.

Exercise 115. Prove that an open ball $V_r(\boldsymbol{x}_0) \in \mathbb{R}^3$ is a convex set. Prove that a convex set is starlike and a starlike set is smoothly contractible.

Let $\Omega \subset \mathbb{R}^n$ be an open set. We introduce the mappings on it

$$j_1 : \Omega \to \Omega \times [0,1]; j_1(\boldsymbol{x}) = (\boldsymbol{x}, 1), \quad j_0 : \Omega \to \Omega \times [0,1]; j_1(\boldsymbol{x}) = (\boldsymbol{x}, 0) \tag{14.63}$$

and their adjoints

$$j_i^* : \Lambda^k(\Omega \times [0,1]) \to \Lambda^k(\Omega) \tag{14.64}$$

by (14.53).

Exercise 116. Prove the following. If

$$\Lambda^k(\Omega \times [0,1]) \ni \omega^k = a(\boldsymbol{x}, t)\mathrm{d}t \wedge \mathrm{d}\boldsymbol{x}_{I_{k-1}},$$

where $I_{k-1} = \{i_1, \cdots, i_{k-1}\}$, then

$$j_i^* \omega^k = 0 \tag{14.65}$$

and if

$$\Lambda^k(\Omega \times [0,1]) \ni \omega^k = a(\boldsymbol{x}, t)\mathrm{d}\boldsymbol{x}_I,$$

where $I = \{i_1, \cdots, i_k\}$, then

$$j_1^* \omega^k(\boldsymbol{x}) = a(\boldsymbol{x}, 1)\mathrm{d}\boldsymbol{x}_I, \quad j_0^* \omega^k(\boldsymbol{x}) = a(\boldsymbol{x}, 0)\mathrm{d}\boldsymbol{x}_I. \tag{14.66}$$

Also let h be as in (14.62). Prove that

$$(h \circ j_1)^* = \iota, \quad (h \circ j_0)^* = 0, \tag{14.67}$$

where ι indicates the identity.

Solution. To prove (14.65), apply Theorem 14.4 with $\frac{\partial(x_{i_1}, \cdots x_{i_{k-1}}, t)}{\partial(x_{j_1}, \cdots, x_{j_k})} = 0$ since $t = 1$ or $t = 0$ identically.

Note that

$$(h \circ j_1)(\boldsymbol{x}) = h(\boldsymbol{x}, 1) = \boldsymbol{x}, \quad (h \circ j_0)(\boldsymbol{x}) = h(\boldsymbol{x}, 0) = \boldsymbol{x}_0$$

or

$$h \circ j_1 = \iota, \quad h \circ j_0 = \text{const.}$$

Since $\langle I s_k, \omega^k \rangle = \langle s_k, \omega^k \rangle = \langle s_k, I\omega^k \rangle$, the first equality in (14.66) follows. The second follows from the vanishing of the Jacobian.

Now we introduce a series of linear maps

$$\varkappa^k : \Lambda^{k+1}(\Omega \times [0,1]) \to \Lambda^k(\Omega) \tag{14.68}$$

by

$$\varkappa^k(a(\boldsymbol{x},t)\mathrm{d}t \wedge \mathrm{d}\boldsymbol{x}_I) = \left(\int_0^1 a(\boldsymbol{x},t)\mathrm{d}t\right)\mathrm{d}\boldsymbol{x}_I, \quad \varkappa^k(a(\boldsymbol{x},t)\mathrm{d}\boldsymbol{x}_I \wedge \mathrm{d}x_{i_{k+1}}) = 0, \tag{14.69}$$

where $\mathrm{d}\boldsymbol{x}_I = \mathrm{d}x_{i_1} \wedge \cdots \wedge \mathrm{d}x_{i_k}$.

Lemma 14.3. *For $\omega^k \in \Lambda^k(\Omega \times [0,1])$ we have*

$$\varkappa^k \mathrm{d}\omega^k + \mathrm{d}\varkappa^{k-1}\omega^k = j_1^*\omega^k - j_0^*\omega^k. \tag{14.70}$$

j_1^* and j_0^* are called algebraically homotopic and $\varkappa$ their algebraic homotopy.

Proof. We abbreviate $\omega = \omega^k$ and $\varkappa = \varkappa^k$. If $\omega = a(\boldsymbol{x},t)\mathrm{d}\boldsymbol{x}_I$, then $\varkappa\omega = 0$ by (14.69), and so $\mathrm{d}\varkappa\omega = 0$. Since

$$\mathrm{d}\omega = \frac{\partial}{\partial t}a(\boldsymbol{x},t)\mathrm{d}t \wedge \mathrm{d}\boldsymbol{x}_I + \sum_{i=1}^n \frac{\partial}{\partial x_i}a(\boldsymbol{x},t)\mathrm{d}x_i \wedge \mathrm{d}\boldsymbol{x}_I, \tag{14.71}$$

it follows from (14.69) that

$$\varkappa\mathrm{d}\omega = \left(\int_0^1 \frac{\partial}{\partial t}a(\boldsymbol{x},t)\mathrm{d}t\right)\mathrm{d}\boldsymbol{x}_I = (a(\boldsymbol{x},1) - a(\boldsymbol{x},0))\mathrm{d}\boldsymbol{x}_I.$$

Hence (14.70) follows from Exercise 116.

If $\omega = a(\boldsymbol{x},t)\mathrm{d}t \wedge \mathrm{d}\boldsymbol{x}_I$, then $j_1^*\omega = 0, j_0^*\omega = 0$ by Exercise 116. By (14.69) and Corollary 11.9,

$$\mathrm{d}\varkappa\omega = \mathrm{d}\left(\int_0^1 a(\boldsymbol{x},t)\mathrm{d}t\right)\mathrm{d}\boldsymbol{x}_I = \sum_{i=1}^n \left(\int_0^1 \frac{\partial}{\partial x_i}a(\boldsymbol{x},t)\mathrm{d}t\right)\mathrm{d}x_i \wedge \mathrm{d}\boldsymbol{x}_I.$$

For the other term, we have similarly to (14.71),

$$\mathrm{d}\omega = -\sum_{i=1}^n \frac{\partial}{\partial x_i}a(\boldsymbol{x},t)\mathrm{d}t \wedge \mathrm{d}x_i \wedge \mathrm{d}\boldsymbol{x}_I, \tag{14.72}$$

and so by (14.69)

$$\varkappa\mathrm{d}\omega = -\sum_{i=1}^n \left(\int_0^1 \frac{\partial}{\partial x_i}a(\boldsymbol{x},t)\mathrm{d}t\right)\mathrm{d}x_i \wedge \mathrm{d}\boldsymbol{x}_I.$$

These two equalities show that the right-hand side of (14.70) is 0 which is equal to the left-hand side, proving the lemma. □

Exercise 117. Check (14.71) and (14.72).

Theorem 14.6. (Poincaré's lemma) *If the domain Ω is smoothly contractible, then $H^k(\Omega) = 0$, i.e. every closed k-form is exact. Or if*

$$d\omega^k = 0, \tag{14.73}$$

then there is a $(k-1)$-form ω^{k-1} such that

$$d\omega^{k-1} = \omega^k. \tag{14.74}$$

Proof. For any $\omega^k \in \Lambda^k(\Omega)$ such that $d\omega^k = 0$, we put

$$\omega^{k-1} = \varkappa^{k-1} h^* \omega^k, \tag{14.75}$$

where h is a homotopy of Ω to $\boldsymbol{x}_0$ in (14.62), $h^* : \Lambda^k(\Omega) \to \Lambda^k(\Omega \times [0,1])$ is its adjoint and $\varkappa^{k-1} : \Lambda^k(\Omega \times [0,1]) \to \Lambda^{k-1}(\Omega)$ in (14.69). In view of this, $\omega^{k-1} \in \Lambda^{k-1}(\Omega)$.

By Theorem 14.5, (iv) and (14.67), we have

$$j_1^* \circ h^* = \iota, \quad j_0^* \circ h^* = 0.$$

Applying this to the right-hand side of (14.70) with ω^k replaced by $h^*\omega^k$, we find that it reduces to

$$\varkappa^k d(h^*\omega^k) + d\omega^{k-1} = \varkappa^k d(h^*\omega^k) + d\varkappa^{k-1}(h^*\omega^k) = \omega^k. \tag{14.76}$$

By Theorem 14.5, (iii), $\varkappa^k dh^*\omega^k = \varkappa^k h^* d\omega^k$, which is 0 by assumption. Hence (14.76) establishes (14.74), completing the proof. □

Corollary 14.1. *If the domain Ω is smoothly contractible and $\omega^k = 0 = d\omega_1^{k-1} = d\omega_2^{k-1}$, then there exists a $(k-2)$ form ω^{k-2} such that*

$$\omega_1^{k-1} - \omega_2^{k-1} = d\omega^{k-2}, \tag{14.77}$$

which means that the solution of (14.73) is unique up to $d\omega^{k-2}$.

For applying the derivation to (14.77), we see that $\omega_1^{k-1} - \omega_2^{k-1}$ is closed and Poincaré's lemma applies.

Now we introduce the dual of the notion of cohomology. In Definition 14.2 the vector space structure over $\mathbb{R}$ is introduced in $V = V^k$ and *a fortiori* on Λ^k, which is done by making $\langle s_k, \omega^k \rangle$ linear in the second variable.

We introduce the addition and scalar multiplication in the first variable by

$$\begin{aligned} \langle s_k^1 + s_k^2, \omega^k \rangle &= \langle s_k^1, \omega^k \rangle + \langle s_k^2, \omega^k \rangle \\ \langle \alpha s_k, \omega^k \rangle &= \alpha \langle s_k, \omega^k \rangle. \end{aligned} \tag{14.78}$$

We are to consider the vector space $C^k(\Omega)$ over $\mathbb{R}$ consisting of all linear combinations of k-singular cubes s_k in Ω with real coefficients, i.e. $C^k(\Omega) = \sum \mathbb{R}s_k$, the vector space generated by all s_k's. Cf. the notion of a direct sum in Definition 1.2.

Definition 14.9. The vector space $C^k(\Omega)$ is called the space of **k-chains** over Ω. Its elements are expressed as

$$c_k = \sum a_j s_k^j, \tag{14.79}$$

finite sums with real coefficients.

Definition 14.10. Define the mappings

$$\begin{aligned} k_m^+ &: \mathbb{I}^k \to \mathbb{I}^{k-1}; k_m^+(t_1, \cdots, t_m, \cdots, t_k) = (t_1, \cdots, b_m, \cdots, t_k), \\ k_m^- &: \mathbb{I}^k \to \mathbb{I}^{k-1}; k_m^+(t_1, \cdots, t_m, \cdots, t_k) = (t_1, \cdots, a_m, \cdots, t_k). \end{aligned} \tag{14.80}$$

Then $\phi \circ k_m^\pm$ are mappings of class C^1 into Ω and we may speak of the $(k-1)$-singular cubes

$$w_m^+ = (k_m^+(\mathbb{I}^k), \phi \circ k_m^+(\mathbb{I}^k)), \quad w_m^- = (k_m^-(\mathbb{I}^k), \phi \circ k_m^-(\mathbb{I}^k)) \tag{14.81}$$

and call them the **walls** of the cube s_k. Define the **boundary** of s_k by

$$\partial s_k = \sum_{m=1}^{k} (-1)^{m-1} \left(w_m^+ - w_m^-\right) \tag{14.82}$$

and extend it to $C^k(\Omega)$ by linearity (14.79):

$$\partial c_k = \sum a_j \partial s_k^j. \tag{14.83}$$

Then the **boundary map** ∂ is

$$\partial : C^k(\Omega) \to C^{k-1}(\Omega). \tag{14.84}$$

It turns out that the derivation defined in (14.31) and the boundary map in Definition 14.10 are adjoint to each other, which is the content of the following theorem.

Theorem 14.7. (Stokes-Poincaré) *For each k-form $\omega^k \in \Lambda^k(\Omega)$ and each chain $c_k \in C^k(\Omega)$ the equality holds true:*

$$\langle \partial c_k, \omega^{k-1} \rangle = \langle c_k, \mathrm{d}\omega^{k-1} \rangle, \tag{14.85}$$

which amounts to the general Stokes theorem, Theorem 14.8.

Proof. Since both sides are linear in the chain part by (14.83) and (14.79), it suffices to prove it for a one cube $s_k = (\mathbb{I}^k, \phi)$, $\phi : \mathbb{I}^k \subset \Omega \subset \mathbb{R}^n$, say. With ι denoting the identity map in $\mathbb{R}^k$, let $r_k = (\mathbb{I}^k, \iota)$ be a singular cube in $\Omega' \subset \mathbb{R}^k$. Since the composite map $\phi \circ \iota$ can be formed, we have $s_k = \phi \circ r_k$.

We define walls $v_m^\pm$ so that those defined by (14.81) are the images $\phi(v_m^\pm)$ by

$$v_m^+ = (k_m^+(\mathbb{I}^k), \iota), \quad v_m^- = (k_m^-(\mathbb{I}^k), \iota).$$

Since (14.82) holds for r_k and $v_m^\pm$:

$$\partial r_k = \sum_{m=1}^{k} (-1)^{m-1} \left(v_m^+ - v_m^-\right), \tag{14.86}$$

it holds for s_k and $w_m^\pm$ in the form

$$\partial s_k = \phi(\partial r_k).$$

Hence the left-hand side of (14.85) with c_k replaced by s_k reads

$$\langle \partial s_k, \omega^{k-1} \rangle = \langle \phi(\partial r_k), \omega^{k-1} \rangle = \langle \partial r_k, \phi^* \omega^{k-1} \rangle \tag{14.87}$$

by (14.53).

On the other hand, by (14.53) again, we have

$$\langle s_k, \omega^k \rangle = \langle \phi r_k, \omega^k \rangle = \langle r_k, \phi^* \omega^k \rangle,$$

whence the right-hand side of (14.85) amounts to

$$\langle s_k, \mathrm{d}\omega^{k-1} \rangle = \langle r_k, \phi^* \mathrm{d}\omega^{k-1} \rangle = \langle r_k, \mathrm{d}(\phi^* \omega^{k-1}) \rangle, \tag{14.88}$$

the last equality being due to (14.60).

It follows from (14.87) and (14.88) that we are to prove

$$\langle \partial r_k, \phi^* \omega^{k-1} \rangle = \langle r_k, \mathrm{d}(\phi^* \omega^{k-1}) \rangle$$

for all $\omega^{k-1} \in \Lambda^{k-1}(\Omega)$. But since $\phi^* \omega^{k-1} \in \Lambda^{k-1}(\Omega)$, it suffices to prove

$$\langle \partial r_k, \omega^{k-1} \rangle = \langle r_k, \mathrm{d}\omega^{k-1} \rangle \tag{14.89}$$

for all $\omega^{k-1} \in \Lambda^{k-1}(\Omega)$, which will be done in the following exercise. □

Exercise 118. Prove (14.89).

Solution. Let $I = I_{k-1}$ be a $(k-1)$-ple index in (14.3). Without loss of generality, we may suppose that

$$I = \{1, \overset{\overset{m}{\vee}}{\cdots}, k-1\}$$

with m missing and we consider a special case of (14.5)

$$\omega^{k-1} = a(x)\mathrm{d}x_1 \wedge \overset{m}{\overset{\vee}{\cdots}} \wedge \mathrm{d}x_{k-1}$$

with $\mathrm{d}x_m$ missing. By (14.32) and rearranging,

$$\mathrm{d}\,\omega^{k-1} = \sum_{i=1}^{k-1} \frac{\partial a}{\partial x_i}\mathrm{d}x_i \wedge \mathrm{d}x_1 \wedge \overset{m}{\overset{\vee}{\cdots}} \wedge \mathrm{d}x_k = (-1)^{m-1}\frac{\partial a}{\partial x_m}\mathrm{d}x_1 \wedge \cdots \wedge \mathrm{d}x_k.$$

Hence by Theorem 14.1 with $\boldsymbol{t} = \boldsymbol{x}$, we have

$$\langle r_k, \mathrm{d}(\omega^{k-1})\rangle = \int_{\mathbb{I}^k} (-1)^{m-1}\frac{\partial a}{\partial x_m}\,\mathrm{d}\boldsymbol{x} \tag{14.90}$$

$$= (-1)^{m-1}\int_{\substack{x_i\in[a_i,b_i]\\ i\neq m}} (a(x_1,\cdots,b_m,\cdots,x_k) - a(x_1,\cdots,a_m,\cdots,x_k))\,\mathrm{d}\boldsymbol{x}_m$$

on integrating in $\boldsymbol{x}_m$ which is permissible by Fubini's theorem, where $\mathrm{d}\boldsymbol{x}_m = \mathrm{d}x_1\wedge\overset{m}{\overset{\vee}{\cdots}}\wedge\mathrm{d}x_k$. By (14.81), the second expression in (14.90) amounts to

$$(-1)^{m-1}\left(\int_{k_m^+(\mathbb{I}^k)} a(x)\,\mathrm{d}x_m - \int_{k_m^-(\mathbb{I}^k)} a(x)\,\mathrm{d}x_m\right) \tag{14.91}$$

$$= (-1)^{m-1}\left(\langle v_m^+, \omega^{k-1}\rangle - \langle v_m^-, \omega^{k-1}\rangle\right)$$

by Theorem 14.1 with $\boldsymbol{t} = \boldsymbol{x}$. By the same theorem with distinct $\boldsymbol{t}, \boldsymbol{x}$, the integral is 0, so that $\langle v_i^\pm, \omega^{k-1}\rangle = 0$ for $i \neq m$. Hence (14.91) may be written as $\sum_{i=1}^k (-1)^{i-1}\left(\langle v_i^+, \omega^{k-1}\rangle - \langle v_i^-, \omega^{k-1}\rangle\right)$, which is $\langle \partial r_k, \omega^{k-1}\rangle$ by (14.86), completing the proof.

14.2 Vector analysis

Definition 14.11. Every vector-valued function $\boldsymbol{f} = \boldsymbol{f}(\boldsymbol{x})$ defined on $X \subset \mathbb{R}^3$ may be thought of as assigning each $\boldsymbol{x} \in X$ a directed line segment with the initial point $\boldsymbol{x}$. In this interpretation, it is termed as a **vector field** or we say that $\boldsymbol{f}$ gives rise to a vector field. Hereafter we will use both terms interchangeably. From a scalar function $f = f(x,y,z)$ there arises the gradient in Theorem 5.1, which we express as a column vector:

$$\mathrm{grad}(f) = \nabla f = \begin{pmatrix} \frac{\partial f}{\partial x} \\ \frac{\partial f}{\partial y} \\ \frac{\partial f}{\partial z} \end{pmatrix}. \tag{14.92}$$

This corresponds to the 1-form $\omega^1 = P\mathrm{d}x + Q\mathrm{d}y + R\mathrm{d}z$ with $P = f_x$, etc.

For a vector-valued function $\boldsymbol{f} = \begin{pmatrix} P \\ Q \\ R \end{pmatrix} \in C^1(X)$ (i.e. $P, Q, R \in C^1(X)$), we define its **curl** denoted $\operatorname{curl} \boldsymbol{f}$ (also called rotation $\operatorname{rot} \boldsymbol{f}$) and **divergence** $\operatorname{div} \boldsymbol{f}$ by

$$\operatorname{curl} \boldsymbol{f} = \begin{pmatrix} \begin{vmatrix} \frac{\partial}{\partial y} & Q \\ \frac{\partial}{\partial z} & R \end{vmatrix} \\ -\begin{vmatrix} \frac{\partial}{\partial x} & P \\ \frac{\partial}{\partial z} & R \end{vmatrix} \\ \begin{vmatrix} \frac{\partial}{\partial x} & P \\ \frac{\partial}{\partial y} & Q \end{vmatrix} \end{pmatrix} = \begin{pmatrix} \frac{\partial R}{\partial y} - \frac{\partial Q}{\partial z} \\ -\left(\frac{\partial R}{\partial x} - \frac{\partial P}{\partial z}\right) \\ \frac{\partial Q}{\partial x} - \frac{\partial P}{\partial y} \end{pmatrix} = \nabla \times \boldsymbol{f} \tag{14.93}$$

and

$$\operatorname{div} \boldsymbol{f} = \frac{\partial P}{\partial x} + \frac{\partial Q}{\partial y} + \frac{\partial R}{\partial z} = \nabla \cdot \boldsymbol{f} \tag{14.94}$$

respectively. Here (14.93) and (14.94) are formal vector product (A.3) and inner product (1.9), respectively.

The formal inner product

$$\Delta = \nabla \cdot \nabla = \frac{\partial^2}{\partial x^2} + \frac{\partial^2}{\partial y^2} + \frac{\partial^2}{\partial z^2} \tag{14.95}$$

is called the **Laplacian** and the solution ψ of the **Laplace equation**

$$\Delta\psi = \nabla \cdot \nabla\psi = \frac{\partial^2 \psi}{\partial x^2} + \frac{\partial^2 \psi}{\partial y^2} + \frac{\partial^2 \psi}{\partial z^2} = 0 \tag{14.96}$$

is called a **harmonic function.**

$\operatorname{curl} \boldsymbol{f}$ corresponds to the 2-form $\omega^2 = \left(\frac{\partial R}{\partial y} - \frac{\partial Q}{\partial z}\right) \mathrm{d}y\mathrm{d}z - \left(\frac{\partial R}{\partial x} - \frac{\partial P}{\partial z}\right) \mathrm{d}z\mathrm{d}x + \left(\frac{\partial Q}{\partial x} - \frac{\partial P}{\partial y}\right) \mathrm{d}x\mathrm{d}y$ while $\operatorname{div} \boldsymbol{f}$ corresponds to the 3-form $\left(\frac{\partial P}{\partial x} + \frac{\partial Q}{\partial y} + \frac{\partial R}{\partial z}\right) \mathrm{d}x\mathrm{d}y\mathrm{d}z$.

Theorem 14.8. (General Stokes' theorem) *Let $\mathcal{M}$ be a $k+1$-dimensional (smooth) manifold, $\partial\mathcal{M}$ its boundary and let ω be a differential form of degree k in n variables. Then the identity*

$$\underbrace{\int \cdots \int}_{k} {}_{\partial\mathcal{M}}\, \omega = \underbrace{\int \cdots \int}_{k+1} {}_{\mathcal{M}}\, \mathrm{d}\omega \tag{14.97}$$

holds true or in the form

$$\omega(\partial\mathcal{M}) = \mathrm{d}\omega(\mathcal{M}). \tag{14.98}$$

name	degree	number of var.	
General Stokes' thm	$k-1$	n	$\langle \partial c_k, \omega^{k-1}\rangle = \langle c_k, \mathrm{d}\omega^{k-1}\rangle$
Green's thm	1	2	$\langle \partial c_2, \omega^1\rangle = \langle c_2, \mathrm{d}\omega^1\rangle$
Gauss divergence thm	2	3	$\langle \partial c_3, \omega^2\rangle = \langle c_3, \mathrm{d}\omega^2\rangle$
Stokes' thm	1	3	$\langle \partial c_2, \omega^1\rangle = \langle c_2, \mathrm{d}\omega^1\rangle$

Table 4. Special cases of general Stokes' theorem

In what follows we apply Theorem 14.3 in conjunction with (14.29) and deduce the important special cases from Theorem 14.8. We omit the wedge symbol and simply write $\mathrm{d}x\mathrm{d}y$ for $\mathrm{d}x \wedge \mathrm{d}y$.

Theorem 14.9. (Green's theorem) *Let $\mathcal{M} = \mathcal{D} \subset \mathbb{R}^2$ be a domain, $\partial\mathcal{M} = \partial\mathcal{D} \subset \mathbb{R}^2$ its boundary and let ω be a C^1 class differential form of degree 1 in 2 variables:*

$$\omega = P\mathrm{d}x + Q\mathrm{d}y. \tag{14.99}$$

Then we have

$$\iint_{\mathcal{D}} \mathrm{d}\omega = \int_{\partial\mathcal{D}} \omega \tag{14.100}$$

or more concretely,

$$\iint_{\mathcal{D}} \left(\frac{\partial Q}{\partial x} - \frac{\partial P}{\partial y}\right) \mathrm{d}x\mathrm{d}y = \int_{\partial\mathcal{D}} P\mathrm{d}x + Q\mathrm{d}y. \tag{14.101}$$

Proof.

$$\mathrm{d}\omega = \mathrm{d}P\mathrm{d}x + \mathrm{d}Q\mathrm{d}y = \left(\frac{\partial P}{\partial x}\mathrm{d}x + \frac{\partial P}{\partial y}\mathrm{d}y\right)\mathrm{d}x + \left(\frac{\partial Q}{\partial x}\mathrm{d}x + \frac{\partial Q}{\partial y}\mathrm{d}y\right)\mathrm{d}y \tag{14.102}$$

$$= \frac{\partial P}{\partial y}\mathrm{d}y\mathrm{d}x + \frac{\partial Q}{\partial x}\mathrm{d}x\mathrm{d}y = \left(\frac{\partial Q}{\partial x} - \frac{\partial P}{\partial y}\right)\mathrm{d}x\mathrm{d}y.$$

□

We shall state this theorem again as Theorem 14.12 and prove it by the fundamental theorem of infinitesimal calculus.

Theorem 14.10. (Gauss divergence theorem) *Let $\mathcal{M} = R \subset \mathbb{R}^3$ be a domain, $\partial\mathcal{M} = \partial R \subset \mathbb{R}^3$ its boundary and let ω be a C^1 class differential form of degree 2 in 3 variables:*

$$\omega = P\mathrm{d}y\mathrm{d}z + Q\mathrm{d}z\mathrm{d}x + R\mathrm{d}x\mathrm{d}y. \tag{14.103}$$

Then we have

$$\iiint_{\mathcal{D}} \mathrm{d}\omega = \iint_{\partial\mathcal{D}} \omega \tag{14.104}$$

or more concretely,

$$\iiint_{\mathcal{D}} \left(\frac{\partial P}{\partial x} + \frac{\partial Q}{\partial y} + \frac{\partial R}{\partial z} \right) \mathrm{d}x\mathrm{d}y\mathrm{d}z = \iint_{\partial\mathcal{D}} P\mathrm{d}y\mathrm{d}z + Q\mathrm{d}z\mathrm{d}x + R\mathrm{d}x\mathrm{d}y. \tag{14.105}$$

Proof.

$$\begin{aligned} \mathrm{d}\omega &= \mathrm{d}P\mathrm{d}y\mathrm{d}z + \mathrm{d}Q\mathrm{d}z\mathrm{d}x + R\mathrm{d}x\mathrm{d}y \\ &= \left(\frac{\partial P}{\partial x}\mathrm{d}x + \frac{\partial P}{\partial y}\mathrm{d}y + \frac{\partial P}{\partial y}\mathrm{d}z \right) \mathrm{d}y\mathrm{d}z + \left(\frac{\partial Q}{\partial x}\mathrm{d}x + \frac{\partial Q}{\partial y}\mathrm{d}y + \frac{\partial Q}{\partial y}\mathrm{d}z \right) \mathrm{d}z\mathrm{d}x \\ &\quad + \left(\frac{\partial R}{\partial x}\mathrm{d}x + \frac{\partial R}{\partial y}\mathrm{d}y + \frac{\partial R}{\partial y}\mathrm{d}z \right) \mathrm{d}x\mathrm{d}y \\ &= \frac{\partial P}{\partial x}\mathrm{d}x\mathrm{d}y\mathrm{d}z + \frac{\partial Q}{\partial y}\mathrm{d}y\mathrm{d}z\mathrm{d}x + \frac{\partial R}{\partial z}\mathrm{d}z\mathrm{d}x\mathrm{d}y. \end{aligned} \tag{14.106}$$

□

The name comes from the fact that the left-hand side of (14.105) is equal to the integral of divergence defined by (14.94).

Theorem 14.11. (Stokes' theorem) *Let $\mathcal{M} = \mathcal{D} \subset \mathbb{R}^3$ be a domain, $\partial\mathcal{M} = \partial\mathcal{D} \subset \mathbb{R}^3$ its boundary and let ω be a C^1 class differential form of degree 1 in 3 variables:*

$$\omega = P\mathrm{d}x + Q\mathrm{d}y + R\mathrm{d}z. \tag{14.107}$$

Then we have

$$\iint_{\mathcal{D}} \mathrm{d}\omega = \int_{\partial\mathcal{D}} \omega \tag{14.108}$$

or more concretely,

$$\begin{aligned} &\iint_{\mathcal{D}} \left(\frac{\partial Q}{\partial x} - \frac{\partial P}{\partial y} \right) \mathrm{d}x\mathrm{d}y + \iint_{\mathcal{D}} \left(\frac{\partial R}{\partial y} - \frac{\partial Q}{\partial z} \right) \mathrm{d}y\mathrm{d}z \\ &+ \iint_{\mathcal{D}} \left(\frac{\partial P}{\partial z} - \frac{\partial R}{\partial x} \right) \mathrm{d}z\mathrm{d}x = \int_{\partial\mathcal{D}} P\mathrm{d}x + Q\mathrm{d}y + R\mathrm{d}z. \end{aligned} \tag{14.109}$$

14.3 Integrals over a curve and a surface

Definition 14.12. A smooth (n-dimensional) curve $\boldsymbol{x} = \boldsymbol{x}(t)$ is a C^1 class map $\boldsymbol{x} : [a, b] \to \mathbb{R}^n$ given in (5.14) and the tangent vector is given in (5.17).

Definition 14.13. A (smooth) surface $S \subset \mathbb{R}^3$ is a (smooth) mapping from $D \subset \mathbb{R}^2 \to \mathbb{R}^3$. As in (14.7), we restrict ourselves to the parametric representation with a 2-dimensional interval $D = \mathbb{I}^2$.

$$\boldsymbol{\phi} : \mathbb{I}^2 \to S \subset \mathbb{R}^3; \quad {}^t(x, y, z) = \boldsymbol{x} = \boldsymbol{\phi}(\boldsymbol{t}), \quad \boldsymbol{t} = {}^t(u, v) \tag{14.110}$$

in component form

$$\begin{cases} x = x(u, v) \\ y = y(u, v) \\ z = z(u, v), \end{cases} \qquad {}^t(u, v) \in \mathbb{I}^2. \tag{14.111}$$

We assume throughout that the functions are C^1 and so the gradient of $\boldsymbol{\phi}$ is

$$\nabla \boldsymbol{\phi} = \begin{pmatrix} \nabla x \\ \nabla y \\ \nabla z \end{pmatrix} = \begin{pmatrix} \frac{\partial x}{\partial u} & \frac{\partial x}{\partial v} \\ \frac{\partial y}{\partial u} & \frac{\partial y}{\partial v} \\ \frac{\partial z}{\partial u} & \frac{\partial z}{\partial v} \end{pmatrix} = (\boldsymbol{x}_u, \boldsymbol{x}_v) \tag{14.112}$$

for which there is no Jacobian and definition like (14.11) does not seem applicable. But the local-global principle applies. We remark that $\boldsymbol{x}_u(u_0, v_0), \boldsymbol{x}_v(u_0, v_0)$ are tangent vectors of the curves $\boldsymbol{x}(u, v_0)$ and $\boldsymbol{x}(u_0, v)$ on S, respectively at the point $\boldsymbol{x}_0 = \boldsymbol{x}_0(u_0, v_0)$ on the surface. Hence the infinitesimal parallelogram formed by them has the area element

$$\mathrm{d}A = |\boldsymbol{x}_u \times \boldsymbol{x}_v| \mathrm{d}u\mathrm{d}v \tag{14.113}$$

by Remark A.1 in Appendix A. Hence for a continuous function defined on S, its integral over S is to be defined as

$$\iiint_S f = \iint_{\mathbb{I}^2} f(\boldsymbol{\phi}(\boldsymbol{t}))|\boldsymbol{x}_u \times \boldsymbol{x}_v| \,\mathrm{d}u\mathrm{d}v, \tag{14.114}$$

where

$$|\boldsymbol{x}_u \times \boldsymbol{x}_v| = \sqrt{\left(\frac{\partial(x, y)}{\partial(u, v)}\right)^2 + \left(\frac{\partial(y, z)}{\partial(u, v)}\right)^2 + \left(\frac{\partial(z, x)}{\partial(u, v)}\right)^2}. \tag{14.115}$$

By Exercise 145 (cf. also Exercise 120), $\boldsymbol{x}_u \times \boldsymbol{x}_v$ is an outer normal perpendicular to $\boldsymbol{x}_u$ and $\boldsymbol{x}_v$. The vector $\boldsymbol{n} = \frac{1}{|\boldsymbol{x}_u \times \boldsymbol{x}_v|}\boldsymbol{x}_u \times \boldsymbol{x}_v$ is called an outer **normal**.

More precisely, one would need to argue as follows. By a division of D into subdomains D_{ij} gives rise to that of S into subsurfaces S_{ij}. Choosing an arbitrary point $p_{ij} \in S_{ij}$ and form the Riemann sum

$$\sum f(p_{ij})A(S_{ij}),$$

where A means the area. Then letting the division smaller and prove the existence of the limit just as in the case of the ordinary Riemann integral. (14.114) allows you compute the integral over a surface as an ordinary double integral.

Exercise 119. A torus T (doughnut) is given by the parametric representation

$$T : \begin{cases} x = (r - \cos v)\cos u \\ y = (r - \cos v)\sin u \\ z = \sin v, \end{cases} \tag{14.116}$$

where $r > 1$ and

$${}^t(u, v) \in [-\pi, \pi] \times [-\pi, \pi].$$

Find the area of T.

Solution. We find that

$$|\boldsymbol{x}_u \times \boldsymbol{x}_v| = r - \cos v, \tag{14.117}$$

so that $A(T) = \int_{-\pi}^{\pi} \mathrm{d}u \int_{-\pi}^{\pi} (r - \cos v)\,\mathrm{d}v = (2\pi)^2 r$.

Corollary 11.3 is the fundamental theorem in calculus giving the way of computing definite integrals as the difference of values at the end points. This may be thought of as a 1-dimensional integral amounting to the 0-dimensional one. In this way, Green's formula, Theorem 14.9, may be thought of as a 2-dimensional integral amounting to the 1-dimensional (line) integral. As its vast generalization, we restate Theorem 14.7 as

Theorem 14.12. (Green) *Let C be a piecewise smooth Jordan curve with D its interior. If functions $P(x, y), Q(x, y)$ are of $C^1(\bar{D})$ and the integrals*

$$\iint_{\bar{D}} \frac{\partial P}{\partial x}\,\mathrm{d}x\mathrm{d}y, \quad \iint_{\bar{D}} \frac{\partial Q}{\partial y}\,\mathrm{d}x\mathrm{d}y$$

exist, then the line integral $\int_C P\,\mathrm{d}x + Q\,\mathrm{d}y$ along the positive direction of C exists and

$$\iint_{\bar{D}} \left(\frac{\partial P}{\partial x} - \frac{\partial Q}{\partial y}\right) \mathrm{d}x\mathrm{d}y = \int_C P\,\mathrm{d}y + Q\,\mathrm{d}x. \tag{14.118}$$

We prove this theorem only in the case of a vertically and horizontally ordinate set in Definition 11.4 where for the boundary may be vertical and horizontal line segments. I.e. we restrict to the case where C is parametrized as a vertically and horizontally ordinate set by

$$\gamma = \gamma(t) = {}^t(x, \psi(x)), \quad a \le x \le b. \tag{14.119}$$

Proof. (Proof of Theorem 14.12.) Suppose $\bar{D}$ is vertically and horizontally ordinate set. Circumscribe $\bar{D}$ by a rectangle that meet at most one point and let vertical tangent points and horizontal tangent points be K, L and M, N, respectively. Then KML and LNK are expressed as one-valued functions $y = \phi_1(x)$ and $y = \phi_2(x)$ on $[a, b]$, respectively. Since $C =$ arc KML + arc LNK, it follows that

$$\begin{aligned} \int_C Q \, \mathrm{d}x &= \int_{\text{arc KML}} Q \, \mathrm{d}x + \int_{\text{arc LNK}} Q \, \mathrm{d}x \\ &= \int_a^b Q(x, \phi_1(x)) \, \mathrm{d}x + \int_b^a Q(x, \phi_2(x)) \, \mathrm{d}x. \end{aligned} \tag{14.120}$$

By Theorem 11.12,

$$\begin{aligned} \iint_{\bar{D}} -\frac{\partial Q}{\partial y} \, \mathrm{d}x\mathrm{d}y &= -\int_a^b \int_{\phi_1(x)}^{\phi_2(x)} \frac{\partial Q}{\partial y} \, \mathrm{d}x\mathrm{d}y \\ &= -\int_a^b [Q(x, y)]_{\phi_1(x)}^{\phi_2(x)} \, \mathrm{d}x = \int_a^b Q(x, \phi_1(x)) \, \mathrm{d}x - \int_a^b Q(x, \phi_2(x)) \, \mathrm{d}x, \end{aligned} \tag{14.121}$$

which is $\int_C Q \, \mathrm{d}x$ by (14.120). Hence

$$\iint_{\bar{D}} \frac{\partial Q}{\partial y} \, \mathrm{d}x\mathrm{d}y = \int_C Q \, \mathrm{d}x. \tag{14.122}$$

Similarly, working with NKM, MLN, we obtain

$$\iint_{\bar{D}} \frac{\partial P}{\partial x} \, \mathrm{d}x\mathrm{d}y = \int_C P \, \mathrm{d}y. \tag{14.123}$$

Adding these completes the proof. □

Generalizing Definition 14.13 to n-dimension, we formulate

Definition 14.14. An n-dimensional curve $S \subset \mathbb{R}^n$ is the image of the composition of mappings $\mathbb{I} \to D$, $D \subset \mathbb{R}^k \to \mathbb{R}^n$. In addition to (14.10), we have

$$\boldsymbol{t} = \boldsymbol{t}(s) = {}^t(t_1, \cdots, t_k), \quad t_j = t_j(s) \tag{14.124}$$

in component form

$$\begin{cases} x_1 = x_1(s) = x_1(t_1, \cdots, t_k) = x_1((t_1(s), \cdots, t_k(s))) \\ \cdots \\ x_n = x_n(s) = x_n(t_1, \cdots, t_k) = x_n((t_1(s), \cdots, t_k(s))), \end{cases} \qquad s \in \mathbb{I}^1 = [a, b]. \tag{14.125}$$

We consider the situation in Definition 14.1 with $k = n - 1$. We assume that the functions are C^1 and so the gradient of ϕ is

$$\nabla\phi = \begin{pmatrix} \nabla x_1 \\ \vdots \\ \nabla x_n \end{pmatrix} = \begin{pmatrix} \frac{\partial x_1}{\partial t_1} & \cdots & \frac{\partial x_1}{\partial t_{n-1}} \\ & \cdots & \\ \frac{\partial x_n}{\partial t_1} & \cdots & \frac{\partial x_n}{\partial t_{n-1}} \end{pmatrix} = (\boldsymbol{\psi}_1, \cdots, \boldsymbol{\psi}_{n-1}), \tag{14.126}$$

where

$$\boldsymbol{\psi}_j = \frac{\partial}{\partial t_j}\boldsymbol{x} = {}^t\left(\frac{\partial x_1}{\partial t_j}, \cdots, \frac{\partial x_n}{\partial t_j}\right), \quad 1 \le j \le n-1. \tag{14.127}$$

Let

$$\boldsymbol{p} = {}^t(p_1, \cdots, p_n), \quad p_i = (-1)^{i-1}\frac{\partial(x_1, \overset{\overset{i}{\vee}}{\cdots}, x_n)}{\partial(t_1, \cdots, t_{n-1})}, \quad 1 \le i \le n \tag{14.128}$$

with x_i missing, which is a special case of (14.12) with $k = n - 1$ and

$$\boldsymbol{n} = \frac{1}{|\boldsymbol{p}|}\boldsymbol{p}. \tag{14.129}$$

Exercise 120. Prove that the vector $\boldsymbol{n}$ in (14.129) is perpendicular to each column of $\nabla\phi$ ("tangent vectors") in (14.126).

Solution. By Theorem 6.1, we have

$$\frac{\mathrm{d}x_i}{\mathrm{d}s} = \sum_{j=1}^{n-1}\frac{\partial x_i}{\partial t_j}\frac{\mathrm{d}t_j}{\mathrm{d}s}.$$

Hence

$$\nabla\phi(s) = \sum_{j=1}^{n-1}\frac{\mathrm{d}t_j}{\mathrm{d}s}\boldsymbol{\psi}_j. \tag{14.130}$$

Hence the inner product of (14.127) and $\boldsymbol{p}$ in (14.128) is

$$(\boldsymbol{p}, \boldsymbol{\psi}_j) = \sum_{i=1}^{n}(-1)^{i-1}\frac{\partial x_i}{\partial t_j}\frac{\partial(x_1, \overset{i}{\cdots}, x_n)}{\partial(t_1, \cdots, t_{n-1})},$$

which is the cofactor expansion of the determinant with respect to the first column of

$$\frac{\partial(x_1, \cdots, x_n)}{\partial(t_j, t_1, \cdots, t_{n-1})}.$$

However, this determinant has the first and the jth columns equal, so that it is zero. Hence $\boldsymbol{n}$ is perpendicular to the tangent vectors in (14.127).

The general Stokes' theorem, Theorem 14.8, may be also stated as

Theorem 14.13. (General Green's theorem) *Let $D \subset \mathbb{R}^n$ be a domain and let $\boldsymbol{a} = \boldsymbol{a}(\boldsymbol{x}) = {}^t(x_1, \cdots, x_n)$ be a vector field. Then the identity*

$$\int_{\partial D} (\boldsymbol{a}, \boldsymbol{n})\, \mathrm{d}S = \int_D \operatorname{div} \boldsymbol{a}\, \mathrm{d}\boldsymbol{x} \tag{14.131}$$

holds true, where $\boldsymbol{n}$ is the outer normal (14.129) *and* $\mathrm{d}S$ *indicates the area element of ∂D.*

Proof. In Lemma 14.1 we set $k = n - 1$ and

$$\omega^{n-1} = \omega^{n-1}(\boldsymbol{x}) = \sum_{i=1}^{n} (-1)^{i-1} a_i \mathrm{d}x_1 \wedge \overset{\overset{i}{\vee}}{\cdots} \wedge \mathrm{d}x_n,$$

where $\boldsymbol{a} = \boldsymbol{a}(\boldsymbol{x}) = {}^t(a_1, \cdots, a_n)$ is a vector field and $\mathrm{d}x_i$ is missing. Then the left-hand side of (14.131) is $\int_{\partial D} \omega^{n-1}$.

On the other hand, as in Exercise 118

$$\begin{aligned} \mathrm{d}\,\omega^{n-1} &= \sum_{i=1}^{n} (-1)^{i-1} \frac{\partial a_i}{\partial x_i} \mathrm{d}x_i \wedge \mathrm{d}x_1 \wedge \overset{\overset{i}{\vee}}{\cdots} \wedge \mathrm{d}x_n \\ &= \sum_{i=1}^{n} \frac{\partial a_i}{\partial x_i} \mathrm{d}x_1 \wedge \cdots \wedge \mathrm{d}x_n = \operatorname{div} \boldsymbol{a}\, \mathrm{d}\,\boldsymbol{x}. \end{aligned} \tag{14.132}$$

Hence the right-hand side of (14.85) is $\int_D \operatorname{div} A\ \mathrm{d}\boldsymbol{x}$. Hence the assertion follows from Theorem 14.8. □

We introduce the notation. Generalizing (14.95), we introduce the **Laplacian**

$$\Delta u = \nabla \cdot \nabla u = \sum_{i=1}^{n} \frac{\partial^2 u}{\partial x_i^2} \tag{14.133}$$

and the normal derivative in the direction of the outer normal (14.129):

$$\frac{\partial u}{\partial \boldsymbol{n}} = (\nabla u, \boldsymbol{n}). \tag{14.134}$$

Corollary 14.2. (Green's formulas) *Let* $u, v \in C^2(D)$, $D \subset \mathbb{R}^n$. *Then*

$$\int_D (\nabla u, \nabla v)\mathrm{d}\boldsymbol{x} + \int_D u \cdot \Delta v \mathrm{d}\boldsymbol{x} = \int_{\partial D} u \cdot \frac{\partial v}{\partial \boldsymbol{n}} \mathrm{d}S \tag{14.135}$$

and

$$\int_D (u \cdot \Delta v - v \cdot \Delta u)\mathrm{d}\boldsymbol{x} = \int_{\partial D} \left(u \cdot \frac{\partial v}{\partial \boldsymbol{n}} - v \cdot \frac{\partial u}{\partial \boldsymbol{n}} \right) \mathrm{d}S. \tag{14.136}$$

Proof. Choosing $\boldsymbol{a} = \boldsymbol{a}(\boldsymbol{x}) = u(\boldsymbol{x})\nabla v(\boldsymbol{x})$ in Theorem 14.13, the integrand on the right-hand side of (14.131) is

$$\operatorname{div} \boldsymbol{a}(\boldsymbol{x}) = \sum_{i=1}^{n} \frac{\partial}{\partial x_i} \left(u(\boldsymbol{x}) \frac{\partial v(\boldsymbol{x})}{\partial x_i} \right) = (\nabla u, \nabla v) + u \cdot \Delta v.$$

The one on left-hand side is

$$(\boldsymbol{a}, \boldsymbol{n}) = u(\boldsymbol{x}) \cdot (\nabla v, \boldsymbol{n})$$

which is the normal derivative in (14.134). Hence (14.131) leads to (14.135).

Subtracting from (14.135), the formula obtained from (14.135) with u and v changed proves (14.136). □

Cf. Example 15.9 below for distribution-theoretic interpretation of one of Green's formulas (14.136).

Exercise 121. Deduce from Theorem 14.13 a form of integration by parts

$$\int_D \frac{\partial u}{\partial x_i} v = \int_{\partial D} u v n_i \,\mathrm{d}S - \int_D u \frac{\partial v}{\partial x_i}, \tag{14.137}$$

where $n_i = (\boldsymbol{n}, \boldsymbol{e}_i)$ is the ith entry of the outer normal (14.129).

Solution. We let

$$\boldsymbol{a} = uv\boldsymbol{e}_i = {}^t(0, \cdots, 0, uv, 0, \cdots, 0).$$

Then $(\boldsymbol{a}, \boldsymbol{n}) = uvn_i$ and

$$\operatorname{div} \boldsymbol{a} = \frac{\partial u}{\partial x_i} v + u \frac{\partial v}{\partial x_i},$$

whence the result.

14.4 Maxwell equations

Schematically the correspondence in Definition 14.11 may be expressed as

$$f \longrightarrow \operatorname{grad}(f) = \nabla f \longrightarrow \operatorname{curl}(\boldsymbol{f}) = \nabla \times \boldsymbol{f} \longrightarrow \operatorname{div}(\boldsymbol{f}) = \nabla \cdot \boldsymbol{f} \quad (14.138)$$

which corresponds to the complex (cf. (14.47))

$$C :\rightarrow \Lambda^0(\Omega) \xrightarrow{\mathrm{d}} \Lambda^1(\Omega) \xrightarrow{\mathrm{d}} \Lambda^2(\Omega) \xrightarrow{\mathrm{d}} \Lambda^3(\Omega). \quad (14.139)$$

Hence Theorem 14.3, (ii), cf. also (14.48), entails

Lemma 14.4. (i)

$$\nabla \times (\nabla f) = \operatorname{curl}(\operatorname{grad} f) = \boldsymbol{o}, \quad (14.140)$$

and
(ii)

$$\nabla \cdot (\nabla \times \boldsymbol{A}) = \operatorname{div}(\operatorname{curl} \boldsymbol{A}) = 0. \quad (14.141)$$

In this setting, Poincaré's lemma, Theorem 14.6, reads for vector fields

Theorem 14.14. (Poincaré's lemma for space) *If the domain Ω is smoothly contractible and $\boldsymbol{A}$ is a vector field of class C^1 on Ω, then*
(i) $\operatorname{curl} \boldsymbol{A} = 0 \Longleftrightarrow$ *there exists a function $f \in C^1(\Omega)$, unique up to a constant, such that $\boldsymbol{A} = \operatorname{grad} f = \nabla f$.*
(ii) $\operatorname{div} \boldsymbol{A} = 0 \Longleftrightarrow$ *there exists a vector field $\boldsymbol{V}$ of class C^1, unique up to ∇g, where g is an arbitrary scalar function such that $\boldsymbol{A} = \operatorname{curl} \boldsymbol{V}$.*

Exercise 122. Prove that the vector potential $\boldsymbol{V}$ of class C^1 in Theorem 14.14, (ii) is unique up to ∇g.

Solution. Suppose $\boldsymbol{V}$ satisfying $\boldsymbol{A} = \operatorname{curl} \boldsymbol{V}$ is shifted by ∇g. Then

$$\nabla \times (\boldsymbol{V} + \nabla g) = \nabla \times \boldsymbol{V} + \nabla \times (\nabla g) = \nabla \times \boldsymbol{V}$$

in view of (14.140) which also follows from

$$\nabla \times \nabla = 0.$$

Definition 14.15. The function f in Theorem 14.14, (i) is called a **potential** for the vector field $\boldsymbol{A}$. In particular, if $\boldsymbol{A}$ is a velocity field of a flow satisfying the condition $\operatorname{curl} \boldsymbol{A} = 0$, then it is called an **irrotational flow** and f is called a **velocity potential**. If $\boldsymbol{A}$ is a force field satisfying the condition curl $\boldsymbol{A} = 0$, then it is called a **conservative** field and $U = -f$ is called a **potential energy**.

On the other hand, a vector field satisfying the condition in Theorem 14.14, (ii) is called **solenoidal** or **divergence-free**. The velocity field of an incompressible fluid is solenoidal.

Let $\boldsymbol{E}$ and $\boldsymbol{H}$ denote the **electric field** and the **magnetic field**, respectively. Additionally, let $\boldsymbol{D} = 4\pi\boldsymbol{E}$ and $\boldsymbol{B} = \mu\boldsymbol{H}$ denote the **electric induction field** and the **magnetic induction field**, respectively. Classical electro-magnetism is concerned with the relationship among current, electrons and electric and magnetic fields.

Thus classical electro-magnetism is governed by the **Maxwell equations** which read:
Faraday's induction law:

$$\operatorname{curl}(\boldsymbol{E}) = -\frac{1}{c}\frac{\partial \boldsymbol{B}}{\partial t} = -\frac{\mu}{c}\frac{\partial \boldsymbol{H}}{\partial t}. \tag{14.142}$$

Absence of magnetic charge:

$$\operatorname{div}(\boldsymbol{B}) = 0. \tag{14.143}$$

Generalized Oersted's law:

$$\operatorname{curl}(\boldsymbol{H}) = \frac{4\pi}{c}\left(j + \frac{\partial E}{\partial t}\right) = \frac{4\pi}{c}j + \frac{1}{c}\frac{\partial \boldsymbol{D}}{\partial t}, \tag{14.144}$$

and

$$\operatorname{div}(\boldsymbol{E}) = \nabla \cdot \boldsymbol{E} = 4\pi\rho, \tag{14.145}$$

where the scalar function ρ describes the distribution of charge and the vector function $\boldsymbol{j}$ the distribution of current. Here the coefficients μ, c or 4π are immaterial and we consider a more concise from ($c = 1$) below.

In special relativity theory, the 4-dimensional expression is known for electro-magnetism. We introduce the 4-dimensional electric current

$$\boldsymbol{J} = {}^t(j_1(\boldsymbol{x}, t), j_2(\boldsymbol{x}, t), j_3(\boldsymbol{x}, t), \rho(\boldsymbol{x}, t)), \quad \boldsymbol{x} = {}^t(x, y, z). \tag{14.146}$$

Then the **equation of continuity** of current reads

$$\operatorname{div} \boldsymbol{J} = \frac{\partial j_1}{\partial x} + \frac{\partial j_2}{\partial y} + \frac{\partial j_3}{\partial z} + \frac{\partial \rho}{\partial t} = 0. \tag{14.147}$$

Indeed, first applying (14.141) to (14.144), we have

$$0 = c \operatorname{div} \operatorname{curl} \boldsymbol{H} = 4\pi \operatorname{div} \boldsymbol{j} + \operatorname{div} \frac{\partial \boldsymbol{D}}{\partial t}. \tag{14.148}$$

Then the differentiated form of (14.145) reads

$$\operatorname{div}\frac{\partial \boldsymbol{D}}{\partial t} = \frac{\partial}{\partial t}\operatorname{div}\boldsymbol{D} = 4\pi\frac{\partial \rho}{\partial t}, \tag{14.149}$$

in view of $\operatorname{div}\frac{\partial}{\partial t} = \frac{\partial}{\partial t}\operatorname{div}$. Substituting (14.149) in (14.148), we obtain

$$0 = \operatorname{div}\boldsymbol{j} + \frac{\partial \rho}{\partial t} = \operatorname{div}\boldsymbol{J}, \tag{14.150}$$

proving (14.147).

Exercise 123. Prove that the equation of continuity (14.147) may be expressed as

$$\mathrm{d}\mathcal{J}^3 = 0, \tag{14.151}$$

where the 3-form $\mathcal{J}$ is defined by

$$\mathcal{J}^3 = j_1\mathrm{d}y\wedge\mathrm{d}z\wedge\mathrm{d}t + j_2\mathrm{d}z\wedge\mathrm{d}t\wedge\mathrm{d}x + j_3\mathrm{d}t\wedge\mathrm{d}x\wedge\mathrm{d}y - \rho\mathrm{d}x\wedge\mathrm{d}y\wedge\mathrm{d}z. \tag{14.152}$$

Solution. By the alternating properties in Lemma 11.2 we have

$$\mathrm{d}j_1\mathrm{d}y\wedge\mathrm{d}z\wedge\mathrm{d}t = \frac{\partial j_1}{\partial x}\mathrm{d}x\wedge\mathrm{d}y\wedge\mathrm{d}z\wedge\mathrm{d}t + \frac{\partial j_1}{\partial y}\mathrm{d}y\wedge\mathrm{d}y\wedge\mathrm{d}z\wedge\mathrm{d}t + \cdots \tag{14.153}$$

$$= \frac{\partial j_1}{\partial x}\mathrm{d}x\wedge\mathrm{d}y\wedge\mathrm{d}z\wedge\mathrm{d}t$$

in view of $\mathrm{d}y\wedge\mathrm{d}y = 0, \cdots$. Hence

$$\mathrm{d}\mathcal{J}^3 = \operatorname{div}(\boldsymbol{J})\mathrm{d}x\wedge\mathrm{d}y\wedge\mathrm{d}z\wedge\mathrm{d}t, \tag{14.154}$$

whence (14.151).

Exercise 124. Define

$$\mathcal{E}^2 = (E_1\mathrm{d}x + E_2\mathrm{d}y + E_3\mathrm{d}z)\wedge\mathrm{d}t + B_1\mathrm{d}y\wedge\mathrm{d}z + B_2\mathrm{d}z\wedge\mathrm{d}x + B_3\mathrm{d}x\wedge\mathrm{d}y \tag{14.155}$$

and

$$\mathcal{H}^2 = -c(H_1\mathrm{d}x + H_2\mathrm{d}y + H_3\mathrm{d}z)\wedge\mathrm{d}t + D_1\mathrm{d}y\wedge\mathrm{d}z + D_2\mathrm{d}z\wedge\mathrm{d}x + D_3\mathrm{d}x\wedge\mathrm{d}y. \tag{14.156}$$

Prove that the equations (14.142) and (14.143) may be expressed as

$$\mathrm{d}\,\mathcal{E}^2 = 0, \tag{14.157}$$

while (14.144) and (14.145) may be by

$$\mathrm{d}\,\mathcal{H}^2 + 4\pi\mathcal{J}^3 = 0. \tag{14.158}$$

Solution. We have

$$d\mathcal{E}^2 = \left(\operatorname{curl}(\boldsymbol{E}) + \frac{\partial \boldsymbol{B}}{\partial t}\right) \cdot {}^t(dy \wedge dz \wedge dt, dz \wedge dx \wedge dt, dx \wedge dy \wedge dt) \tag{14.159}$$
$$+ \operatorname{div}(\boldsymbol{B}) dx \wedge dy \wedge dz.$$

Hence (14.157) implies the coefficients are 0, which amounts to (14.142) and (14.143).

In the same way as we prove (14.159), we have

$$d\mathcal{H}^2 = \left(-c\operatorname{curl}(\boldsymbol{H}) + \frac{\partial \boldsymbol{D}}{\partial t}\right) \cdot {}^t(dy \wedge dz \wedge dt, dz \wedge dx \wedge dt, dx \wedge dy \wedge dt) \tag{14.160}$$
$$+ \operatorname{div}(\boldsymbol{D}) dx \wedge dy \wedge dz.$$

Hence the left-hand side of (14.158) reads

$$\left(-c\operatorname{curl}(\boldsymbol{H}) + 4\pi \boldsymbol{j} + \frac{\partial \boldsymbol{D}}{\partial t}\right) \cdot {}^t(dy \wedge dz \wedge dt, dz \wedge dx \wedge dt, dx \wedge dy \wedge dt) \tag{14.161}$$
$$+ (\operatorname{div}(\boldsymbol{D}) - 4\pi\rho)\, dx \wedge dy \wedge dz.$$

Hence (14.158) implies (14.144) and (14.145), completing the proof.

Supposing that (14.157) holds on a smoothly contractible set, say on the whole $\mathbb{R}^4$, it follows from Poincaré's lemma, Theorem 14.6 that there exists a 1-form

$$\omega^1 = A_1 dx + A_2 dy + A_3 dz + A_0 dt \tag{14.162}$$

such that

$$\mathcal{E}^2 = d\,\omega^1. \tag{14.163}$$

In the terminology of vector analysis, this means that there exists a vector potential $\boldsymbol{V} = {}^t(A_1, A_2, A_3)$ such that

$$\boldsymbol{H} = \operatorname{curl} \boldsymbol{V} \tag{14.164}$$

and

$$\boldsymbol{E} = \nabla A_0 - \frac{\partial}{\partial t}\boldsymbol{V}. \tag{14.165}$$

Exercise 125. Prove (14.164) and (14.165) by applying Theorem 14.14.

Solution. Applying Theorem 14.14, (ii) to (14.167), we find a vector potential $\boldsymbol{V}$ of class C^1 satisfying (14.164). Substituting this in (14.166), we find

$$\boldsymbol{o} = \nabla \times \boldsymbol{E} + \frac{\partial}{\partial t}\operatorname{curl}\boldsymbol{V} = \nabla \times \left(\boldsymbol{E} + \frac{\partial}{\partial t}\boldsymbol{V}\right),$$

whence, by Theorem 14.14, (i), there exists a function A_0, unique up to a constant satisfying (14.164).

We shall study the following slightly modified form (with $c = 1$).

$$\operatorname{curl}(\boldsymbol{E}) = \nabla \times \boldsymbol{E} = -\frac{\partial \boldsymbol{H}}{\partial t}, \tag{14.166}$$

$$\operatorname{div}(\boldsymbol{H}) = \nabla \cdot \boldsymbol{H} = 0, \tag{14.167}$$

$$\operatorname{curl}(\boldsymbol{H}) = \nabla \times \boldsymbol{H} = 4\pi\left(\boldsymbol{j} + \frac{\partial \boldsymbol{E}}{\partial t}\right), \tag{14.168}$$

$$\operatorname{div}(\boldsymbol{E}) = \nabla \cdot \boldsymbol{E} = \rho. \tag{14.169}$$

We shall prove the following

Theorem 14.15. *Suppose the domains of all relevant vector fields are smoothly contractible. Then under the condition*

$$\frac{1}{4\pi}\nabla \cdot \boldsymbol{V} - \frac{\partial f}{\partial t} = 0 \tag{14.170}$$

the solutions of the Maxwell equations amount to those of

$$\nabla^2 f - 4\pi\frac{\partial^2 f}{\partial t^2} = \rho \tag{14.171}$$

and

$$\nabla^2 \boldsymbol{V} - \frac{\partial^2 \boldsymbol{V}}{\partial t^2} = -\boldsymbol{j} \tag{14.172}$$

plus the shift ψ which are solutions of the inhomogeneous **wave equation**

$$\nabla^2 \psi - 4\pi\frac{\partial^2 \psi}{\partial t^2} = -h, \tag{14.173}$$

or in coordinate form

$$\frac{\partial^2 \psi}{\partial x^2} + \frac{\partial^2 \psi}{\partial y^2} + \frac{\partial^2 \psi}{\partial z^2} - 4\pi\frac{\partial^2 \psi}{\partial t^2} + h = 0, \tag{14.174}$$

where h is a constant given by

$$-h = \nabla \cdot \boldsymbol{V} - 4\pi\frac{\partial f}{\partial t}. \tag{14.175}$$

Here f and $\boldsymbol{V}$ are particular solutions of (14.171) *and* (14.172), *respectively.*

Proof. We denote A_0 by $f \in C^1(\Omega)$ and restate (14.165) as

$$\boldsymbol{E} = \nabla f - \frac{\partial}{\partial t}\boldsymbol{V}. \tag{14.176}$$

Substituting (14.176), (14.169) reads

$$\rho = \nabla \cdot (\nabla f - \frac{\partial}{\partial t}\boldsymbol{V}) = \nabla^2 f - \frac{\partial}{\partial t}(\nabla \cdot \boldsymbol{V}) \tag{14.177}$$

and (14.168) reads on substituting (14.164)

$$\boldsymbol{j} = \frac{1}{4\pi}\nabla \times (\nabla \times \boldsymbol{V}) - \frac{\partial}{\partial t}\left(\nabla f - \frac{\partial \boldsymbol{V}}{\partial t}\right),$$

which becomes by (14.180)

$$\boldsymbol{j} = \frac{1}{4\pi}\nabla(\nabla \cdot \boldsymbol{V}) - \nabla^2 \boldsymbol{V} - \nabla\frac{\partial f}{\partial t} + \frac{\partial^2 \boldsymbol{V}}{\partial t^2}, \tag{14.178}$$

and which in turn becomes

$$\nabla^2 \boldsymbol{V} - \frac{\partial^2 \boldsymbol{V}}{\partial t^2} = -\boldsymbol{j} + \nabla\left(\frac{1}{4\pi}\nabla \cdot \boldsymbol{V} - \frac{\partial f}{\partial t}\right). \tag{14.179}$$

Under (14.170), (14.177), which takes a particularly simple form, and (14.179) amount to (14.171) and (14.172), respectively.

If $\boldsymbol{V}$ is shifted $\boldsymbol{V} + \mathrm{grad} g$, then Exercise 126 shows $f + \frac{\partial g}{\partial t}$ also satisfies (14.176). Once the shift function g is obtained, then all the solution of the respective equations are determined. Hence Theorem 14.15 holds true. □

In addition to the formulas in Lemma 14.4, which are useful in vector analysis, we also appeal to as an analogue of (D.2)

$$\mathrm{curl}(\mathrm{curl}(\boldsymbol{V})) = \nabla \times (\nabla \times \boldsymbol{V}) = \nabla(\nabla \cdot \boldsymbol{V}) - \nabla^2 \boldsymbol{V} = \mathrm{grad}\,\mathrm{div}(\boldsymbol{V}) - \nabla^2 \boldsymbol{V}. \tag{14.180}$$

Exercise 126. Suppose $\boldsymbol{V}$ is shifted as in Theorem 14.14, (ii): $\boldsymbol{V} + \mathrm{grad} g$. Then the shift $f + \frac{\partial g}{\partial t}$ also satisfies (14.176) if f does so.

Solution.

$$\nabla\left(f + \frac{\partial g}{\partial t}\right) - \frac{\partial}{\partial t}(\boldsymbol{V} + \mathrm{grad} g) = \nabla f - \frac{\partial}{\partial t}\boldsymbol{V} + \nabla\frac{\partial}{\partial t}g - \frac{\partial}{\partial t}\nabla g = \boldsymbol{E}.$$

We need to choose g so as to satisfy (14.170), i.e.

$$\frac{1}{4\pi}\nabla \cdot (\boldsymbol{V} + \mathrm{grad} g) - \frac{\partial}{\partial t}(f + \frac{\partial g}{\partial t}) = 0,$$

or

$$\nabla^2 g - 4\pi\frac{\partial^2 g}{\partial t^2} = -\nabla \cdot \boldsymbol{V} + 4\pi\frac{\partial f}{\partial t}, \tag{14.181}$$

which is of the form (14.173), where h is given by (14.175).

14.5 Transition to a Hilbert space

In this section we shall encounter possible warping from classical real analysis to modern analysis. The following theorem is a warp from the Bolzano-Weierstrass theorem by way of Cantor diagonal argument.

Theorem 14.16. *Every bounded sequence in a Hilbert space contains a weakly convergent subsequence.*

Proof. Suppose $\{t_n\}$ is bounded, $\|t_n\| \leq c$, $c > 0$. For the dense sequence $\{y_m\}$ described above, the double sequence (y_m, t_n) is bounded for each m since $|(y_m, t_n)| \leq c\|y_m\|$. Hence by Lemma 14.5, we may choose a subsequence (y_m, t_{ν_n}) such that $\lim_{n\to\infty}(y_m, t_{\nu_n}) = (y_m, c)$ for each y_m. □

Lemma 14.5. (Cantor diagonal argument) *Suppose the double sequence $\{a_{mn}\}$ is bounded. Then one can choose a strictly increasing subsequence $\{\nu_n\} \subset \mathbb{N}$ such that $\{a_{m\nu_n}\}$ is convergent for each m.*

Proof. By the Bolzano-Weierstrass theorem, one can choose a subsequence $\{\nu_{1n}\}$ of $\{n\}$ so that $\{a_{m\nu_{1n}}\}$ is convergent for each m. Inductively one can choose a subsequence $\{\nu_{mn}\}$ of $\{\nu_{m-1,n}\}$ such that $\{a_{m\nu_{mn}}\}$ is convergent for each m. Now put $\nu_n = \nu_{nn}$. Then save for a finite number of terms, $\{\nu_n\}$ is a subsequence of $\{\nu_{mn}\}$ and so $\{a_{m\nu_n}\}$ is convergent for each m. □

Chapter 15

Theory of distributions

The results in this chapter are partly modifications of results from [Chakraborty *et al.* (2016)] and Vista II [Chakraborty *et al.* (2016), pp. 192–197, 214–223] which in turn depend on good references including [Donoghue (1969)], [Mizohata (1965), Chapter 2], [Papoulis (1962)], [Yoshida (1956), Chapter 12]. We use the basic results in these references freely without mentioning their sources. Papoulis must be popular among the engineers in the US as it is cited in [Doss (1994)] together with J. Kitchener's "Centennial". The aim is to get readers familiar with the use of distributions rather than giving robust foundations of the theory.

15.1 Some facts from measure theory and Fourier transforms

Definition 15.1. Let X be a locally compact Haustoria space. Let $\mathcal{B}$ be the smallest family of subsets of X satisfying

(i) $\mathcal{B} \supset \mathcal{A}(X)$, the family of all closed subsets of X ($\mathcal{A}$ is for "Abgeschlossene"–closed),

(ii) $\mathcal{B}$ is closed under formation of countable union,

(iii) $\mathcal{B}$ is closed under complementation A^c.

Each element of $\mathcal{B}$ is called a **Borel set** of X.

By (iii), we may take the family $\mathcal{O}(X)$ of all open sets of X for $\mathcal{A}(X)$ in Definition 15.1.

A non-negative **measure** μ is a **set function** $\mathcal{B} \to \mathbb{R}_+ \cup \{0\}$ (non-negative reals) which is **completely additive**, i.e. if $E_i \cap E_j = \emptyset$ for $i \neq j$

$$\mu\left(\cup_{k=1}^{\infty} E_k\right) = \sum_{k=1}^{\infty} \mu(E_k).$$

A measure μ on X is called a **Borel measure** if the Borel sets in X are measurable, which is equivalent to the assertion that all open sets are measurable.

$f : X \to \mathbb{C}$ is called a **Borel function** if $f^{-1}(U)$ is a Borel set for all $U \in \mathcal{O}(X)$.

Let μ be a measure on X and let $\int_X f \, \mathrm{d}\mu$ denote a bounded linear functional defined for bounded continuous functions f with compact support (for these notions, cf. §15.2). It is called the **integral** of f over X with respect to μ. It is known that a Borel measurable function and a Lebesgue measurable function coincide a.e. and so we are dealing with Lebesgue integrals.

For $0 < p < \infty$, let

$$\|f\|_p = \left(\int_X |f|^p \, \mathrm{d}\mu\right)^{1/p}.$$

This $\|f\|_p$ satisfies the conditions for a norm and is called the p-norm. $L^p(\mu) = \{f | \ \|f\|_p < \infty\}$ forms a **Banach space** and especially, $L^2(\mu)$ forms a **Hilbert space** with respect to the inner product

$$(f, g) = \int_X f\bar{g} \, \mathrm{d}\mu.$$

$L^\infty(\mu)$ is the space of all bounded Borel functions normed by

$$\|f\|_\infty = \operatorname{ess\,sup}_{x \in X} |f(x)|,$$

where

$$\operatorname{ess\,sup} = \min_{\lambda}\{\mu(x \,|\, |f(x)| > \lambda) = 0\}.$$

Let $T > 0$ and f be a periodic function of period $2T$ ($f(t) = f(t + 2T)$). The most commonly used cases are $2T = 2\pi$ or $2T = 1$. The nth **Fourier coefficient** c_n is defined by

$$c_n = \frac{1}{2T} \int_{-T}^{T} e^{-i\lambda_n t} f(t) \, \mathrm{d}t, \quad \lambda_n = \frac{2\pi}{2T} n. \tag{15.1}$$

Then we have the Fourier expansion

$$f(t) \sim \sum_{n=-\infty}^{\infty} c_n \, e^{i\lambda_n t}, \tag{15.2}$$

where the equality holds under some conditions, e.g. smoothness. Cf. [Zygmund (2003)] for more details. For a class of smooth functions f for $f \in L^2(\mathbb{R})$ in the sense of l.i.m., the **Fourier transform** $\hat{f} = \hat{f}_T$ with parameters $T > 0$ and $c > 0$ is defined by

$$\hat{f}_T(z) = \hat{f}_{T,c}(z) = \frac{1}{c}\int_{-\infty}^{\infty} e^{-i\frac{\pi}{T}zt} f(t)\,\mathrm{d}t. \tag{15.3}$$

Then we compute $\hat{\hat{f}}(-t) \approx \int_{-\infty}^{\infty} e^{i\frac{\pi}{T}zt}\hat{f}(z)\,\mathrm{d}z$ by substituting (15.3): Changing the order of integration we deduce that

$$\begin{aligned}\int_{-\infty}^{\infty} e^{i\frac{\pi}{T}zt}\hat{f}(z)\,\mathrm{d}z &= \frac{1}{c}\int_{-\infty}^{\infty} e^{i\frac{\pi}{T}zt}\,\mathrm{d}z\int_{-\infty}^{\infty} e^{-i\frac{\pi}{T}zu} f(u)\,\mathrm{d}u \\ &= \frac{1}{c}\int_{-\infty}^{\infty} f(u)I(t-u)\,\mathrm{d}u, \end{aligned}\tag{15.4}$$

where

$$I(w) = \int_{-\infty}^{\infty} e^{\frac{\pi}{T}zw}\,\mathrm{d}z = 2T\delta(w). \tag{15.5}$$

This follows from (15.36) by putting $\frac{\pi}{T}z = 2\pi y$. Substituting (15.5) in (15.4), we obtain

$$\hat{\hat{f}}(-t) := \int_{-\infty}^{\infty} e^{i\frac{\pi}{T}zt}\hat{f}(z)\,\mathrm{d}z = \frac{2T}{c}f(t). \tag{15.6}$$

Hence we have the Fourier integral theorem

$$f(t) = \frac{c}{2T}\int_{-\infty}^{\infty} e^{i\frac{\pi}{T}zt}\hat{f}(z)\,\mathrm{d}z = \frac{c}{2T}\hat{\hat{f}}(-t) \tag{15.7}$$

and we have a few commonly used choices

c	$2T$	1	$\sqrt{2T}$
$\frac{1}{c}$	$\frac{1}{2T}$	1	$\frac{1}{\sqrt{2T}}$
$\frac{c}{2T}$	1	$\frac{1}{2T}$	$\frac{1}{\sqrt{2T}}$

Table 5. Coefficients of the FT

Convention. In what follows we fix the notation. The nth Fourier coefficient c_n is defined by (15.1) and the Fourier expansion (15.2) with equality is supposed to hold.

The Fourier transform $\hat{f}$ is defined by

$$\hat{f}(z) = \hat{f}_{\pi,1}(z) = \int_{-\infty}^{\infty} e^{-izt} f(t)\,\mathrm{d}t. \tag{15.8}$$

If e.g. f, f' are piecewise smooth and $f \in L^1$, then the Fourier inversion formula (15.7) with $c = 1, T = \pi$ holds:

$$f(t) = \frac{1}{2\pi} \int_{-\infty}^{\infty} e^{itz} \hat{f}(z)\, \mathrm{d}z, \tag{15.9}$$

where the left-hand side at jump discontinuities is to mean the arithmetic mean $\frac{1}{2}(f(t+) + f(t-))$. Hence we have the Fourier transform pair $f(t) \leftrightarrow \frac{1}{2\pi}\hat{f}(z)$ and its reflection $\hat{f}(t) \leftrightarrow 2\pi f(-z)$.

Exercise 127. By viewing λ_n in (15.1) as an infinitesimal as $T \to \infty$ and by heuristic argument, prove an analogue of (15.6).

Solution. We may view (15.1) as

$$2Tc_n \to \int_{-\infty}^{\infty} e^{-i\lambda t} f(t)\, \mathrm{d}t = \hat{f}(\lambda_n), \quad T \to \infty. \tag{15.10}$$

Hence we may view (15.2) as

$$\begin{aligned} f(t) &= \frac{1}{2\pi} \sum_{n=-\infty}^{\infty} 2Tc_n\, e^{i\lambda_n t} \Delta\lambda_n \to \frac{1}{2\pi} \sum_{n=-\infty}^{\infty} \hat{f}(\lambda_n)\, e^{i\lambda_n t} \Delta\lambda_n \\ &\to \frac{1}{2\pi} \int_{-\infty}^{\infty} e^{izt} \hat{f}(z)\, \mathrm{d}z = \frac{1}{2\pi} \hat{\hat{f}}(-t) \end{aligned} \tag{15.11}$$

on using (15.10). This amounts to (15.7) with $T = \pi$, $c = 1$.

15.2 Basics of the theory of distributions

The objective of introducing distributions (or generalized functions) is, as is stated in Preface, to generalize the notion of functions through "integration by parts," which we explain below. It should not be confused with distribution function in statistics.

We think of a "functional" as a distribution, which assigns the complex number $\langle f, \varphi \rangle = \int f(x)\varphi(x)\, \mathrm{d}x$ to each test function argument φ. Here a **test function** is an infinitely differentiable function which vanishes outside of a compact set. The test functions φ work as a variable in calculus and we may equate two distributions by their values at all test functions, as in Examples 15.7, 15.8. Since test functions vanish in the far-end of the interval, we may just ignore the boundary values. E.g. if we would like to know the effect of a possible derivative $T_{f'}$ of a non-differentiable f on the test function φ, we may argue as follows. By integration by parts,

$$T_{f'}(\varphi) = \int_{-\infty}^{\infty} f'(x)\varphi(x)\, \mathrm{d}x = [f\varphi]_{-\infty}^{\infty} - \int f(x)\varphi'(x)\, \mathrm{d}x = T_f(-\varphi'). \tag{15.12}$$

Viewing T_f as f as in Example 15.3 below, this amounts to

$$f'(\varphi) = f(-\varphi'), \tag{15.13}$$

i.e. "the value of the derivative f'" is the value of the functional f at $-\varphi'$. Cf. Examples 15.7, 15.8 for concrete calculations. We begin with

Definition 15.2. Any measurable function on $\mathbb{R}^n$ which is integrable in a neighborhood of each point (in the sense of Lebesgue) is called **locally integrable** (and sometimes denoted L^1_{loc}). According to [Mizohata (1965), p. 26], each point x is classified into two categories: (1) in a sufficiently small neighborhood, the function is 0 a.e. or (2) for an arbitrarily small neighborhood V, $\int_V |f(x)|\,dx > 0$. The set of all points of the second category forms a closed set, called a **support** of the function. For an infinitely differentiable function f on an open set $\mathfrak{U} \subset \mathbb{R}^n$, $f \in C^\infty(\mathfrak{U})$, we define

$$\operatorname{supp} f = \overline{\{x \in \mathbb{R}^n | \, |f(x)| > 0\}}.$$

Any function $f \in C^\infty(\mathfrak{U})$ with compact support $\subset \mathfrak{U}$, $f \in C_0^\infty(\mathfrak{U})$, is called a **test function**. We denote the totality of test functions on $\mathfrak{U}$ by $\mathfrak{D}$:

$$\mathfrak{D} = \mathfrak{D}(\mathfrak{U}) = \{f : \text{test-function} | \operatorname{supp} f \subset \mathfrak{U}\},$$

which forms a linear space over $\mathbb{R}$ or $\mathbb{C}$.

For any n-dimensional index vector $\alpha = (m_1, m_2, \cdots, m_n)$, we define

$$D^\alpha \varphi = \frac{\partial^{|\alpha|}}{\partial x_1^{m_1} \cdots \partial x_n^{m_n}} \varphi,$$

where $|\alpha| = m_1 + \cdots + m_n$.

We introduce a family of (semi-) norms on $\mathfrak{D}$:

$$\|\varphi\|_N = \sum_{|\alpha| \le N} \|D^\alpha \varphi\|_\infty, \quad N \in \mathbb{N} \cup \{0\}, \tag{15.14}$$

where $\|f\|_\infty = \sup|f(x)|$ for a continuous function f.

Definition 15.3. A linear functional (a linear mapping which assigns a complex number to each function in its domain of definition)

$$T : \mathfrak{D}(\mathfrak{U}) \to \mathbb{C},$$

is called a **distribution** if there exist a compact set $K \subset \mathfrak{U}$ and a constant $C > 0$ and $N \in \mathbb{N}$ such that

$$|T(\varphi)| \le C\|\varphi\|_N \tag{15.15}$$

for all test-functions φ, with $\operatorname{supp} \varphi \subset K$. The set of all distributions is denoted by $T \in \mathfrak{D}'(\mathfrak{U})$, the space of distributions.

If N can be chosen independently of K and N is indeed chosen to be the smallest such, then T is said to be **of order** N where 0 is also allowed.

Remark 15.1. Instead of (15.15), one often adopts the continuity of T as a defining condition in the sense that $\lim_{\varphi_j \to 0} T\varphi_j = 0$ in the norm defined by (15.14).

Definition 15.4. Let X be a locally compact space and μ be a Borel measure on X. Then if μ takes on only finite values on all compact subsets of X, then μ is called a **Radon measure**.

Theorem 15.1. ([Donoghue (1969), p. 25]) *Let X be a locally compact space and let $\mathcal{C}_0(X)$ be the linear space of all continuous functions on X with compact support. If $T(f)$ is a functional on $\mathcal{C}_0(X)$ having the property that $T(f) \geq 0$ whenever $f(x) \geq 0$, then there exists a Radon measure on X such that*

$$T(f) = \int_X f(x)\,\mathrm{d}\mu(x). \tag{15.16}$$

Remark 15.2. Since the Euclidean space $\mathbb{R}^n$ is a locally compact Hausdorff complete metric space, all that precedes apply to it and we may think of the Radon measure as Lebesgue measure or more simply as Jordan measure providing the Riemann integral. We may equate two distributions by their values at all test functions, thus, the test functions φ work as a variable in calculus, as in Examples 15.7, 15.8. Since test functions are 0 outside of a compact set, we may explain away by saying "by integration by parts and boundary conditions."

Example 15.1. Let $\mathrm{d}\mu$ be a Radon measure on $\mathfrak{U} \subset \mathbb{R}^n$. Then the functional $T = T_\mu$ defined by (15.16)

$$T(\varphi) = T_\mu(\varphi) = \int \varphi(x)\,\mathrm{d}\mu(x) \tag{15.17}$$

is a distribution of order 0, **where the integration is performed over the subdomain $\mathfrak{U}$ of $\mathbb{R}^n$, which is not mentioned at each occurrence. Also the variable is often written x rather than x where there is no fear of confusion.**

Indeed, let $K \subset \mathfrak{U}$ be compact and let $C = \int_K |\mathrm{d}\mu(x)|$ = total mass of $\mathrm{d}\mu$ on K. Then

$$|T(\varphi)| \leq C\|\varphi\|_\infty = C\|\varphi\|_0$$

for all $\varphi \in \mathcal{D}(\mathfrak{U})$ with $\operatorname{supp}\varphi \subset K$.

Example 15.2. (Dirac distribution) The **Dirac δ-distribution** is a Radon measure consisting of positive unit mass at the origin,

$$\delta(\varphi) = T_\delta(\varphi) = \int \varphi(x)\,\mathrm{d}\delta(x) = \int \varphi(x)\delta(x)\,\mathrm{d}x = \varphi(0) = \varphi(0,\cdots,0) \tag{15.18}$$

for test functions φ continuous at the origin. This may be thought of as an extensive generalization of the Kronecker delta in (1.84). The Dirac distribution δ_a supported at a, is such that

$$\delta_a(\varphi) = \int \varphi(x)\delta(x-a)\,\mathrm{d}x = \varphi(a) \tag{15.19}$$

for test functions φ continuous at the point a. We need only assume φ is continuous at the respective points in view of Theorem 15.1.

Example 15.3. If f is locally integrable on $\mathfrak{U}$, then $f(x)\,\mathrm{d}x$ is a Radon measure on $\mathfrak{U}$, so that the functional $T = T_f$ defined by

$$T_f(\varphi) = \int \varphi(x)f(x)\,\mathrm{d}x \tag{15.20}$$

is a distribution, **which will be used as the definition and is often identified with the function itself and we often speak of distributions which are a polynomial, a C^∞-function or the characteristic function of a set**. In this sense we may express (15.20) as

$$f(\varphi) = \int \varphi(x)f(x)\,\mathrm{d}x,$$

whereby generalizing the meaning of a function. It can be proved that

Theorem 15.2. *T is a distribution of order 0 if and only if T is a Radon measure.*

Example 15.4. For $T, S \in \mathcal{D}'(\mathfrak{U})$, we define their sum and the scalar multiplication by

$$(T+S)(\varphi) = T(\varphi) + S(\varphi),$$

$$(\alpha T)(\varphi) = T(\alpha\varphi),\ \alpha \in \mathbb{C}.$$

Then

$$T_f + T_g = T_{f+g},\ \alpha T_f = T_{\alpha f}.$$

Also for $\alpha = \alpha(x) \in C^\infty$, define

$$T(\alpha\varphi) = (T_\alpha T)(\varphi).$$

Then $T_\alpha T_f = T_{\alpha f}$.

Let $T = T_f$ be a distribution on $\mathfrak{U}$ and let x_k be a coordinate. Then by (15.20)

$$T_f\left(\frac{\partial\varphi}{\partial x_k}\right) = \int \frac{\partial\varphi}{\partial x_k} f(x)\,\mathrm{d}x = [\varphi f] - \int \varphi(x)\frac{\partial f}{\partial x_k}(x)\,\mathrm{d}x = -T_{\frac{\partial f}{\partial x_k}}(\varphi) \tag{15.21}$$

by integration by parts and boundary conditions. In order that we should have "distributional derivative = ordinary derivative":

$$\frac{\partial T_f}{\partial x_k} = T_{\frac{\partial f}{\partial x_k}} \tag{15.22}$$

we should define

$$\frac{\partial T_f}{\partial x_k}(\varphi) = -T\left(\frac{\partial\varphi}{\partial x_k}\right). \tag{15.23}$$

Indeed, since $\frac{\partial\varphi}{\partial x_k}$ is a function $\mathfrak{D} \to \mathbb{C}$ and if

$$T(\varphi) \leq C\|\varphi\|_N,$$

then

$$T\left(\frac{\partial\varphi}{\partial x_k}\right) \leq C\|\varphi\|_{N+1},$$

whence

$$\frac{\partial T}{\partial x_k} \in \mathfrak{D}'(\mathfrak{U}).$$

Example 15.5. We may prove that

$$\frac{\partial^{|\alpha|}T(\varphi)}{\partial x_1^{m_1}\cdots\partial x_n^{m_n}} = D^\alpha T(\varphi) = (-1)^{|\alpha|}T(D^\alpha\varphi). \tag{15.24}$$

In particular, if T is C^2, and $\Delta = \frac{\partial^2}{\partial x_1^2} + \cdots + \frac{\partial^2}{\partial x_n^2}$ is the Laplacian in (14.133), then

$$\Delta T(\varphi) = T(\Delta\varphi).$$

Example 15.6. If $\mathfrak{U} \ni 0$, then we may prove that

$$(D^\alpha\delta)(\varphi)\left(= (-1)^{|\alpha|}\delta(D^\alpha\varphi)\right) = (-1)^{|\alpha|}D^\alpha\varphi(0).$$

And more generally, if $\mathfrak{U} \ni a$, then

$$(D^\alpha\delta_a)(\varphi) = \left((-1)^{|\alpha|}\delta_a(D^\alpha\varphi)\right) = (-1)^{|\alpha|}D^\alpha\varphi(a). \tag{15.25}$$

(15.25) reads in the form of an integral

$$\int_{\mathbb{R}} \delta^{(\alpha)}(x-y)\varphi(y)\mathrm{d}y = (-1)^{|\alpha|}\varphi^{(\alpha)}(x).$$

The $\alpha = 0$ case is (15.18).

Example 15.7. Let $\mathfrak{U} = \mathbb{R}$ and let H be the Heaviside function (or the unit step function) in Definition 11.3 which has the value 0 for $x \le 0$ and 1 for $x > 0$. Then $H' = \delta$.

Indeed,

$$H'(\varphi) = -\int_{\mathbb{R}} \varphi' H \, \mathrm{d}x = -\int_0^\infty \varphi' \, \mathrm{d}x = [\varphi(x)]_0^\infty = \varphi(0) = \delta(\varphi).$$

Example 15.8. Let $f(x) = |x|$ on $\mathbb{R}$. Then $T_f{}'' = 2\delta$. Indeed,

$$T_f{}''(\varphi) = T_f(\varphi'') = \int \varphi'' f \, \mathrm{d}x = [\varphi' f]_{-\infty}^\infty - \int \varphi' f' \, \mathrm{d}x.$$

Now noting that $f'(x) = (H(x) - H(-x))$, $x \neq 0$, it follows that

$$\begin{aligned} T_f{}''(\varphi) &= -\int \varphi' \left(H(x) - H(-x)\right) \mathrm{d}x = -\int_0^\infty \varphi' \, \mathrm{d}x + \int_{-\infty}^0 \varphi' \, \mathrm{d}x \\ &= 2\varphi(0) = 2\delta(\varphi). \end{aligned}$$

Example 15.9. Suppose $D \subset \mathbb{R}^n$ be a bounded domain with smooth boundary ∂D and f is $C^\infty(\mathbb{R}^n)$. Let $[f; D] = [f(x); D]$ be the restriction of f to D, i.e. $[f(x); D] = f(x)$ if $x \in D$ and $[f(x); D] = 0$ if $x \notin D$. Then we have a distribution-theoretic interpretation of Green's formula (14.136):

$$\Delta T_{[f;D]}(\varphi) - T_{[\Delta f;D]}(\varphi) = T_{[f;\partial D]}\left(\frac{\partial \varphi}{\partial \boldsymbol{n}}\right) - T_{[\frac{\partial f}{\partial \boldsymbol{n}};\partial D]}(\varphi). \tag{15.26}$$

Proof depends essentially on twice application of (15.21) with some modifications. It suffices to consider the kth entry

$$\frac{\partial^2}{\partial x_k^2} T_{[f;D]}(\varphi) = T_{[f;D]}\left(\frac{\partial^2 \varphi}{\partial x_k^2}\right).$$

By (15.20),

$$\begin{aligned} T_{[f;D]}\left(\frac{\partial^2 \varphi}{\partial x_k^2}\right) &= \int_D f(x) \frac{\partial^2 \varphi}{\partial x_k^2} \, \mathrm{d}x = \int_{\partial D} f(x) \frac{\partial \varphi}{\partial x_k} \, \mathrm{d}x_1 \cdot \overset{k}{\cdots} \cdot \mathrm{d}x_n - \int_D \frac{\partial f}{\partial x_k} \frac{\partial \varphi}{\partial x_k} \, \mathrm{d}x \\ &= \int_{\partial D} f(x) \frac{\partial \varphi}{\partial x_k} \, \mathrm{d}x_1 \cdot \overset{k}{\cdots} \cdot \mathrm{d}x_n - \int_{\partial D} \frac{\partial f}{\partial x_k} \varphi(x) \, \mathrm{d}x_1 \cdot \overset{k}{\cdots} \cdot \mathrm{d}x_n + \int_D \frac{\partial^2 f}{\partial x_k^2} \varphi(x) \, \mathrm{d}x. \end{aligned} \tag{15.27}$$

If $\theta_k(x)$ denotes the angle formed by the outer normal $\boldsymbol{n}$ and the x_k-axis at x, then $\cos\theta_k(x) = \frac{\partial x_k}{\partial \boldsymbol{n}}$. Hence $\mathrm{d}x_1 \cdot \overset{k}{\cdots} \cdot \mathrm{d}x_n = \cos\theta_k(x) \mathrm{d}S$ and so (15.27) reads

$$\begin{aligned} &T_{[f;D]}\left(\frac{\partial^2 \varphi}{\partial x_k^2}\right) \\ &= \int_{\partial D} f(x) \frac{\partial \varphi}{\partial x_k} \cos\theta_k(x) \mathrm{d}S - \int_{\partial D} \frac{\partial f}{\partial x_k} \varphi(x) \cos\theta_k(x) \mathrm{d}S + \int_D \frac{\partial^2 f}{\partial x_k^2} \varphi(x) \, \mathrm{d}x. \end{aligned} \tag{15.28}$$

Adding (15.28) over $k = 1, \cdots, n$ and noting that

$$\sum_{k=1}^{n} \frac{\partial f}{\partial x_k} \cos\theta_k(x) = \frac{\partial f}{\partial \boldsymbol{n}}, \quad \sum_{k=1}^{n} \frac{\partial \varphi}{\partial x_k} \cos\theta_k(x) = \frac{\partial \varphi}{\partial \boldsymbol{n}}, \tag{15.29}$$

we conclude (15.26).

Definition 15.5. For any function f defined on $\mathbb{R}^n$ and any vector $h \in \mathbb{R}^n$ we define the translation operator $\mathcal{T}$ by

$$(\mathcal{T}_h f)(x) = f(x - h) \tag{15.30}$$

and extend it to the space of distributions.
For any test-functions φ, ψ their convolution is defined by

$$(\varphi * \psi)(x) = \int \varphi(x - y)\psi(y)\,\mathrm{d}y, \tag{15.31}$$

the integral being taken over a compact set F, say. We express F as a finite union of disjoint measurable sets F_j each of whose diameter being less than $\varepsilon > 0$. Choosing a point $y_j \in F_j$, the approximating Riemann sum is

$$\sum_j \varphi(x - y_j) \int_{F_j} \psi(y)\,\mathrm{d}y = \sum_j m_j \mathcal{T}_{y_j}\varphi(x). \tag{15.32}$$

Definition 15.6. Let $\{\varphi_\varepsilon\}$ be a family of functions which are positive, C^∞-functions vanishing outside the sphere of radius ε such that $\int_{\mathbb{R}^n} \varphi_\varepsilon(x)\,\mathrm{d}x = 1$. For a locally integrable function f, we form the **regularization** of f by

$$f_\varepsilon(x) = \int_{\mathbb{R}^n} \varphi_\varepsilon(x - y) f(y)\,\mathrm{d}y = \frac{1}{\varepsilon^n} \int_{\mathbb{R}^n} \varphi\left(\frac{x - y}{\varepsilon}\right) f\left(\frac{y}{\varepsilon}\right) \mathrm{d}y, \tag{15.33}$$

where $\varphi(x)$ is a positive, C^∞-function vanishing outside the unit sphere with the integral equal to 1.

Theorem 15.3. *The space* $\mathfrak{D} = \mathfrak{D}(\mathbb{R}^n)$ *is separable with respect to distance* $d(\varphi, \psi) = |\varphi - \psi| = \max_{x \in \mathbb{R}^n} |\varphi(x) - \psi(x)|$.

Proof. For any test-function f its regularization $(f * \varphi_\varepsilon)(x)$ converges to f in $\mathfrak{D}$ as ε tends to 0 through a countable set of values. Combining above definitions, we see that $f_\varepsilon(x)$ being a convolution of φ_ε and f, can be approximated by the Riemann sum (15.32):

$$(f * \varphi_\varepsilon)(x) = \lim \sum_j m_j \mathcal{T}_{y_j}\varphi(x). \tag{15.34}$$

The Riemann sums are text-functions converging to f in $\mathfrak{D}$. It follows that the system of functions $\sum_j m_j \mathcal{T}_{y_j}\varphi(x)$ forms a countable dense subset of $\mathfrak{D}$ as the coefficients m_j run through all rational numbers, the y_j's through countable dense subset of $\mathbb{R}^n$ and the ε through a sequence tending to 0. □

15.3 The Dirac delta-function

In this section we follow the argument of Papoulis [Papoulis (1962)] and Yoshida [Yoshida (1956)] to prove some basic results on the Dirac delta-function and Fourier transforms, following the notation of [Kanemitsu and Tsukada (2007), Chapter 7]. We base our argument on (15.18) and

$$\int_{-\infty}^{\infty} e^{2\pi i(z-x)t}\,\mathrm{d}t = \delta(z-x) \tag{15.35}$$

which is an equivalent to the Dirichlet integral (15.36) below.

First we interpret the Dirichlet integral ([Kanemitsu and Tsukada (2007), p. 155]) as a distribution.

Theorem 15.4. (Dirichlet integral) *The equality for the Dirichlet integral*

$$\int_{-\infty}^{\infty} e^{2\pi i x t}\,\mathrm{d}t = \delta(x) \tag{15.36}$$

is true as distributions, or what amounts to the same thing,

$$2\int_{0}^{\infty} \cos 2\pi x t\,\mathrm{d}t = \delta(x). \tag{15.37}$$

Putting

$$f_t(x) := \frac{1}{\pi x}\sin 2\pi t x = \int_{-t}^{t} e^{2\pi i x u}\,\mathrm{d}u = 2\int_{0}^{t}\cos 2\pi x u\,\mathrm{d}u, \tag{15.38}$$

the limit $\lim_{t\to\infty} f_t(x) = \lim_{t\to\infty}\int_{-t}^{t} e^{2\pi i x u}\,du$ does not exist in ordinary sense. However, we may prove the following

Lemma 15.1. *For* $\varphi \in \mathfrak{D} = \mathfrak{D}(\mathbb{R})$, *prove that* $\lim_{t\to\infty} T_{f_t}(\varphi)$ *exists and is equal to*

$$T_{f_\infty}(\varphi) := 2DT_g(\varphi) + \int_{-\infty}^{\infty}\left(\int_{-1}^{1} e^{2\pi i x u}\,\mathrm{d}u\right)\varphi(x)\mathrm{d}x, \tag{15.39}$$

where

$$g(x) = \int_{1}^{\infty} -\frac{\cos 2\pi u x}{(2\pi u)^2}\,\mathrm{d}u = -\lim_{t\to\infty} g_t(x), \tag{15.40}$$

and where

$$g_t(x) = \int_{1}^{t} -\frac{\cos 2\pi u x}{(2\pi u)^2}\,\mathrm{d}u.$$

Proof. By the convergence theorem, (15.40) leads to

$$T_g(\varphi) = \lim_{t\to\infty} T_{g_t}(\varphi).$$

Hence, by the theorem on termwise differentiation with $D^2 = \frac{\mathrm{d}^2}{\mathrm{d}t^2}$, we have

$$(DT_g)(\varphi) = \lim_{t\to\infty} DT_{g_t}(\varphi).$$

Here $D^2 T_{g_t} = T_{D^2 g_t}$, so that

$$\begin{aligned} D^2 g_t(x) &= \frac{\mathrm{d}^2}{\mathrm{d}t^2}\int_1^t -\frac{\cos 2\pi ux}{(2\pi u)^2}\,\mathrm{d}u = \int_1^t \cos 2\pi xu\mathrm{d}u \\ &= \frac{\sin 2\pi xt}{2\pi x} - \int_0^1 \cos 2\pi xu\mathrm{d}u = \frac{1}{2}f_t(x) - \int_0^1 \cos 2\pi xu\mathrm{d}u, \end{aligned}$$

whence (15.39) follows. □

Lemma 15.2. *The evaluation of the Dirichlet integral reads*

$$\lim_{t\to\infty}\int_{-\infty}^{\infty}\frac{\sin 2\pi tx}{\pi x}\varphi(x)\,\mathrm{d}x = \varphi(0) \tag{15.41}$$

provided that φ is continuous at the origin.

Proof. Divide the integral into three parts $(-\infty,-\varepsilon),(-\varepsilon,\varepsilon),(\varepsilon,\infty)$ with $\varepsilon > 0$. Then since the function $\frac{\varphi(x)}{x}$ is integrable over two infinite intervals, it follows from the Riemann-Lebesgue lemma [Kanemitsu and Tsukada (2007), Proposition 7.1, p. 137] that the corresponding integrals tend to 0 as $t \to \infty$. On the interval $(-\varepsilon, \varepsilon)$, $\varphi(x)$ can be approximated by $\varphi(0)$ and so

$$\begin{aligned} \int_{-\varepsilon}^{\varepsilon}\frac{\sin 2\pi tx}{\pi x}\varphi(x)\,\mathrm{d}x &= \varphi(0)\int_{-\varepsilon}^{\varepsilon}\frac{\sin 2\pi tx}{\pi}\frac{\mathrm{d}x}{x} + o(1) \\ &= \frac{\varphi(0)}{\pi}\int_{-2\pi\varepsilon t}^{2\pi\varepsilon t}\frac{\sin x}{x}\,\mathrm{d}x + o(1), \end{aligned} \tag{15.42}$$

which tends to $\varphi(0)$ in view of

$$\int_{-\infty}^{\infty}\frac{\sin x}{x}\,\mathrm{d}x = \pi. \tag{15.43}$$

For this value, cf. e.g. [Chakraborty *et al.* (2016), p. 23]. □

Proof of Theorem 15.4. Equation (15.41) means

$$T_{f_\infty}(\varphi) = \lim_{t\to\infty} T_{f_t}(\varphi) = \varphi(0) = T_\delta(\varphi). \tag{15.44}$$

Symbolically, (15.44) amounts to $f_\infty(x) = \delta(x)$, or in view of (15.38),

$$\lim_{t\to\infty} \frac{1}{\pi x}\sin 2\pi t x = \int_{-\infty}^{\infty} e^{2\pi i x t}\,\mathrm{d}t = 2\int_0^{\infty} \cos 2\pi i x t\,\mathrm{d}t = f_\infty(x) = \delta(x), \tag{15.45}$$

which is (15.36), completing the proof.

If r_{2T} signifies the rectangular pulse function in Example 15.10, then

$$r_{2T}(t-t_0) \leftrightarrow 2\frac{\sin Tz}{z}e^{-it_0 z}, \quad \frac{T}{\pi}\frac{\sin(Tt-n\pi)}{Tt-n\pi} \leftrightarrow r_{2T}(z)e^{-in\frac{\pi}{T}z}. \tag{15.46}$$

By (15.36), $\hat{\delta}(t) = 1$ or more generally

$$\hat{\delta}(t-t_0) = e^{-it_0 z}. \tag{15.47}$$

This suggests the validity of the following lemma which has been extensively used in our investigations on functional equations (cf. [Kanemitsu and Tsukada (2007), p. 75]). Let $(T > 0)$

$$p_{2T}(t) = \sum_{n=-\infty}^{\infty} \delta(t-2nT) \tag{15.48}$$

be the **pulse train** consisting of a sequence of equidistant pulses $\delta(t-2nT)$ distance $2T$ apart.

Lemma 15.3. *The Fourier transform of the pulse train is again a pulse train*

$$\hat{p}_{2T}(t) = \frac{\pi}{T}\sum_{n=-\infty}^{\infty} \delta\left(t-\frac{\pi}{T}n\right) \tag{15.49}$$

which in the range $-T < t < T$ amounts to the Fourier expansion of the Dirac delta-function

$$\frac{1}{2T}\sum_{n=-\infty}^{\infty} e^{i\frac{\pi}{T}nt} = \frac{1}{2T}\lim_{N\to\infty} D_N(t) = \delta(t), \tag{15.50}$$

where $D_N(t) = \sum_{n=-N}^{N} e^{i\frac{\pi}{T}nt}$.

Proof. The Dirichlet kernel $D_N(t)$ is expressed by [Kanemitsu and Tsukada (2007), (7.13)] as

$$\frac{1}{2T}D_N(t) = \frac{1}{2T}\frac{\sin\left(2\pi\frac{N+\frac{1}{2}}{2T}t\right)}{\pi t}\frac{\pi t}{\sin\frac{\pi}{2T}t}. \tag{15.51}$$

Now the second factor of the right-hand side of (15.51) is bounded on $(-T, T)$ and the first factor tends to $\delta(t)$ by (15.45), we obtain

$$\frac{1}{2T}\lim_{N\to\infty} D_N(t) = \frac{\pi t}{2T}\left(\sin\frac{\pi t}{2T}\right)^{-1}\delta(t) = \delta(t), \tag{15.52}$$

where we used the equality valid for a function φ continuous at the origin

$$\delta(\varphi) = \int \delta(t)\varphi(t)\mathrm{d}t = \varphi(0). \tag{15.53}$$

(15.52) now leads to (15.50), which in turn leads to the Fourier inversion for (15.49) in view of the periodicity of $p_T(t)$, completing the proof. □

Now we may state a generalization of the above lemma.

Theorem 15.5. *The Fourier transform $\hat{f}$ of a periodic function f ($f(t) = f(t+2T)$) is given by a sequence of equidistant pulses*

$$\hat{f}(z) = \pi \sum_{n=-\infty}^{\infty} c_n \delta\left(z - \frac{\pi}{T}n\right), \tag{15.54}$$

distance $\frac{\pi}{T}$ apart, where c_n is the Fourier coefficient given by (15.1) *and conversely.*

Definition 15.7. For $f, g \in L(\mathbb{R})$, we define their **convolution** $f * g$ by

$$(f * g)(x) = \int_{-\infty}^{\infty} f(x-t)g(t)\,\mathrm{d}t. \tag{15.55}$$

Then $f * g \in L(\mathbb{R})$ and $f * g = g * f$. Further we have

Exercise 128. Prove that

$$\widehat{f * g} = \hat{f}\hat{g}. \tag{15.56}$$

Example 15.10. (i) If H is the Heaviside function, then

$$f * H(x) = \int_0^{\infty} f(x-t)\,\mathrm{d}t = \int_{-\infty}^{x} f(t)\,\mathrm{d}t.$$

(ii) If $r_{2T}(t) = H(t+T) - H(t-T)$ is the **rectangular pulse function**, then

$$(f * r_{2T})(x) = \int_{x-T}^{x+T} f(t)\,\mathrm{d}t \tag{15.57}$$

known as the **smoothing**.
(iii) Let

$$f_0(t) = \begin{cases} f(t), & |t| < T, \\ 0, & |t| > T. \end{cases} \tag{15.58}$$

Then

$$f = f_0 * p_{2T}. \tag{15.59}$$

Proof. Equation (15.59) follows from

$$f(t) = \sum_{n=-\infty}^{\infty} f_0(t + 2nT),$$

the definition (15.48) and the evenness of the delta-function. □

We are now ready to prove Theorem 15.5.

Proof. (*Proof of Theorem 15.5*) By (15.59), (15.56) and (15.49),

$$\hat{f}(z) = \hat{f}_0(z)\widehat{p_{2T}}(z) = \hat{f}_0(z)\frac{\pi}{T}\sum_{n=-\infty}^{\infty} \delta\left(z - \frac{\pi}{T}n\right)$$

$$= 2\pi \sum_{n=-\infty}^{\infty} \frac{1}{2T}\hat{f}_0\left(\frac{\pi}{T}n\right)\delta\left(z - \frac{\pi}{T}n\right). \tag{15.60}$$

(15.8) reads

$$\hat{f}_0(z) = \int_{-\infty}^{\infty} e^{-izt} f_0(t)\,\mathrm{d}t = \int_{-T}^{T} e^{-izt} f(t)\,\mathrm{d}t,$$

and so we have

$$\frac{1}{2T}\hat{f}_0\left(\frac{\pi n}{T}\right) = c_n. \tag{15.61}$$

Substituting this in (15.60) proves (15.54). The reverse implication being trivial, this completes the proof. □

By the above proof, the Fourier series (cf. e.g. [Kanemitsu and Tsukada (2007), Theorem 7.2, p. 141]) for a periodic function $f(t)$ may be written as

$$\sum_{n=-\infty}^{\infty} f_0(t + 2nT) = \frac{1}{2T}\sum_{n=-\infty}^{\infty} e^{in\frac{\pi}{T}t}\hat{f}_0\left(\frac{\pi}{T}n\right). \tag{15.62}$$

The following generalization of (15.62) is known as the **Poisson summation formula** which is very useful in vast areas of science (cf. e.g. [Rademacher (1973), pp. 71–79]).

Theorem 15.6. *If* $f \in C^2(-\infty, \infty)$ *and* $f \in L^1 \cap L^2$, *then*

$$\sum_{n=-\infty}^{\infty} f(t + 2nT) = \frac{1}{2T}\sum_{n=-\infty}^{\infty} e^{in\frac{\pi}{T}t}\hat{f}\left(\frac{\pi}{T}n\right). \tag{15.63}$$

Proof. Proof is similar to that of Theorem 15.5. First we write, correspondingly to (15.59),

$$\sum_{n=-\infty}^{\infty} f(t + 2nT) = f * p_{2T}(t). \tag{15.64}$$

Hence the Fourier transform of the right-hand side of (15.64) is

$$\hat{f}(z)\,\hat{p_{2T}}(z) = \hat{f}(z)\frac{\pi}{T}\sum_{n=-\infty}^{\infty}\delta\left(z-\frac{\pi}{T}n\right) = \frac{\pi}{T}\sum_{n=-\infty}^{\infty}\hat{f}_0\left(\frac{\pi}{T}n\right)\delta\left(z-\frac{\pi}{T}n\right). \tag{15.65}$$

Hence the Fourier inversion of the right-hand side of (15.65) is the right-hand side of (15.63), completing the proof. □

Corollary 15.1. *Under the same conditions on f, we have*

$$S(u) := \sum_{n=-\infty}^{\infty} f(n+u) = \sum_{n=-\infty}^{\infty}\hat{f}(2\pi n)e^{2\pi inu} \tag{15.66}$$

uniformly in u in any finite interval, and in particular

$$S(0) = \sum_{n=-\infty}^{\infty} f(n) = \sum_{n=-\infty}^{\infty}\hat{f}(2\pi n), \tag{15.67}$$

where

$$\hat{f}(2\pi n) = \int_{-\infty}^{\infty} e^{-2\pi int} f(t)\,\mathrm{d}t$$

and the left-hand side is to mean the mean at discontinuities.

This is a special case of Theorem 15.6 with $T=\frac{1}{2}$.

Cf. Exercise 130 for another proof.

Other conditions on f may be like, $f \in C(\mathbb{R}) \cap \mathrm{L}^1(\mathbb{R})$ (cf. e.g. [Berndt (1975)] for a generalization).

Exercise 129. Recall the first periodic Bernoulli polynomial $\bar{B}_1(x) = x - [x] - \frac{1}{2}$ in (13.79). Prove that for $x \notin \mathbb{Z}$ and $1 < N \in \mathbb{N}$, the approximation

$$\bar{B}_1(x) = -\sum_{n=1}^{N}\frac{\sin 2\pi nx}{\pi n} + O\left(\min\left(1, \frac{1}{N\,\|x\|}\right)\right) \tag{15.68}$$

holds true, where $\|x\| = \min\{1-\{x\},\{x\}\}$ is the distance from x to the nearest integer. Hence the Fourier expansion (13.79) follows which is uniformly convergent in any interval not containing an integer.

Solution. First we check oddness

$$\bar{B}_1(1-x) = -\bar{B}_1(x). \tag{15.69}$$

This follows from $[1-x] = [-[x]+1-\{x\}] = -[x]$. By periodicity and oddness, it suffices to consider the case $0 < x \le \frac{1}{2}$. Note that for $m \neq 0$

$$\int_{\frac{1}{2}}^{x} e^{2\pi imt}\,\mathrm{d}t = \frac{1}{2\pi im}\left(e^{2\pi imx} - e^{\pi im}\right).$$

Summing this over $m = -N, \cdots, N, m \neq 0$, we deduce that

$$\int_{\frac{1}{2}}^{x} \sum_{\substack{m=-N \\ m\neq 0}}^{N} e^{2\pi imt}\,\mathrm{d}t = \sum_{\substack{m=-N \\ m\neq 0}}^{N} \frac{1}{2\pi im} e^{2\pi imx}. \tag{15.70}$$

We may modify the formula in Exercise 77, (ii) into

$$\sum_{k=-N}^{N} e^{2\pi ikt} = \frac{\sin(2N+1)\pi t}{\sin \pi t}. \tag{15.71}$$

Adding $x - \frac{1}{2}$ to both sides of (15.70) and using this formula, we derive that

$$\sum_{\substack{m=-N \\ m\neq 0}}^{N} \frac{1}{2\pi im} e^{2\pi imx} + x - \frac{1}{2} = R(x) \tag{15.72}$$

where $R(x) = \int_{\frac{1}{2}}^{x} \frac{\sin(2N+1)\pi t}{\sin \pi t}\,\mathrm{d}t$. Since $\frac{1}{\sin \pi t}$ is decreasing on $[x, \frac{1}{2}]$, Corollary 11.7 applies to give

$$|R(x)| = \left| \frac{1}{\sin \pi x} \int_{\frac{1}{2}}^{\xi} \sin(2N+1)\pi t\,\mathrm{d}t \right| \le \frac{1}{2x} \frac{1}{(2N+1)\pi}, \tag{15.73}$$

where we have used Jordan's inequality (7.12) in the last step. Taking the range of x into account, we conclude that (15.73) leads to the estimate in (15.68) and so (15.72) proves the assertion.

Exercise 130. Prove the Poisson summation formula in the following form for $f \in C^1([a,b])$:

$$\sum_{a<n\le b} f(n) = \sum_{n=-N}^{N} \int_a^b f(x) \cos 2\pi nx\,\mathrm{d}x + O\left(\frac{M \log N}{N}(b-a)\right),$$

where a,b are half-integers, whence as the limiting case

$$\sum_{a\le n<b} f(n) = \lim_{N\to\infty} \sum_{n=-N}^{N} \int_a^b f(x)\, e^{2\pi inx}\,\mathrm{d}x.$$

Solution. Since $\bar{B}_1(a) = \bar{B}_1(b) = 0$, Euler's summation formula (Corollary 10.2) reads

$$\sum_{a<n\le b} f(n) = \int_a^b f(x)\,\mathrm{d}x + \int_a^b \bar{B}_1(x) f'(x)\,\mathrm{d}x \tag{15.74}$$

the second term on the right of which becomes, by (15.68),

$$-\int_a^b \bar{B}_1(x) f'(x)\,\mathrm{d}x = \sum_{n=1}^{N} \frac{1}{\pi n}\int_a^b f'(x)\sin 2\pi n x\,\mathrm{d}x + R, \tag{15.75}$$

where

$$R = O\left(M\int_a^b \min\left(1, \frac{1}{N\|x\|}\right)\mathrm{d}x\right), \quad M = \max_{[a,b]}|f'(x)|.$$

To estimate the error term R, we entrap all integers in $[a,b]$ (which are $\sim [b-a]$ in number) in the neighborhood of length $\frac{1}{N}$, in which |integrand| ≤ 1 to obtain the estimate $[b-a]\cdot M\frac{1}{N}$. In the remaining intervals, we use $\frac{1}{N}\frac{1}{\|x\|}$ to obtain $\frac{1}{N}\int_{\frac{1}{N}}^{1}\frac{1}{x}\,\mathrm{d}x = \frac{\log N}{N}$. Hence altogether, the error term $R = O\left(\frac{M\log N}{N}(b-a)\right)$.

Now the first term on the right of (15.75) becomes $-2\sum_{n=1}^{N}\int_a^b f(x)\cos 2\pi n x\,\mathrm{d}x$, whence (15.75) leads to

$$\int_a^b \bar{B}_1(x) f'(x)\,\mathrm{d}x = \sum_{\substack{n=-N\\ n\ne 0}}^{N}\int_a^b f(x)\cos 2\pi n x\,\mathrm{d}x + R.$$

Substituting this in (15.74) completes the solution.

The following example is from Papoulis [Papoulis (1962), pp. 50–52]

Example 15.11. (Sampling theorem, Shannon 1949) If the Fourier transform $\hat{f}(z)$ is zero for $|z| \ge T > 0$, then f is given by

$$f(t) = \sum_{n=-\infty}^{\infty} f_n \frac{\sin(Tt - \pi n)}{Tt - \pi n}, \tag{15.76}$$

where

$$f_n = f\left(n\frac{\pi}{T}\right). \tag{15.77}$$

In particular, f is uniquely determined by its values (samples) at a sequence of equi-distant points, distance $\frac{\pi}{T}$ apart.

Proof. Proof is similar to that of Theorem 15.5. Since the Fourier inversion formula (15.9) reads

$$f(t) = \frac{1}{2\pi}\int_{-T}^{T} e^{-izt}\hat{f}(z)\,\mathrm{d}z,$$

it follows that

$$f_n = f\left(n\frac{\pi}{T}\right) = \frac{T}{\pi}\frac{1}{2T}\int_{-T}^{T} e^{-iz\frac{T}{\pi}n}\hat{f}(z)\,\mathrm{d}z = \frac{T}{\pi}\hat{c}_n,$$

where $\hat{c}_n$ is the nth Fourier coefficient of $\hat{f}$ (to be more precise, $\hat{f}(-t)$) and so $\hat{c}_n = \frac{\pi}{T}f_n$ correspondingly to (15.61). Hence the Fourier expansion (15.2) of $\hat{f}$ in the interval $(-T, T)$ is

$$\hat{f}(z) = \frac{\pi}{T}\sum_{n=-\infty}^{\infty} f_n e^{-in\frac{\pi}{T}z},$$

so that we may write

$$\hat{f}(z) = r_{2T}(z)\frac{\pi}{T}\sum_{n=-\infty}^{\infty} f_n e^{-in\frac{\pi}{T}z}$$

with the rectangular pulse function r_{2T}. Recalling the pair in (15.46), we immediately conclude (15.76). □

Example 15.12. (**Theta transformation formula**) Let $\theta(t)$ denote the **Jacobi elliptic theta function**

$$\theta(t) = \sum_{n=-\infty}^{\infty} e^{-\pi n^2 t}, \quad t > 0. \tag{15.78}$$

Then it satisfies the transformation formula

$$\theta\left(\frac{1}{t}\right) = t^{1/2}\theta(t). \tag{15.79}$$

Proof. In Corollary 15.1 we choose $f(x) = e^{-\pi x^2 t}$ and evaluate the Fourier coefficients

$$\hat{f}(n) = \int_{-\infty}^{\infty} e^{-\pi x^2 t}e^{-2\pi inx}\mathrm{d}x. \tag{15.80}$$

The integrand may be expressed as $e^{-\pi n^2/t}e^{-\pi(x\sqrt{t}+in/\sqrt{t})^2}$. Hence it amounts to establishing

$$\int_{-\infty}^{\infty} e^{-\pi(x+iu)^2}\mathrm{d}x = 1 \tag{15.81}$$

for any $u \in \mathbb{R}$, which is done in Exercise 131 below.

Under (15.81), we have $\hat{f}(n) = t^{-1/2}e^{-\pi n^2/t}$. Hence (15.67) leads to (15.79), completing the proof. □

Exercise 131. Prove (15.81).

Solution. Using the value of the probability integral in (11.158)

$$\int_{-\infty}^{\infty} e^{-\pi x^2}\mathrm{d}x = 1 \tag{15.82}$$

we show that the value of the integral in (15.81) is independent of the value of u.

Differentiating the integral in (15.81) under the integral sign, which we may because of absolute convergence, we see that the result is

$$\int_{-\infty}^{\infty} -2\pi i(x+iu)e^{-\pi(x+iu)^2}\mathrm{d}x, \tag{15.83}$$

which we may view as

$$i\int_{-\infty}^{\infty} \frac{\partial}{\partial x}e^{-\pi(x+iu)^2}\mathrm{d}x. \tag{15.84}$$

Since this is simply $\left[ie^{-\pi(x+iu)^2}\right]_{-\infty}^{\infty}$, which is 0, and we have shown that the integral in (15.81) is independent of the value of u. Choosing $u = 0$ gives the result.

Exercise 132. Use Corollary 15.1 to prove the **general theta-transformation formula**

$$\sum_{n=-\infty}^{\infty} e^{-\pi(n+\alpha)^2/x} = x^{1/2}\sum_{n=-\infty}^{\infty} e^{-\pi n^2 x+2\pi i n\alpha} \tag{15.85}$$

for $\mathrm{Re}\, x > 0$, $|\arg x^{1/2}| < \frac{\pi}{4}$.

Solution. In Corollary 15.1 we choose $f(x) = e^{-\pi t^2/x}$, then

$$\hat{f}(n) = \int_{-\infty}^{\infty} e^{-\pi t^2/x}e^{-2\pi i n t}\mathrm{d}t. \tag{15.86}$$

And the same argument applies

$$\hat{f}(n) = x^{\frac{1}{2}}e^{-\pi n^2 x}. \tag{15.87}$$

By (15.66),

$$\begin{aligned}\sum_{n=-\infty}^{\infty} e^{-\pi(n+\alpha)^2/x} &= \sum_{n=-\infty}^{\infty} f(n+\alpha) = \sum_{n=-\infty}^{\infty} \hat{f}(n)e^{2\pi i n\alpha}\\ &= \sum_{n=-\infty}^{\infty} x^{\frac{1}{2}}e^{-\pi n^2 x}e^{2\pi i n\alpha}.\end{aligned}$$

Chapter 16

Topological spaces

In this chapter we assemble some basics of (general) topology part of which has been used in many places without much notice. There are many standard references on topology including [Engelking (1977)], [Kelley (1955)], [Sugawara (1966)] etc. We partially follow [Sugawara (1966)] in the presentation of results and naturally we cannot be exhaustive. In §§16.5–16.7 we prove two things, i.e. we construct the real number system and prove that it is a Hilbert space and on the other hand we construct a completion of a number field and the proof works for a general metric spaces. This type of local-global view has been a fashion since Iwasawa and Tate in the mid 1960's (it seems that Hensel was the first to use the local method), cf. e.g. [Goldstein (1971)] and references therein.

16.1 Open sets and neighborhoods

Definition 16.1. A set $X \neq \emptyset$ is called a **topological space** if there exists a family $\mathcal{O} \subset 2^X$ of **open sets** satisfying
O_1: $X, \emptyset \in \mathcal{O}$
O_2: $U_1, \cdots, U_n \in \mathcal{O} \Longrightarrow \cap_{k=1}^{n} U_k \in \mathcal{O}$
O_3: $U_\lambda \in \mathcal{O}(\lambda \in \Lambda) \Longrightarrow \cup_{\lambda\in\Lambda} U_\lambda \in \mathcal{O}$.
More precisely, X is said to be a topological space with open sets in $\mathcal{O}$ and $\mathcal{O}$ is said to determine the topology of X. Sometimes, one writes $X = (X, \mathcal{O})$ and speak of topology τ of the topological space $(X, \mathcal{O})$. Every element of X is called a point of X. For different topologies on the same space X, we may introduce order relation. Suppose $(X, \mathcal{O}_1)$ and $(X, \mathcal{O}_2)$ be topological spaces and that $\mathcal{O}_1 \supset \mathcal{O}_2$, then the topology τ_1 of $(X, \mathcal{O}_1)$ is stronger than the topology τ_2 of $(X, \mathcal{O}_2)$ and write $\tau_1 \geq \tau_2$ or τ_2 is weaker than τ_1, denoted $\tau_2 \leq \tau_1$.

Definition 16.2. Let X be a topological space. For each $x \in X$, any open set containing x is called an **open neighborhood**. The family of all subsets of X that contains an open set containing x is called the system of neighborhoods and denoted $\mathcal{U}(x)$. The family of all open neighborhoods of x is called the system of open neighborhoods and denoted $\mathcal{U}(x)$. A subfamily $\mathcal{U}^*(x)$ is called a **fundamental system of neighborhoods** if for any $U \in \mathcal{U}(x)$ there is a $V \in \mathcal{U}^*(x)$ contained in U. To express an open neighborhood, one often uses the symbol U coming from the German word "Umgebung" or V coming from "vicinity". It is often the case that U is used to indicate an open set and V a neighborhood.

Proposition 16.1. *Suppose in a topological space X, there is given a fundamental system of neighborhoods $\mathcal{U}^*(x)$ for each $x \in X$. Then $U \subset X$ is an open set if and only if for any $x \in U$ there is a $V_x \in \mathcal{U}^*(x)$ such that $V_x \subset U$.*

Proof. ($\Longrightarrow$). If U is an open set, then for every $x \in U$,U is an open neighborhood of x. Hence there is a $V_x \in \mathcal{U}^*(x)$ such that $V_x \subset U$.
($\Longleftarrow$). Since $U = \cup_{x \in U} V_x$, it follows from O_3 that U is an open set. □

It is often the case that one provides a fundamental system of neighborhoods to endow topology to a set X based on Proposition 16.1 which introduces the family of open sets, thus the topology in X. Any point x satisfying the condition that there is a $V_x \in \mathcal{U}^*(x)$ such that $V_x \subset U$ may be called an **inner point** and the set of all inner points is called the interior of U, denoted $\operatorname{Int} U$. Thus $U \in \mathcal{O} \Longleftrightarrow U = \operatorname{Int} U$. For a metrical space in §16.4, it is customary to define an open set by this condition, i.e. for every point x of U there is a δ-neighborhood $V_\delta(x)$ which is $\subset U$.

Exercise 133. Prove the following.
(i) Any open interval $(a, b) \subset \mathbb{R}$ is an open set.
(ii) Any union of open intervals $\in \mathbb{R}$ is an open set and conversely, any open set $\in \mathbb{R}$ is given as a union of open intervals.

Solution. (i) For any $x \in I = (a, b)$ choose $\delta = \min\{x - a, b - x\} > 0$. Then $V_\delta(x) = (x - \delta, x + \delta) \subset I$. Hence $x \in \operatorname{Int} I$.
(ii) By (i) and O_3, any union is open. Conversely, for any open set $U \subset \mathbb{R}$ and any $x \in U$, choose a δ-neighborhood $V(x) = V_\delta(x) \subset U$. Then $U = \cup\{V(x) | x \in U\}$.

Theorem 16.1. *The fundamental system of neighborhoods $\mathcal{U}^*(x)$ satisfies U_1^*: For all $V \in \mathcal{U}^*(x)$, $V \ni x$.*

U_2^*: *For any* $V_1, V_2 \in \mathcal{U}^*(x)$, *there exists a* $V_3 \in \mathcal{U}^*(x)$ *such that* $V_3 \subset V_1 \cap V_2$.
U_3^*: *For any* $V \in \mathcal{U}^*(x)$ *and any* $y \in V$, *there exists a* $W \in \mathcal{U}^*(y)$ *such that* $W \subset V$.
Conversely, suppose for any $x \in X$ *there is given the fundamental system of neighborhoods* $\mathcal{U}^*(x)$ *satisfying* U_1^*-U_3^*. *Then the family* $\mathcal{O}$ *of* U*'s satisfying the condition in Proposition 16.1 satisfies* O_1-O_3, *introducing topology on* X. *The family* $\mathcal{U}^*(x)$ *defined with respect to this topology is a fundamental system of neighborhoods.*

Proof. U_2^*: By U_1^*, $x \in V_1$, $x \in V_2$, so that $x \in V_1 \cap V_2$ which is an open set in X by O_2. Hence $V_1 \cap V_2 \in \mathcal{U}(x)$ and so there exists a $V_3 \in \mathcal{U}^*(x)$ contained in $V_1 \cap V_2$.
U_3^*: Since $V \in \mathcal{U}(y)$, there exists a $W \in \mathcal{U}^*(y)$ contained in V.
Conversely,
O_1: For any $x \in X$, there exists a $V \in \mathcal{U}^*(x)$ such that $x \in V \subset X$. Hence x is an inner point of X, so that $X \in \mathcal{O}(X)$. Since $\operatorname{Int} \emptyset \subset \emptyset$, $\emptyset \in \mathcal{O}(X)$.
O_2: Suppose $U_j \in \mathcal{O}(X)$, $1 \leq j \leq n$. Then for any $x \in U_1 \cap \cdots \cap U_n$, $U_j \in \mathcal{U}(x)$. Hence by U_2^*, there exists a $V \in \mathcal{U}^*(x)$ such that $V \subset U_j$, whence $V \subset U_1 \cap \cdots \cap U_n$ and so $U_1 \cap \cdots \cap U_n \in \mathcal{O}(X)$.
O_3: Suppose $O_\lambda \in \mathcal{O}(X)$, $\lambda \in \Lambda$. Then for any $x \in \cup_{\lambda \in \Lambda} U_\lambda$, there is a $\mu \in \Lambda$ such that $U_\mu \in \mathcal{U}(x)$. Hence by U_2^*, there exists a $V \in \mathcal{U}^*(x)$ such that $V \subset U_\mu$, whence $V \subset \cup_{\lambda \in \Lambda} U_\lambda$, and so the union is $\in \mathcal{O}(X)$. □

Proposition 16.2. *For a subfamily* $\mathcal{O}^*$ *of open sets in* X *the following are equivalent.*
(i) *For any* $x \in X$ *the family* $\mathcal{U}^*(x) = \{V | x \in V \in \mathcal{O}^*\}$ *is a fundamental system of neighborhoods of* x.
(ii) *For any* $U \in \mathcal{O}$ *and any* $x \in U$ *there exists a* $V \in \mathcal{O}^*$ *such that* $x \in V \subset U$.
(iii) *Any* $U \in \mathcal{O}$ *is expressed as a union of sets in* $\mathcal{O}^*$:

$$\mathcal{O} = \{\cup_{\lambda \in \Lambda} V_\lambda \mid V_\lambda \in \mathcal{O}^* \quad (\lambda \in \Lambda)\}. \tag{16.1}$$

Proof. (i)$\Longleftrightarrow$(ii) follows from the fact that $U \in \mathcal{U}^*(x)$. From (ii), U is the union of all those V's for which $\mathcal{O}^* \ni V \subset U$ whence (iii). Conversely, if $x \in \cup_{\lambda \in \Lambda} V_\lambda$, then $x \in V_\mu \subset U$ for some μ, i.e. (ii). □

Any family $\mathcal{O}^*$ satisfying the conditions in Proposition 16.2 is called an **open base**. It is clear that given a fundamental system of neighborhoods $\mathcal{U}^*(x)$ at each $x \in X$, the union $\{\mathcal{U}^*(x) \mid x \in X\} = \mathcal{O}^*$ is an open base of X.

Thus the notions of a fundamental system of neighborhoods and an open base are parallel in that the former is point-wise and the latter set-wise.

Theorem 16.2. *An open base $\mathcal{O}^*(x)$ satisfies*
O_1^*: *For any $x \in X$ there exists a $V \in \mathcal{O}^*(x)$ such that $V \ni x$.*
O_2^*: *For any $V_1, V_2 \in \mathcal{O}^*$ and any $x \in V_1 \cap V_2$, there exists a $V_3 \in \mathcal{O}^*$ such that $x \in V_3 \subset V_1 \cap V_2$.*
Conversely, suppose for any X there is given a family $\mathcal{O}^*$ satisfying O_1^* and O_2^*. Then the family $\mathcal{O}$ defined by (16.1) satisfies O_1–O_3, introducing topology on X. The family $\mathcal{O}^*$ defined with respect to this topology is an open base.

Proof. U_2^*: By U_1^*, $x \in V_1$, $x \in V_2$, so that $x \in V_1 \cap V_2$ which is an open set in X by O_2. Hence $V_1 \cap V_2 \in \mathcal{U}^*(x)$ and so there exists a $V_3 \in \mathcal{U}^*(x)$ contained in $V_1 \cap V_2$.
U_3^*: Since $V \in \mathcal{U}(y)$, there exists a $W \in \mathcal{U}^*(y)$ contained in V.
Conversely,
O_1: For any $x \in X$, there exists a $V \in \mathcal{U}^*(x)$ such that $x \in V \subset X$. Hence x is an inner point of X, so that $X \in \mathcal{O}(X)$. Since $\mathrm{Int}\,\emptyset \subset \emptyset$, $\emptyset \in \mathcal{O}(X)$.
O_2: Suppose $O_j \in \mathcal{O}(X)$, $1 \le j \le n$. Then for any $x \in U_1 \cap \cdots \cap U_n$, $U_j \in \mathcal{U}^*(x)$. Hence by U_2^*, there exists a $V \in \mathcal{U}^*(x)$ such that $V \subset U_j$, whence $V \subset U_1 \cap \cdots \cap U_n$ and so $U_1 \cap \cdots \cap U_n \in \mathcal{O}(X)$.
O_3: Suppose $O_\lambda \in \mathcal{O}(X)$, $\lambda \in \Lambda$. Then for any $x \in \cup_{\lambda \in \Lambda} U_\lambda$, there is a $\mu \in \Lambda$ such that $U_\mu \in \mathcal{U}(x)$. Hence by U_2^*, there exists a $V \in \mathcal{U}^*(x)$ such that $V \subset U_\mu$, whence $V \subset \cup_{\lambda \in \Lambda} U_\lambda$, and so the union is $\in \mathcal{O}(X)$. □

There are several ways known to introduce topology into a given set which we shall state subsequently, corresponding to some of the axioms of continuity of real numbers. It is therefore important to distinguish in the first stage which axiom is essential for general theory.

Definition 16.3. If a topological space X has at most a countable system of fundamental neighborhoods, it is said to satisfy the **first axiom of choice**. In such a space, the convergence of a sequence may be used to introduce topology on X. A sequence $\{x_n | n \in \mathbb{N}\} \subset X$ is said to converge to $x \in X$, denoted $\lim_{n\to\infty} x_n = x$ if for every neighborhood U of x there exists an $N \in \mathbb{N}$ such that $\{x_n | n > N\} \subset U$.

Theorem 16.3. *Suppose X is a topological space satisfying the first axiom of choice. Then $U (\subset X)$ is an open set if and only if for every $x \in U$ and* *every sequence* $\{x_n | n \in \mathbb{N}\} \subset X$ *converging to* x, *there exists an* $N \in \mathbb{N}$ *such that* $\{x_n | n > N\} \subset U$.

Proof. Necessity is clear from definition of convergence. To prove sufficiency, for each $x \in X$ choose an at most countable $\mathcal{U}^*(x) = \{U_1, \cdots, U_n\}$ and put

$$V_n = U_1 \cup \cdots \cup U_n \tag{16.2}$$

(if $\mathcal{U}^*(x)$ is a finite set with n_0 elements, then we put $V_n = V_{n_0}$ for $n > n_0$). Then $\mathcal{U}_V^*(x) = \{V_n | n \in \mathbb{N}\}$ is a fundamental neighborhood of x, which is decreasing in the sense of inclusion. Suppose U is not open. Then there exists an $x \in U$ such that no $V_n \in \mathcal{U}_V^*(x)$ is contained in U. Hence we may choose an $x_n \in U \setminus V_n$, finding a sequence $\{x_n\} \not\subset U$. Then for any $U' \in \mathcal{U}(x)$ we may choose an $N \in \mathbb{N}$ such that $V_N \subset U'$. Hence for $n > N$, we have $x_n \in V_n \subset V_N \subset U'$, so that $\lim_{n\to\infty} x_n = x$. However, this sequence is not contained in U, invalidating the condition in the theorem, whence sufficiency follows. □

Remark 16.1. The sufficiency condition in Theorem 16.3 does not necessarily hold in a topological space not satisfying the first axiom of choice.

16.2 Other ways of introducing topology

We shall state a few more ways of introducing a topology on a set.

Definition 16.4. Let $X = (X, \mathcal{O})$ be a topological space. Any $F \subset X$ is called a **closed set** if the complement is an open set: $X - F \in \mathcal{O}$. Denote the family of all closed sets in X by $\mathcal{A} = \mathcal{A}(X)$:

$$\mathcal{A} = \{F \subset X | X - F \in \mathcal{O}\}.$$

Then $\mathcal{A}$ satisfies
A_1: $X, \emptyset \in \mathcal{A}$
A_2: $F_1, \cdots, F_n \in \mathcal{A} \Longrightarrow \cup_{k=1}^n F_k \in \mathcal{A}$
A_3: $F_\lambda \in \mathcal{A}(\lambda \in \Lambda) \Longrightarrow \cap_{\lambda\in\Lambda} F_k \in \mathcal{A}$.
Here "A" is from a German word "abgeschlossene" and "F" from a French word "fermé". Correspondingly to the interior, we define the intersection of all closed sets containing A to be the **closure** of A denoted $\bar{A}$:

$$\bar{A} = \cap_{A\subset F\in\mathcal{A}} F.$$

It is the smallest closed set containing A. Using the derived set in Definition 2.2, we have

$$\bar{A} = A \cup A^d.$$

A subset $D \subset A$ such that $\bar{D} = A$ is called **dense** in A.

Theorem 16.4. *Given a family $\mathcal{A}$ of closed sets in a set X, the family $\mathcal{O} = \{U \subset X | X - U \in \mathcal{A}\}$ satisfies the conditions O_1–O_3 providing a topology on X; thus a topological space with $\mathcal{A}$ as the family of closed sets: $(X, \mathcal{A})$.*

Definition 16.5. For a topological space X, a family $\mathcal{U} \subset 2^X$ such that $\cup\mathcal{U} \supset X$ is said to be a **covering** of X or cover X. A covering with a countably [finitely] many sets is called a countable [finite] covering. A covering consisting of open sets is called an **open covering**.
If any open covering of X contains a finite sub-covering, X is said to be **compact**.
A family $\mathcal{F} \subset 2^X$ of subsets of X is said to have the **finite intersection property** if any finitely many sets $F_1, \cdots, F_n$ has an intersection, $F_1 \cap \cdots \cap F_n \neq \emptyset$.

Theorem 16.5. *A topological space is compact if and only if any family $\mathcal{F}$ of closed subsets of X having the finite intersection property satisfies*

$$\cap_{\lambda \in \Lambda} F_\lambda \neq \emptyset. \tag{16.3}$$

Proof. To prove necessity, suppose $\cap\mathcal{F} = \emptyset$. Then by the de Morgan law, $\mathcal{U} := \{X - F | F \in \mathcal{F}\}$ is an open covering of X. Hence there exists a finite covering $\mathcal{U}_1 \subset \mathcal{U}$. Then $\{F | X - F \in \mathcal{U}_1\} = \{F_1, \cdots, F_n\}$, say and $F_1 \cap \cdots \cap F_n = \emptyset$. A contradiction.
Sufficiency may proved almost verbatim. Suppose $\mathcal{U}$ is an open covering of X. Then $\mathcal{F} := \{F | X - F \in \mathcal{U}\}$ is a family of closed subsets of X satisfying (16.3). Hence there exist finitely many sets $\{F_1, \cdots, F_n\}$ which are disjoint, i.e. $F_1 \cap \cdots \cap F_n = \emptyset$ or $(X - F_1) \cup \cdots \cup (X - F_n) = X$ is a finite sub-covering $\subset \mathcal{U}$ of X. □

Theorem 16.6. (Cantor intersection theorem) *If X is compact, then any sequence of its closed subsets $\{F_j\}$ satisfying $F_n \supset F_{n+1}$ has an intersection, i.e. $\cap F_n \neq \emptyset$.*

This is a generalization of R_4 (Nested set property by Cantor). Proof follows immediately from Theorem 16.5 since $F_k \cap \cdots \cap F_{k+n} = F_{k+n} \neq \emptyset$.

Definition 16.6. If in a topological space X there are no open sets $U_1 \neq \emptyset$, $U_2 \neq \emptyset$ satisfying

$$X = U_1 \cup U_2, \quad U_1 \cap U_2 = \emptyset, \tag{16.4}$$

X is called a **connected set**.

For the definition of a continuous map, cf. Theorem 3.2 and Remark 3.1 above.

Theorem 16.7. *For topological spaces X, Y and a map $f : Y \to X$, the family*

$$\mathcal{O}(Y) = \{f^{-1}(U) \mid U \in \mathcal{O}(X)\} \tag{16.5}$$

defines a topology on Y that is the weakest topology regarding which f is continuous, called the **induced topology** *by f.*

Proof that $\mathcal{O}(Y)$ defines a topology follows from O_1–O_3 of X and Lemma 1.1. f is clearly continuous. For any family $\mathcal{O}_1(Y)$ of open sets regarding which f is continuous, we have $f^{-1}(U) \in \mathcal{O}(Y)$ i.e. $\mathcal{O}(Y) \subset \mathcal{O}_1(Y)$.

Lemma 16.1. *Given a family $\{\tau_\lambda | \lambda \in \Lambda\}$ of topologies on a set X with $\mathcal{O}_\lambda$ the open sets of τ_λ, we have*

$$\mathcal{O}_o := \bigcap_{\lambda \in \Lambda} \mathcal{O}_\lambda = \inf\{\tau_\lambda | \lambda \in \Lambda\}, \tag{16.6}$$

i.e. the strongest of all topologies weaker than any τ_λ.

Theorem 16.8. *Given an arbitrary family $\mathcal{O}^o$ of a set X, there exists the weakest one among those topologies which have all the sets in $\mathcal{O}^o$ as their open sets. Indeed, the topology τ which has as its open base*

$$\mathcal{O}^* = \{V_1 \cap \cdots \cap V_n \mid V_i \in \mathcal{O}^o, 1 \le i \le n\} \cup \{X\}. \tag{16.7}$$

Proof. Let

$$\mathcal{O}_o := \bigcap_{\lambda \in \Lambda} \mathcal{O}_\lambda, \tag{16.8}$$

where $\mathcal{O}_\lambda$ rings through all family of open sets containing $\mathcal{O}^*$. By Lemma 16.1, $\mathcal{O}_o$ defines the required topology on X. We show that the family of open sets $\mathcal{O}$ defined by (16.1) coincides with $\mathcal{O}_o$. □

16.3 Relative topology, product topology

Let $X = (X, \mathcal{O})$ be a topological space and $\emptyset \ne A \subset X$. Then the **inclusion map** is defined by

$$i = i_A : A \subset X. \tag{16.9}$$

For any $U \in \mathcal{O}(X)$,

$$i^{-1}(U) = U \cap A. \tag{16.10}$$

The **relative topology** on A is the induced topology by the inclusion map:

Corollary 16.1. *Let A be a subset of a topological space X. Then the relative topology τ is the weakest topology for which i is continuous. It is given by the family*

$$\mathcal{O}(A) = \{U \cap A | U \in \mathcal{O}\}. \tag{16.11}$$

This is a special case of Theorem 16.7. It is instructive to give a direct proof.

The Cartesian product $\prod_{\lambda \in \Lambda} X_\lambda$ is discussed in §1.3. We use the notation given there. The mapping

$$p_\lambda : \prod_{\lambda \in \Lambda} X_\lambda \to X_\lambda; (a_\lambda) \mapsto a_\lambda \tag{16.12}$$

is called a (natural) **projection**. The product topology is defined to be the weakest one for which all the projections are continuous mappings.

One of the highlights is the celebrated Tychonoff theorem [Bourbaki (1989)], [Kelley (1955)], [Willard (1978)] to the effect that any Cartesian product of compact sets is again compact. The Cartan-Bourbaki proof depends on multiple use of Zorn's lemma (or the Axiom of choice, its equivalent). Tsukada [Tsukada (2015)] is the first who gave a proof which uses Zorn's lemma only once.

Theorem 16.9. *If X_λ $(\lambda \in \Lambda)$ are all compact, then their Cartesian product $\prod_{\lambda \in \Lambda} X_\lambda$ is also compact.*

16.4 Metric spaces

In this section we collect basic properties of a metric space which do not hold true in a general topological space. These properties hold true for the n-dimensional Euclidean space $\mathbb{R}^n$ in view of Exercise 5 which we extend to a general metric space.

Definition 16.7. Let $X \neq \emptyset$ be a set and $d : X \to \mathbb{R}_{\geq 0}$ be a function satisfying the defining conditions of the distance function:
(i) $d(x, y) \geq 0$ and $d(x, y) = 0$ if and only if $x = y$.
(ii) $d(x, y) = d(y, x)$.
(iii) (the triangular inequality) $d(x, z) \leq d(x, y) + d(y, z)$.

The second condition in (i) is a uniqueness condition. If we do not assume it, we call d a pseudo-distance (or pseudo-metric). We may work with pseudo-metric by appealing to [Kelley (1955), Lemma 4.5, p. 116, Theorem 4.15, p. 123].

Cf. the passages after Definition 16.11.

Definition 16.8. A metric space X is said to be **sequentially compact** if each sequence $\{x_n\} \subset X$ contains a subsequence converging to a point in X. $X \subset \mathbb{R}^n$ is said to be **bounded** if there exists a neighborhood $V_n(o)$ of the zero containing X.

Theorem 16.10. *A metric subspace of $\mathbb{R}^n$ is sequentially compact if and only if it is a bounded and closed set.*

Theorem 16.11. *For a metric space X the following are equivalent.*
(i) *X is compact.*
(ii) *X is sequentially compact.*
(iii) *X is paracompact and complete.*

Corollary 16.2. *For a metric subspace X of $\mathbb{R}^n$ to be compact it is necessary and sufficient that X is a bounded and closed subset of $\mathbb{R}^n$.*

Exercise 134. Prove that any closed interval $I = [a, b] \subset \mathbb{R}$ is compact.

Solution. Let $\mathcal{U}$ be an open covering of I. Suppose I is not compact. Then $[a, \frac{a+b}{2}]$ or $[\frac{a+b}{2}, b]$ does not have the finite covering. We name such interval $I_1 = [a_1, b_1]$ (i.e. $a_1 = a$, $b_1 = \frac{a+b}{2}$ or $a_1 = \frac{a+b}{2}$, $b_1 = b$). And we define the intervals $I_{n+1} = [a_{n+1}, b_{n+1}] \subset I_n, n = 1, 2, \cdots$ such that $a_{n+1} = a_n$, $b_{n+1} = \frac{a_n+b_n}{2}$ or $a_{n+1} = \frac{a_n+b_n}{2}$, $b_{n+1} = b_n$ and there exist no finite covering of I_{n+1}. Take $\alpha \in \cap_{n=0}^{\infty} I_n$, then there exists $U_\alpha \in \mathcal{U}$. However $\alpha \in U_\alpha \in \mathcal{O}$. For some $\varepsilon > 0, (\alpha - 2\varepsilon, \alpha + 2\varepsilon) \subset U_\alpha$. Then for $N \gg 0$ $I_N \subset [\alpha - \varepsilon, \alpha + \varepsilon] \subset U_\alpha$. It is a contradiction.

Exercise 135. Prove that an upward [downward] bounded closed set ($\subset \mathbb{R}$) has its maximum [minimum].

Solution. Suppose $F \subset \mathbb{R}$ is bounded from above. Then by R_2, $a \in \sup F$ exists. If $a \neq F$, then $a \in X - F \in \mathcal{O}(X)$, so that there exists an ε-neighborhood $V_\varepsilon(a) \subset X - F$. I.e. $(a - \varepsilon, a + \varepsilon) \subset X - F$. However, since $a - \varepsilon$ is not an upper bound of F, there exists an $x \in F$ such that $x \in (a - \varepsilon, a) \subset X - F$, a contradiction.

Definition 16.9. Let $(X, d), (Y, \rho)$ be metric spaces and let $\{T_n\}$ be a sequence of maps from X to Y. If there is a map $T_0 : X \to Y$ such that

$$T_n(x) \to T_0(x), \quad n \to \infty \tag{16.13}$$

for all $x \in X$. Then we say that $\{T_n\}$ is convergent to T_0 and write:

A map $T : X \to Y$ is said to be **bounded** if there exist $r > 0$ and $y_0 \in Y$ such that

$$\rho(T(x), y_0) \le r \quad \text{or} \quad T(x) \in V_r(y_0) \tag{16.14}$$

for all $x \in X$.

We denote the set of all bounded maps from X to Y by $B(X, Y)$. For $T_1, T_2 \in B(X, Y)$ define

$$\delta(T_1, T_2) = \sup_{x \in X} \rho(T_1(x), T_2(x)). \tag{16.15}$$

Exercise 136. Show that $(B(X, Y), \delta)$ is a metric space.

16.5 Completion of a number field

Let k be an algebraic number field whose special case is the rational number field $\mathbb{Q}$.

Definition 16.10. If a non-negative-valued function $v : k \to \mathbb{R}_+ \cup \{0\}$ satisfies the following conditions, it is called a **valuation** or an **absolute value** on k and also denoted $|\cdot|$:

(i) $v(0) = 0$; $v(x) = 0 \iff x = 0$

(ii) $v(xy) = v(x)v(y)$

(iii) There exists a $C > 0$ such that $v(x + y) \le C \max\{v(x), v(y)\}$.

If (iii) holds with $C = 1$, then v is said to be **non-Archimedean** (null-A) or finite. Otherwise, it is said to be **Archimedean** or infinite.

Exercise 137. Suppose $C \le 2$ in (iii). Then prove that

$$v\left(\sum_{j=1}^{k} x_j\right) \le k \max\{v(x_j) | 1 \le j \le k\}, \tag{16.16}$$

in particular, $v(n) = v(n \cdot 1) \le n$.

Solution. We first note the following. If $k \leq 2^s$, $s = s(n) \in \mathbb{N}$, then (16.16) holds in the form

$$v\left(\sum_{j=1}^{k} x_j\right) \leq C^s M, \tag{16.17}$$

where for simplicity we put $M = \max\{v(x_j) | 1 \leq j \leq k\}$.

This can be proved by induction.

Then for any $t \in \mathbb{N}$ choose $s = s(k)$ such that $2^{s-1} < k^t \leq 2^s$ and consider the expression $\left(\sum_{j=1}^{k} x_j\right)^t$. We have

$$v\left(\sum_{j=1}^{k} x_j\right)^t = v\left(\left(\sum_{j=1}^{k} x_j\right)^t\right) \leq C^s M_1, \tag{16.18}$$

where $M_1 = \max\{v(x_{i_1}) \cdots v(x_{i_t}) | 1 \leq i_j \leq k\}$ and this is $\leq M^t$. Hence (16.18) amounts to

$$v\left(\sum_{j=1}^{k} x_j\right) \leq C^{\frac{s}{t}} M \leq \lim_{t\to\infty} 2^{\frac{s}{t}} M = 2^{\frac{\log k}{\log 2}} M = kM. \tag{16.19}$$

Exercise 138. Prove that $C \leq 2$ in (iii) implies the **trigonometric inequality**

$$v(x + y) \leq v(x) + v(y) \tag{16.20}$$

and conversely, if (16.20) holds, we may take $C = 2$.

Solution. We argue in the spirit of Exercise 137 and consider $(x + y)^n$. Then the counterpart of (16.18) reads in view of (16.16)

$$v(x + y)^n = v\left((x + y)^n\right) \leq nM_2, \tag{16.21}$$

where M_2 is the maximum of $n + 1$ terms $\binom{n}{r} v(x)^r v(y)^{n-r}$. Hence

$$v(x + y)^n \leq n \sum_{r=0}^{n} \binom{n}{r} v(x)^r v(y)^{n-r} = nM_3^n \tag{16.22}$$

where now $M_3 = v(x) + v(y)$. Hence similarly to (16.19),

$$v(x + y) \leq (n + 1)^{\frac{1}{n}} M_3. \tag{16.23}$$

Letting $n \to \infty$ in (16.23), leads to (16.20).

Hereafter, we assume (16.20) and write $|x+y| \le |x| + |y|$. Needless to say, it induces a metric

$$d(x, y) = |x - y| = v(x - y) \tag{16.24}$$

on k. If we do not assume the uniqueness condition (i) then it introduces a pseudo-metric in Definition 16.7. We sometimes use both notations. We also exclude the trivial valuation $v(0) = 0, v(x) = 1, x \neq 0$. There exist infinitely many non-Archimedean valuations corresponding to infinitely many prime ideals in k and $[k : \mathbb{Q}]$ Archimedean valuations corresponding to $[k : \mathbb{Q}]$ embeddings of k into $\mathbb{C}$. Non-Archimedean valuations define the "nearness" by the degree of divisibility of two elements by a power of the prime ideal.

Example 16.1. For an $x \in \mathbb{Q}$ and a prime p, we define the **p-adic valuation** $v_p(x)$ of x (also called p-order) by

$$x = p^{v_p(x)} \frac{a}{b}, \quad a, b \in \mathbb{Z},\ (a, p) = (b, p) = 1. \tag{16.25}$$

Then the p-adic metric φ_p is defined by $\varphi_p(x) = \rho^{v_p(x)}$ for some $0 < \rho < 1$. We usually make the normalization $\rho = \frac{1}{p}$ so that the product formula holds true. Cf. [Goldstein (1971), p. 36].

Definition 16.11. $\tilde{k}$ is called a **completion** of k with respect to the valuation v if

(i) $\tilde{k}$ is complete.

(ii) There exists an isometry $\rho : k \to \tilde{k}$ such that $\overline{\rho(k)} = \tilde{k}$.

Our objective is to prove the following theorem in two stages: First we construct the real number system $\mathbb{R}$ and prove that it is complete §16.7. Then secondly using completeness of the reals, we prove completeness of other fields. For the former purpose we restrict the range of the valuation $v = v_{\mathbb{Q}}$ in Definition 16.10 to $\mathbb{Q}_+ \cup \{0\}$ and use the fact that it is an Archimedean ordered field. If we restrict to pseudo-metric, then we do not need to consider the Cauchy sequences modulo null sequences and we may directly appeal to the argument e.g. in [Kelley (1955), Lemma 4.5, p. 116, Theorem 4.15, p. 123].

Or we may suppose k is such, cf. §16.7.

Theorem 16.12. *For every (non-trivial) valuation v of k there exists a unique (up to isometry) completion k_v of k.*

16.6 Construction of the field

Recall the definition of a Cauchy sequence in Definition 2.1. In case of $v_{\mathbb{Q}}$, $\varepsilon > 0$ is understood as a given rational number.

Exercise 139. Let $\mathcal{R} = \{\{a_n\}\}$ be the set of all Cauchy sequences in k. We define addition $+$ and multiplication $\cdot$ componentwise as in Example 1.2, (ii). Then prove that $(\mathcal{R}, +, \cdot)$ forms a commutative ring.

Solution. We show that under operations $\{a_n\} + \{b_n\} = \{a_n + b_n\}$, $\{a_n\}\{b_n\} = \{a_n b_n\}$, $\mathcal{R}$ satisfies the defining conditions for a ring in Definition 1.4.

We need to show that $\{a_n + b_n\}$ and $\{a_n b_n\}$ are Cauchy sequences. For two Cauchy sequences $\{a_n\}, \{b_n\}$, we note that $|a_n + b_n - a_m - b_m| \le |a_n - a_m| + |b_n - b_m| < 2\varepsilon$ and that $|a_n b_n - a_m b_m| = |a_n b_n - a_n b_m + a_n b_m - a_m b_m| \le (|a_n| + |b_m|)\varepsilon = O(\varepsilon)$.
And $\{1\} = \{1, 1, \cdots\} \in \mathcal{R}$ is the identity since a convergent sequence is a Cauchy sequence.

Lemma 16.2. *Let $\mathcal{I} = \{\{a_n\} | a_n \to 0\,(n \to \infty)\} \subset \mathcal{R}$ be the set of sequences converging to 0 (called the* **null sequences***), which are Cauchy sequences by the remark above. Then $\mathcal{I}$ is a maximal ideal of $\mathcal{R}$ and so the quotient ring $\tilde{k} = \mathcal{R}/\mathcal{I}$ is a field.*

Proof. Since the subtraction $\{x_n\} - \{y_n\} = \{x_n - y_n\} \in \mathcal{I}$, $\mathcal{I}$ is plainly an additive subgroup of $\mathcal{R}$. Since a Cauchy sequence is bounded, multiplication of any element of $\mathcal{I}$ by an element of $\mathcal{R}$ remains within $\mathcal{I}$, whence $\mathcal{I}$ is an ideal.

Next let $y = \{y_n\} \in \mathcal{R} \backslash \mathcal{I}$. Then we show that y^{-1} exists in $\mathcal{R}$. For this we show that $\{y_n\}$ is bounded from below, i.e. $|y_n| \ge \delta$ for all $n \in \mathbb{N}$. We use (2.7) in the form with a fixed $\varepsilon > 0$, $\exists n_0(\varepsilon) \in \mathbb{N}$ such that $|y_n - y_{n_0}| < \varepsilon$. Choose $0 < \varepsilon < \delta$. Then for $n \ge n_0$, $||y_n| - |y_{n_0}|| \le |y_n - y_{n_0}| < \varepsilon$, whence $|y_n| \ge |y_{n_0}| - \varepsilon \ge \delta - \varepsilon > 0$. We may add e.g. $\{1, 1, \cdots, 1, 0, 0, \cdots\} \in \mathcal{I}$ to $\{y_n\}$ to ensure that all the terms of $\{y_n\}$ are non-zero. Hence we obtain $\{{y_n}^{-1}\} \in \mathcal{R}$. Then the ideal $\langle \mathcal{I}, y \rangle$ generated by $\mathcal{I}$ and y contains 1 and must coincide with $\mathcal{R}$. Since any ideal $\mathcal{J}$ containing $\mathcal{I}$ contains $\langle \mathcal{I}, y \rangle$, it follows that $\mathcal{J} = \mathcal{R}$, whence $\mathcal{I}$ is maximal. ($\mathcal{J} \supsetneq \mathcal{I}$.)

The last assertion follows e.g. from [Li *et al.* (2013), Proposition 1.8]. □

Exercise 140. Prove the maximality of $\mathcal{I}$ by the following reasoning. If $\mathfrak{a} \supsetneq \mathcal{I}$ and suppose there is an element $a = \{a_n\} \in \mathfrak{a} \backslash \mathcal{I}$, then $\mathfrak{a} = \mathcal{R}$.

Solution. We must have $\lim_{n\to\infty} v(a_n) > 0$, so that there exists a $\delta > 0$ such that $v(a_n) \geq \delta$ for n large enough. Let $\{b_n\}$ be defined by

$$b_n = \begin{cases} 1 & a_n = 0 \\ \frac{1}{a_n} & a_n \neq 0. \end{cases}$$

Then

$$v(b_n - b_m) \leq \frac{v(a_m - a_n)}{\delta^2} \to 0,$$

and so $\{b_n\} \in \mathcal{R}$. Since $\lim_{n\to\infty} v(a_n b_n) = 1$, it follows that $c = \{c_n\} = \{a_n b_n - 1\} \in \mathcal{I}$. Hence $1 = ab - c \in \mathfrak{a}$ and $\mathfrak{a} = \mathcal{R}$.

16.7 Proof of completeness

We introduce the embedding

$$\rho : k \to \tilde{k}; \quad x \to \{x, x, \cdots\} + \mathcal{I}, \tag{16.26}$$

which is a field injection and we may view k as a subfield of $\tilde{k}$.

We now assume *k is an Archimedean ordered field (in particular, the rational field $\mathbb{Q}$) and the range of valuation is, i.e. the Archimedean principle holds true in k: For any $k \ni a > 0$, $k \ni \varepsilon > 0$, there is an $n_0 \in \mathbb{N}$ such that*

$$n_0 \varepsilon > a.$$

We introduce the ordering on $\tilde{k}$ by defining $\tilde{x} > \tilde{0}$ if there exists an $0 < \varepsilon \in k$ and a representative $\{x_n\} \subset k$ ($\tilde{x} = \{x_n\} + \mathcal{I}$) such that $x_n \geq \varepsilon$. By this $\tilde{k}$ is an ordered field.

Then we may introduce the absolute value on $\tilde{k}$ and therewith a distance $\tilde{d}$ on $\tilde{k}$

$$\tilde{d}(\tilde{x}, \tilde{y}) = |\tilde{x} - \tilde{y}| = |\tilde{x} - \tilde{y}|_{\tilde{k}} = \begin{cases} \tilde{x} - \tilde{y} & \tilde{x} \geq \tilde{y} \\ \tilde{y} - \tilde{x} & \tilde{x} \leq \tilde{y}. \end{cases} \tag{16.27}$$

We may define the limit notion and Cauchy sequences etc. in $\tilde{k}$ as in §2.1: E.g. a sequence $\{\tilde{x}_n\} \subset \tilde{k}$ is a Cauchy sequence if and only if for any $k \ni \varepsilon > 0$, there exists an $n_0 \in \mathbb{N}$ such that for any $\mathbb{N} \ni m, n \geq n_0$, we have

$$\tilde{d}(\tilde{x}_m, \tilde{x}_n) < \varepsilon. \tag{16.28}$$

We note that since $d(x_m, x_n) = \tilde{d}(\rho(x_m), \rho(x_n)) < \varepsilon$ for $m, n \geq n_0$, $\tilde{x} = \{x_n\} + \mathcal{I}$ is equivalent to

$$\tilde{d}(\tilde{x}, \rho(x_n)) < \varepsilon, \tag{16.29}$$

i.e. $x_n = \rho(x_n) \to \tilde{x}$. Hence any element $\tilde{x} \in \tilde{k}$ is the limit of a sequence of k.

Theorem 16.13. *$\tilde{k}$ is a complete metric space and k is its dense subset. In particular, $\mathbb{R} = \mathbb{Q}/\mathcal{I}$ is a complete metric space and $\mathbb{Q}$ is its dense subset.*

Proof. Let $\{\tilde{x}_n\} \subset \tilde{k}$ be a Cauchy sequence. Then by (16.29), we conclude that for each $n \in \mathbb{N}$, there is a $y_n \in k$ such that

$$\tilde{d}(\tilde{x}_n, \rho(y_n)) < \frac{1}{n}. \tag{16.30}$$

Then $\{y_n\}$ is a Cauchy sequence. For choosing n_0 so large that $n_0\varepsilon \ge 1$, we have for $m, n \ge n_0$

$$\begin{aligned} d(y_m, y_n) &= \tilde{d}(\rho(y_m), \rho(y_n)) \\ &\le \tilde{d}(\rho(y_m), \tilde{x}_m) + \tilde{d}(\tilde{x}_m, \tilde{x}_n) + \tilde{d}(\tilde{x}_n, \rho(y_n)) \\ &< \frac{1}{n} + \varepsilon + \frac{1}{n} \le 3\varepsilon. \end{aligned} \tag{16.31}$$

Hence $\{y_n\} \in \mathcal{R}$ and it defines a class $\tilde{y} = \{y_n\} + \mathcal{I}$. Since $\tilde{d}(\rho(y_n), \tilde{y}) \to 0$ as $n \to \infty$, it follows that

$$\tilde{d}(\tilde{x}_n, \tilde{y}) \le \tilde{d}(\tilde{x}_n, \rho(y_n)) + \tilde{d}(\rho(y_n), \tilde{y}) < \frac{1}{n} + \varepsilon \le 2\varepsilon, \tag{16.32}$$

whence $\tilde{x}_n \to \tilde{y}$, i.e. every Cauchy sequence is convergent. By this we establish the particular case of the real number system, the completion $\mathbb{R} = \mathcal{R}/\mathcal{I}$ is a maximal Archimedean ordered field.

In the general case, we let the range of the valuation be the reals as in Definition 16.10. Then we introduce the valuation on $\tilde{k}$ as in Exercise and establish (16.29). Then the argument is verbatim to that given above. □

Remark 16.2. The completion $\mathbb{Q}_p$ of $\mathbb{Q}$ with respect to the p-adic valuation called the p-adic number field, is a locally compact space and the set $\mathbb{Z}_p$ consisting of elements x for which $v_p(x) \ge 0$ called the ring of p-adic integers, is its open subring. The set $U = U_p$ of all invertible elements in $\mathbb{Z}_p$ are called p-adic units for which $v_p(x) = 0$.

Exercise 141. Use completeness of $\mathbb{R}$ to deduce (16.29).

Solution. For $\{x_n\} \subset \mathcal{R}$, $\{v(x_n)\} \subset \mathbb{R}$ is a Cauchy sequence, so that $\lim_{n\to\infty} v(x_n)$ exists. We define

$$\tilde{v}(\{x_n\} + \mathcal{I}) = \lim_{n\to\infty} v(x_n). \tag{16.33}$$

Then recalling (16.24), we define the metric similarly to (16.27)

$$\tilde{d}(\tilde{x}, \tilde{y}) = v(\tilde{x} - \tilde{y}). \tag{16.34}$$

Hence

$$\begin{aligned}\tilde{d}(\tilde{x}, \rho(x_m)) &= \tilde{v}\left(\{x_n\} + \mathcal{I} - (\{x_m\} + \mathcal{I})\right) = \tilde{v}(\{x_n - x_m\} + \mathcal{I}) \\ &= \lim_{n\to\infty} v(x_n - x_m) < \varepsilon.\end{aligned}$$

16.8 Generalized sequences and their limits

Definition 16.12. If on a set $X \neq \emptyset$ the order relation $\leqslant$ is defined such that
(i) (Reflexive law) $x \prec x$
(ii) (Transitive law) $x \prec y$ and $y \prec z$ imply $x \prec z$
(iii) (Anti-reflexive law) $x \prec y$ and $y \prec x$ imply $x = y$.
Then $(X, \prec)$ is called a partially ordered set.

If in $(X, \prec)$, for any $x, y \in X$, either $x \prec y$ or $y \prec x$ holds, then X is called a totally ordered set.

Let $\Lambda = (\Lambda, \prec)$ be a partially ordered set with respect to the ordering relation $\prec$. Λ is called a **directed set** if it satisfies:
(iv) Given two elements μ_1, μ_2, there exists an element ν such that $\mu_1 \prec \nu, \mu_2 \prec \nu$.

Any subset $\Lambda' \subset \Lambda$ is said to **have the common end** with Λ if for any $\lambda \in \Lambda$ there exists a $\lambda' \in \Lambda'$ such that $\lambda \prec \lambda'$.

A mapping $(\Lambda, \prec) \to X$ is called a **generalized sequence**, denoted $\{x_\lambda\}_{\lambda\in\Lambda}$ or simply $\{x_\lambda\}$. In case X is a metric space or a Hausdorff space, we may speak of the convergence of generalized sequences.

Definition 16.13. A generalized sequence (x_λ) in a topological space X is said to converge to x_0 if for any $U \in \mathcal{U}(x_0)$ there exists a $\lambda_0 \in \Lambda$ such that $\lambda \prec \lambda_0$ implies $x_\lambda \in U$. In a metric space, we may use the ε-neighborhood for U.

Exercise 142. In a topological space X, let $\mathcal{U}(x)$ be the system of neighborhoods of $x \in X$. For $U_1, U_2 \in \mathcal{U}(x)$ we write $U_1 \prec U_2$ if $U_2 \subset U_2$. Then prove that this is an order relation.

Solution. Enough to check (iv). We put $U_3 = U_1 \cap U_2$. Then $U_3 \in \mathcal{U}(x)$ and is contained in both U_1, U_2 whence (iv) follows.

Exercise 143. Let $\Lambda = \{\mathbb{N} - A | \mathbb{N} \supset A : \text{ finite set}\}$ and define the relation by inclusion: writing $\mathbb{N} - A_j = U_j$, we define $U_1 \prec U_2$ if $U_1 \subset U_2$, i.e.

$A_2 \subset A_1$. Then check that this is an order relation and that endowing discrete topology in $\mathbb{N}$, we may view Λ as a neighborhood of ∞.

Cf. Example 3.3.

Definition 16.14. A mapping $f : X \to Y$ between topological spaces is continuous at $x \in X$ if the directed set $f(\mathcal{U}(x))$ has the common end with the directed set $\mathcal{U}(f(x))$.

This means that for any $\mathcal{O}_Y \ni U \in \mathcal{U}_Y(f(x))$ there exists a $\mathcal{O}_X \ni V \in \mathcal{U}_X(x)$ such that $f(V) \subseteq U$ (or $f(V) \subset U$). By Theorem 3.2, this amounts to the ordinary definition of continuity.

Theorem 16.14. *Let X be a metric space and $x \in X$. Then the family of $1/n$-neighborhoods $\mathcal{U}^*(x) := \{V_{1/n}(x) \mid n \in \mathbb{N}\}$ has the common end with the system of neighborhoods $\mathcal{U}(x)$ of x, i.e. $\mathcal{U}^*(x)$ forms a fundamental system of neighborhoods of x.*

Proof. For any $U_x \in \mathcal{U}(x)$ there exists an ε-neighborhood $V_\varepsilon(x)$ of x. By Archimedes principle A (p. 36) there exists an $n \in \mathbb{N}$ such that $n > \frac{1}{\varepsilon}$ whence $V_{1/n}(x) \subset V_\varepsilon(x)$ or $V_\varepsilon(x) \prec V_{1/n}(x)$. $\square$

Fig. 16.1: Poincaré

Appendix A

Basics from linear algebra

In this appendix we assemble basic knowledge of linear algebra which is used in this book.

We continue from Exercise 4: An $m \times n$ matrix has a few expressions including the entries, column vector representation and row vector representation.

Exercise 144. With fundamental unit vectors $\boldsymbol{e}_j$ whose entries are 0 save for the jth one which is 1, we have

$$A\boldsymbol{e}_j = \boldsymbol{a}_j, \quad (A\boldsymbol{e}_j, \boldsymbol{e}_i) = a_{ij}. \tag{A.1}$$

Definition A.1. For two vectors $\boldsymbol{a} = {}^t(a_1, \cdots, a_n), \boldsymbol{b} = {}^t(b_1, \cdots, b_n) \in \mathbb{R}^n$, their **scalar product** (or inner product) is defined by

$$(\boldsymbol{a}, \boldsymbol{b}) = \boldsymbol{a} \cdot \boldsymbol{b} = \sum_{k=1}^{n} a_k b_k. \tag{A.2}$$

For two vectors $\boldsymbol{a} = {}^t(a_1, a_2, a_3), \boldsymbol{b} = {}^t(b_1, b_2, b_3) \in \mathbb{R}^3$ we define their **vector product** $\boldsymbol{a} \times \boldsymbol{b}$ by

$$\boldsymbol{a} \times \boldsymbol{b} = \begin{pmatrix} a_1 \\ a_2 \\ a_3 \end{pmatrix} \times \begin{pmatrix} b_1 \\ b_2 \\ b_3 \end{pmatrix} = \begin{pmatrix} \begin{vmatrix} a_2 & b_2 \\ a_3 & b_3 \end{vmatrix} \\ -\begin{vmatrix} a_1 & b_1 \\ a_3 & b_3 \end{vmatrix} \\ \begin{vmatrix} a_1 & b_1 \\ a_2 & b_2 \end{vmatrix} \end{pmatrix} \doteq \begin{pmatrix} a_2 b_3 - a_3 b_2 \\ -(a_1 b_3 - a_3 b_1) \\ a_1 b_2 - a_2 b_1 \end{pmatrix}. \tag{A.3}$$

Exercise 145. (i) For the scalar product of two vectors $\boldsymbol{a}, \boldsymbol{b} \in \mathbb{R}^n$, prove the following formula holds true:

$$(\boldsymbol{a}, \boldsymbol{b}) = |\boldsymbol{a}||\boldsymbol{b}| \cos\theta, \tag{A.4}$$

where θ is the angle formed by the two vectors, $0 \leq \theta \leq \pi$. For $n = 2, 3$, this amounts to the cosine theorem. What about for $n \geq 4$?
(ii) For two vectors $\boldsymbol{a}, \boldsymbol{b} \in \mathbb{R}^3$ we still have the cosine theorem (A.4). Prove that if the vector product $\boldsymbol{a} \times \boldsymbol{b} \neq \boldsymbol{o}$, then it is perpendicular to both $\boldsymbol{a}$ and $\boldsymbol{b}$ and that its length is the area of the parallelogram formed by $\boldsymbol{a}$ and $\boldsymbol{b}$:

$$|\boldsymbol{a} \times \boldsymbol{b}| = |\boldsymbol{a}||\boldsymbol{b}| \sin \theta, \tag{A.5}$$

where θ is the angle subtended by two vectors. Given another vector $\boldsymbol{c}$, prove that the volume formed by these three vectors is the absolute value of the determinant $\det(\boldsymbol{a}, \boldsymbol{b}, \boldsymbol{a})$ or the scalar triple product (D.1)

$$\boldsymbol{a} \times \boldsymbol{b} \cdot \boldsymbol{c}. \tag{A.6}$$

Confirm that if the sign of (A.6) is plus, then $\boldsymbol{a}, \boldsymbol{b}, \boldsymbol{c}$ form a right-handed system, i.e. the direction of $\boldsymbol{c}$ is the direction of a right-handed screw moves when $\boldsymbol{a}$ is rotated to $\boldsymbol{b}$ in counter-clockwise direction. If it is minus, then we have the left-handed system.

Solution. (ii) That $\boldsymbol{a} \times \boldsymbol{b} \cdot \boldsymbol{a} = 0$ and $\boldsymbol{a} \times \boldsymbol{b} \cdot \boldsymbol{b} = 0$ follow from direct computation. Or anteceding that they are determinants with two columns being equal, we may view this as the alternating property of a determinant. Denote the plane formed by them by P. If they are colinear, then the area is 0.

Substituting (A.4) in the Lagrange formulas (1.10), we find that

$$\begin{aligned} \sin \theta &= \frac{1}{|\boldsymbol{a}||\boldsymbol{b}|} \sqrt{(a_2 b_3 - a_3 b_2)^2 + (a_1 b_3 - a_3 b_1)^2 + (a_1 b_2 - a_2 b_1)^2} \\ &= \frac{1}{|\boldsymbol{a}||\boldsymbol{b}|} |\boldsymbol{a} \times \boldsymbol{b}|. \end{aligned} \tag{A.7}$$

As in spherical coordinates (Example 5.2, (ii)), let φ denote the angle which $\boldsymbol{c}$ forms with the positive direction of z-axis measured from the latter. Then $0 \leq \varphi \leq \pi$. There are two cases $0 \leq \varphi \leq \frac{\pi}{2}$ and $\frac{\pi}{2} \leq \varphi \leq \pi$. In the former case, $\boldsymbol{c}$ lies in the same side of P as the positive direction of z-axis and $\boldsymbol{a}, \boldsymbol{b}, \boldsymbol{c}$ form a right-handed system. In the latter case, they form a left-handed system and the value of $\cos \varphi < 0$. Provided that $\boldsymbol{a}, \boldsymbol{b}, \boldsymbol{c}$ form a parallelopiped, then since the volume is the area $\times$ the height $=$ $|\boldsymbol{a} \times \boldsymbol{b}||\boldsymbol{c}| \cos \varphi$ in the case $0 \leq \varphi \leq \frac{\pi}{2}$. This is nothing other than the inner product and (A.6) follows. In the other case, we deduce (A.6) in the same way, the only difference being that the value is negative.

Remark A.1. Let D be a rectangle with vertices at $(0, 0)$, $(a_1, 0)$, $(0, a_2)$, and (a_1, a_2) $(a_1, a_2 > 0)$. Let $A = (\boldsymbol{a}_1, \boldsymbol{a}_2)$, $\boldsymbol{a}_j = \begin{pmatrix} a_{1j} \\ a_{2j} \\ a_{3j} \end{pmatrix}$. Then by the linear map $f(\boldsymbol{x}) = A^t\boldsymbol{x}$, D is mapped onto the parallelogram $f(D)$ with adjacent sides $a_j\boldsymbol{a}_i$ which make the angle θ, say, so that the area of $f(D)$ is

$$|a_1\boldsymbol{a}_1||a_2\boldsymbol{a}_2|\sin\theta = c|a_1a_2|,$$

where

$$c = |\boldsymbol{a}_1||\boldsymbol{a}_2|\sin\theta = |\boldsymbol{a}_1 \times \boldsymbol{a}_2| \tag{A.8}$$

by (A.5).

Exercise 146. In (A.4) with $n = 3$ choose $\boldsymbol{o} \neq \boldsymbol{a} = {}^t(a_1, a_2, a_3)$ and $\boldsymbol{b} = \boldsymbol{e}_j$, $j = 1, 2, 3$. Suppose the angle subtended by $\boldsymbol{a}$ and the axes $\boldsymbol{e}_j$ be $\alpha_1, \alpha_2, \alpha_3$. Then show that

$$\cos\alpha_j = \frac{a_j}{|\boldsymbol{a}|} \tag{A.9}$$

and $\sum_{k=1}^{3}\cos^2\alpha_j = 1$. The vector ${}^t(\cos\alpha_1, \cos\alpha_2, \cos\alpha_3)$ is called the **directional cosines.** Also show that the case $n = 2$ leads to the polar coordinates: ${}^t(\cos\alpha_1, \cos\alpha_2) = {}^t\left(\cos\alpha_1, \sin\left(\frac{\pi}{2} - \alpha_2\right)\right)$.

We consider a square matrix $A = (a_{ij})$ of degree n, i.e. an $n \times n$ matrix. If there exists a matrix A' such that $AA' = A'A = E$, then A is unique (by Exercise 16) and is called the inverse of A, denoted A^{-1}. A matrix with the inverse is called a **regular matrix.**

We refer to Exercise 114 for the definition of a determinant $|A|$. We define the (i, j)**-cofactor** by

$$\tilde{a}_{ij} = (-1)^{i+j}D_{ij} \tag{A.10}$$

where D_{ij} is the **minor** of degree $n - 1$ which is obtained from the determinant $|A|$ of A by deleting the ith row and jth column. The cofactor expansion is the main ingredient for computation of the value of a determinant. Indeed, one may conventionally take the cofactor expansion as its definition.

Theorem A.1. *The* **cofactor expansion** *holds true.*
(i) *(expansion w.r.t. the jth column)*

$$|A| = \sum_{i=1}^{n} a_{ij}\tilde{a}_{ij}, \quad 1 \le j \le n \tag{A.11}$$

(ii) *(expansion w.r.t. the ith row)*

$$|A| = \sum_{j=1}^{n} a_{ij}\tilde{a}_{ij}, \quad 1 \le i \le n. \tag{A.12}$$

Let $\tilde{A} = {}^t(\tilde{a}_{ij})$ be the **cofactor matrix** of A.

Theorem A.2. *We have*

$$\tilde{A}A = A\tilde{A} = |A|E, \tag{A.13}$$

where $E = (\delta_{ij})$ is the identity matrix of degree n in (1.85).

This follows from Theorem A.1.

By Theorem A.2, we may give a one-line proof of Cramér's formula:

Theorem A.3. *Consider the system of equations*

$$A\boldsymbol{x} = \boldsymbol{b}, \tag{A.14}$$

where $\boldsymbol{x} = {}^t(x_1, \cdots, x_n), \boldsymbol{b} = {}^t(b_1, \cdots, b_n)$. Suppose $|A| \neq 0$. Then we have **Cramér's formula**

$$x_j = \frac{1}{|A|} \det(\boldsymbol{a}_1, \cdots, \boldsymbol{b}, \cdots, \boldsymbol{a}_n), \quad 1 \leq j \leq n \tag{A.15}$$

where $\boldsymbol{b}$ is in the jth column.

Proof. Multiplying (A.14) by $\tilde{A}$ from the left and using (A.13) leads to

$$\boldsymbol{x} = \frac{1}{|A|}\tilde{A}\boldsymbol{b}. \tag{A.16}$$

Indeed, $\frac{1}{|A|}\tilde{A} = A^{-1}$. □

Indeed, we may further simplify the proof by the following

Lemma A.1. *We have*

$$\tilde{A}\boldsymbol{b} = \begin{pmatrix} \det(\boldsymbol{b}, \cdots, \boldsymbol{a}_n) \\ \vdots \\ \det(\boldsymbol{a}_1, \cdots, \boldsymbol{b}) \end{pmatrix} \tag{A.17}$$

and in particular,

$$\tilde{A}\boldsymbol{a}_k = |A|\boldsymbol{e}_k. \tag{A.18}$$

Similarly, we denote

$$A = \begin{pmatrix} \boldsymbol{a}'_1 \\ \vdots \\ \boldsymbol{a}'_n \end{pmatrix}, \quad \text{i.e. } {}^t\boldsymbol{e}_i A = \boldsymbol{a}'_i.$$

Then

$${}^t\boldsymbol{b}\tilde{A} \;= \det\left(\begin{pmatrix} {}^t\boldsymbol{b} \\ \vdots \\ \boldsymbol{a}'_n \end{pmatrix}, \cdots, \det\begin{pmatrix} \boldsymbol{a}_1 \\ \vdots \\ {}^t\boldsymbol{b} \end{pmatrix}\right) \tag{A.19}$$

and in particular,

$$\boldsymbol{a}'_k\tilde{A} = |A|\boldsymbol{e}'_k. \tag{A.20}$$

Proof. The jth row of $\tilde{A}\boldsymbol{b}$ is

$$\sum_{i=1}^{n} b_i \tilde{a}_{ij} \tag{A.21}$$

which is the cofactor expansion w.r.t. the jth column of $\det(\boldsymbol{a}_1, \cdots, \boldsymbol{b}, \cdots, \boldsymbol{a}_n)$, $1 \le j \le n$ with $\boldsymbol{b}$ in the jth column.

The ith column of ${}^t\boldsymbol{b}\tilde{A} = {}^t({}^t\tilde{A}\boldsymbol{b})$ is

$$\sum_{j=1}^{n} b_j \tilde{a}_{ij} \tag{A.22}$$

which is the cofactor expansion w.r.t. the ith row of $\det \begin{pmatrix} \boldsymbol{a}'_1 \\ \vdots \\ {}^t\boldsymbol{b} \\ \vdots \\ \boldsymbol{a}'_n \end{pmatrix}$, $1 \le i \le n$ with $\boldsymbol{b}$ in the ith row. □

Now Theorem A.3 is the same as (A.17). Theorem A.2 follows from (A.18) and (A.20).

Theorem A.4. *For a square matrix A of degree n to be regular it is necessary and sufficient that*
(i) $|A| \neq 0$.
(ii) *The system of equations $A\boldsymbol{x} = \boldsymbol{o}$ has only the trivial solution $\boldsymbol{x} = \boldsymbol{o}$.*
(iii) $\operatorname{rank} A = n$.
(iv) *All n rows of A are linearly independent.*
(v) *All n columns of A are linearly independent.*

Definition A.2. Let $A = (a_{ij})$ be a square matrix of degree n with entries in $\mathbb{R}$ or $\mathbb{C}$, which we denote by K. A complex number $\lambda \in \mathbb{C}$ is called an **eigenvalue** of A if there exists a $\boldsymbol{o} \neq \boldsymbol{x} \in \mathbb{C}^n$ (occasionally we may choose $\lambda \in \mathbb{R}, \boldsymbol{x} \in \mathbb{R}^n$) such that

$$A\boldsymbol{x} = \lambda \boldsymbol{x}. \tag{A.23}$$

All the non-zero vectors satisfying (A.23) are called **eigenvectors** belonging to λ. Since (A.16) amounts to $(A - \lambda E)\boldsymbol{x} = \boldsymbol{o}$, it follows from Cramér's rule that eigenvalues are the solutions of the **eigenequation**

$$|A - \lambda E| = 0. \tag{A.24}$$

The set of all eigenvectors belonging to λ and $\boldsymbol{o}$, being a **solution space** of the system (A.23) of equations, form a subspace $E_A(\lambda)$ of K^n called the eigenspace belonging to λ:

$$E_A(\lambda) = \{\boldsymbol{x} | (A - \lambda E)\boldsymbol{x} = \boldsymbol{o}\}. \tag{A.25}$$

Theorem A.5. *Suppose the eigen polynomial in* (A.24) *decomposes as*

$$|A - \lambda E| = (-1)^n (\lambda - \lambda_1)^{m_1} . \cdots (\lambda - \lambda_k)^{m_k}, \quad \sum_{j=1}^{k} m_j = n \tag{A.26}$$

and that the dimension of the corresponding eigenspaces are

$$\dim E_A(\lambda_j) = m_i, \quad 1 \le j \le k. \tag{A.27}$$

Then A is diagonizable, i.e. there exists a regular matrix $P \in \mathrm{GL}_n(K)$ such that

$$P^{-1}AP = \begin{pmatrix} \lambda_1 & 0 & \cdots & 0 \\ 0 & \lambda_1 & \cdots & 0 \\ 0 & \cdots & \cdots & 0 \\ 0 & \cdots & \lambda_k & 0 \\ 0 & \cdots & 0 & \lambda_k \end{pmatrix}. \tag{A.28}$$

Theorem A.6. *Let A be a real symmetric matrix (i.e. ${}^tA = A$) of degree n: $A = (a_{ij})_{1 \le i,j \le n}$. For any $\boldsymbol{x} \in \mathbb{R}^n$ the following are equivalent.*
(i) *$A[\boldsymbol{x}] = {}^t\boldsymbol{x}A\boldsymbol{x}$ is positive [negative] definite, i.e. for any $\boldsymbol{o} \ne \boldsymbol{x} \in \mathbb{R}^n$, $A[\boldsymbol{x}] > 0$ $[A[\boldsymbol{x}] < 0]$.*
(ii) *All eigenvalues are positive [negative].*
(iii) *Let $|A_r|$ denote the principal minor of degree r:*

$$|A_r| = \begin{vmatrix} a_{11} & \cdots & a_{1r} \\ \vdots & \vdots & \vdots \\ a_{r1} & \cdots & a_{rr} \end{vmatrix}. \tag{A.29}$$

Then $|A_r| > 0$ $[(-1)^r |A_r| > 0\,]$ for all $1 \le r \le n$.

Appendix B

Diagonalization of matrices

Example B.1. Let $A_1 = \frac{1}{2}A$, where

$$A = \begin{pmatrix} 0 & 1 & 1 \\ 1 & 0 & 1 \\ 1 & 1 & 0 \end{pmatrix}, \tag{B.1}$$

which is the topological matrix of cyclopropenyl radical [Kitajima and Kanemitsu (2012)]. Given the sequence $\boldsymbol{x}_n$ by

$$\boldsymbol{x}_n = \begin{pmatrix} a_n \\ b_n \\ c_n \end{pmatrix} = A_1^n \boldsymbol{x}_0$$

we find an explicit expression for it. Taking the limit as $n \to \infty$, we find that

$$\lim_{n\to\infty} \boldsymbol{x}_n = \frac{1}{3}(a_0 + b_0 + c_0) \begin{pmatrix} 1 \\ 1 \\ 1 \end{pmatrix}. \tag{B.2}$$

First we find eigenvalues of A.

$$
\begin{aligned}
0 = |A - \lambda E| &= \begin{vmatrix} -\lambda & 1 & 1 \\ 1 & -\lambda & 1 \\ 1 & 1 & -\lambda \end{vmatrix} \\
&\overset{1st\ row+2nd\ row+3rd\ row}{=} \begin{vmatrix} 2-\lambda & 2-\lambda & 2-\lambda \\ 1 & -\lambda & 1 \\ 1 & 1 & -\lambda \end{vmatrix} \\
&= (2-\lambda) \begin{vmatrix} 1 & 1 & 1 \\ 1 & -\lambda & 1 \\ 1 & 1 & -\lambda \end{vmatrix} \\
&\overset{2nd\ row+1st\ row\times(-1)}{=} (2-\lambda) \begin{vmatrix} 1 & 1 & 1 \\ 0 & -1-\lambda & 0 \\ 1 & 1 & -\lambda \end{vmatrix} \\
&= -(\lambda-2)(\lambda+1)^2,
\end{aligned}
$$

whence $\lambda = 2, -1, -1$. We may easily find the eigenspaces

$$
E_A(-1) = \mathbb{R}\begin{pmatrix} -1 \\ 1 \\ 0 \end{pmatrix} \oplus \mathbb{R}\begin{pmatrix} -1 \\ 0 \\ 1 \end{pmatrix}, \quad E_A(2) = \mathbb{R}\begin{pmatrix} 1 \\ 1 \\ 1 \end{pmatrix}, \tag{B.3}
$$

whence by $P = \begin{pmatrix} -1 & -1 & 1 \\ 1 & 0 & 1 \\ 0 & 1 & 1 \end{pmatrix}$, we may diagonalize A as

$$
P^{-1}AP = \begin{pmatrix} -1 & 0 & 0 \\ 0 & -1 & 0 \\ 0 & 0 & 2 \end{pmatrix} =: D. \tag{B.4}
$$

Hence for $A_1 = \frac{1}{2}A$

$$
P^{-1}A_1P = \begin{pmatrix} -\frac{1}{2} & 0 & 0 \\ 0 & -\frac{1}{2} & 0 \\ 0 & 0 & 1 \end{pmatrix} \tag{B.5}
$$

and

$$
A_1^n = P \begin{pmatrix} \left(-\frac{1}{2}\right)^n & 0 & 0 \\ 0 & \left(-\frac{1}{2}\right)^n & 0 \\ 0 & 0 & 1 \end{pmatrix} P^{-1}. \tag{B.6}
$$

Since

$$P^{-1} = \frac{1}{3}\begin{pmatrix} -1 & 2 & -1 \\ -1 & -1 & 2 \\ 1 & 1 & 1 \end{pmatrix}, \tag{B.7}$$

we may express the nth term explicitly as a function in n.

Just to calculate the limits, one may take the limit in (B.6) to deduce that

$$\lim_{n\to\infty} A_1^n = P\begin{pmatrix} 0 & 0 & 0 \\ 0 & 0 & 0 \\ 0 & 0 & 1 \end{pmatrix} P^{-1} = \frac{1}{3}\begin{pmatrix} 1 & 1 & 1 \\ 1 & 1 & 1 \\ 1 & 1 & 1 \end{pmatrix}, \tag{B.8}$$

whence (B.2) follows.

However, since A is a circulant, we may also apply the theory of circulants and use the Fourier matrix to diagonalize it. By Theorem C.1, we may use the Fourier matrix $F = \begin{pmatrix} 1 & 1 & 1 \\ 1 & \omega^2 & \omega \\ 1 & \omega & \omega^2 \end{pmatrix}$ to get another diagonalization

$$F^{-1}AF = \begin{pmatrix} p_\gamma(1) & 0 & 0 \\ 0 & p_\gamma(\omega) & 0 \\ 0 & 0 & p_\gamma(\omega^2) \end{pmatrix}, \tag{B.9}$$

where $p_\gamma(z) = z + z^2$. Hence $p_\gamma(1) = 2, p_\gamma(\omega) = -1, p_\gamma(\omega^2) = -1$ and (B.4) and (B.9) are equivalent.

Transition to (B.6) may be explained as follows.

Eigenvalues are $-1, -1, 2$ (-1 being a double root). $\dim E_A(-1) = 2$ multiplicity of -1 and $\dim E_A(2) = 1$ multiplicity of 2. Hence it is diagonalizable. Put

$$\boldsymbol{p}_1 = \begin{pmatrix} -1 \\ 1 \\ 0 \end{pmatrix}, \quad \boldsymbol{p}_2 = \begin{pmatrix} -1 \\ 0 \\ 1 \end{pmatrix} \in E_A(-1), \quad \boldsymbol{p}_1 = \begin{pmatrix} 1 \\ 1 \\ 1 \end{pmatrix} \in E_A(2)$$

and $P = (\boldsymbol{p}_1, \boldsymbol{p}_2, \boldsymbol{p}_3)$ as above. Then by block multiplication we have

$$\begin{aligned} AP &= A(\boldsymbol{p}_1, \boldsymbol{p}_2, \boldsymbol{p}_3) = (A\boldsymbol{p}_1, A\boldsymbol{p}_2, A\boldsymbol{p}_3) = (-\boldsymbol{p}_1, -\boldsymbol{p}_2, 2\boldsymbol{p}_3) \\ &= (\boldsymbol{p}_1, \boldsymbol{p}_2, \boldsymbol{p}_3)\begin{pmatrix} -1 & 0 & 0 \\ 0 & -1 & 0 \\ 0 & 0 & 2 \end{pmatrix} = P\begin{pmatrix} -1 & 0 & 0 \\ 0 & -1 & 0 \\ 0 & 0 & 2 \end{pmatrix}. \end{aligned}$$

Since P is regular (and indeed, it is easy to find the inverse matrix in (B.7)), multiplying P^{-1} from the left, we deduce (B.6).

Remark B.1. Consider the extremal values of the function $f(\boldsymbol{x}) = A_1[\boldsymbol{x}] = (A_1\boldsymbol{x}, \boldsymbol{x}) = xy + yz + zx$. The Hess matrix is A. Since the eigenvalues are $-1, -1, 2$ and there is no extremal values. We may also confirm this by putting $A[\boldsymbol{x}]$ into a canonical form. For this we orthogonalize P into $P_1 \in O_3$ and we still have ${}^tP_1AP_1 = P_1^{-1}AP_1 = D$. Hence

$$(A\boldsymbol{x}, \boldsymbol{x}) = {}^t\boldsymbol{x}A\boldsymbol{x} = {}^t\boldsymbol{x}P_1D({}^tP_1\boldsymbol{x}) = (D\boldsymbol{y}, \boldsymbol{y}) = -y_1^2 - y_2^2 + 2y_3^2 \qquad \text{(B.10)}$$

is indefinite, where ${}^tP_1\boldsymbol{x} = \boldsymbol{y} = {}^t(y_1, y_2, y_3)$.

Appendix C

Circulants

The following is a concise version of [Davis (1979)].

Definition C.1. For a germ

$$\gamma = (c_1, \ldots, c_N) \in \mathbb{C}^N, \tag{C.1}$$

we call

$$C = \operatorname{circ}\gamma = \operatorname{circ}(c_1, \ldots, c_N) = \begin{pmatrix} c_1 & c_2 & \cdots & c_N \\ c_N & c_1 & \cdots & c_{N-1} \\ \cdots & \cdots & \cdots & \cdots \\ c_2 & c_3 & \cdots & c_1 \end{pmatrix} \tag{C.2}$$

a **circulant matrix** (or a **circulant**). Also, putting

$$\pi = \begin{pmatrix} e'_2 \\ e'_3 \\ \vdots \\ e'_1 \end{pmatrix},$$

we call it the **shift forward matrix** (which plays a fundamental role in the theory of circulant matrices), where

$$e'_k = (\delta_{k1}, \cdots, \delta_{kn}), \quad 1 \le k \le n \tag{C.3}$$

are basic unit *row* vectors (π is for **push** forward). Using this, we conclude that $C = c_1 E + c_2 \pi + \cdots + c_N \pi^{N-1}$. Viewing this as a polynomial, we call

$$p_\gamma(z) = c_1 + c_2 z + \cdots + c_N z^{N-1} \tag{C.4}$$

a **representor** of C.

Note that $n \times n$ circulant matrices are **matrix representations of the group ring** over $\mathbb{C}$ or $\mathrm{GF}(q)$ as the case may be, of the underlying cyclic group ([Willard (1978)]). Here the matrix representation means a homomorphism into the multiplicative group of square matrices.

Letting $\zeta = \zeta_N$ be a primitive Nth root of 1, we define the **Fourier matrix** F by means of its conjugate transpose F^*:

$$F^* = \frac{1}{\sqrt{N}}\left(\zeta^{(i-1)(j-1)}\right) = \frac{1}{\sqrt{N}}\begin{pmatrix} 1 & 1 & \cdots & 1 \\ 1 & \zeta & \cdots & \zeta^{N-1} \\ \cdots & \cdots & \cdots & \cdots \\ 1 & \zeta^{N-1} & \cdots & \zeta^{(N-1)(N-1)} \end{pmatrix}, \tag{C.5}$$

whence

$$F = \frac{1}{\sqrt{N}}\left(\zeta^{-(i-1)(j-1)}\right) = \frac{1}{\sqrt{N}}\begin{pmatrix} 1 & 1 & \cdots & 1 \\ 1 & \zeta^{N-1} & \cdots & \zeta \\ \cdots & \cdots & \cdots & \cdots \\ 1 & \zeta & \cdots & \zeta^{N-1} \end{pmatrix}. \tag{C.6}$$

Theorem C.1. ([Davis (1979), Theorem 3.2.2, p. 72]) *Any circulant matrix C can be diagonalized as*

$$C = F^*\Lambda F \tag{C.7}$$

by the Fourier matrix F, where

$$\Lambda = \Lambda_C = \begin{pmatrix} p_\gamma(1) & 0 & \cdots & 0 \\ 0 & p_\gamma(\zeta) & \cdots & 0 \\ \cdots & \cdots & \cdots & \cdots \\ 0 & \cdots & 0 & p_\gamma\left(\zeta^{N-1}\right) \end{pmatrix}. \tag{C.8}$$

Thus, in particular, the eigenvalues of C are $p_\gamma(1)$, $p_\gamma(\zeta)$, $\dots$, $p_\gamma(\zeta^{n-1})$.

Corollary C.1. ([Davis (1979), (3.2.14), p. 75])

$$\det C = \det(\mathrm{circ}\,\gamma) = \prod_{j=0}^{N-1} p_\gamma(\zeta^j), \tag{C.9}$$

where $p_\gamma(z)$ is the representor of $\mathrm{circ}\,\gamma$ *defined by* (C.4).

Proof. Proof of Theorem C.1 follows verbatim to that of Theorem 3.2.2 in [Davis (1979)] since it is a consequence of Theorem 3.2.1 asserting the diagonalization of π:

$$\pi = F^*\Omega F, \tag{C.10}$$

where

$$\Omega = \Omega_C = \begin{pmatrix} 1 & 0 & \cdots & 0 \\ 0 & \zeta & \cdots & 0 \\ \cdots & \cdots & \cdots & \cdots \\ 0 & \cdots & 0 & \zeta^{N-1} \end{pmatrix}. \tag{C.11}$$

□

Fig. C.1: Riemann

Appendix D

Formulas for vector operations

Lemma D.1. *In the scalar triple product, the dot and the cross may be changed, i.e.*

$$(\boldsymbol{a} \times \boldsymbol{b}) \cdot \boldsymbol{c} = \boldsymbol{a} \cdot (\boldsymbol{b} \times \boldsymbol{c}). \tag{D.1}$$

This is because the both sides are $\det(\boldsymbol{a}, \boldsymbol{b}, \boldsymbol{c})$. It is also written as $[\boldsymbol{a}, \boldsymbol{b}, \boldsymbol{c}]$ or simply as $[\boldsymbol{abc}]$.

Theorem D.1. *The vector triple product satisfies the relation*

$$(\boldsymbol{a} \times \boldsymbol{b}) \times \boldsymbol{c} = (\boldsymbol{a} \cdot \boldsymbol{c})\boldsymbol{b} - (\boldsymbol{b} \cdot \boldsymbol{c})\boldsymbol{a} = \boldsymbol{b}(\boldsymbol{a} \cdot \boldsymbol{c}) - \boldsymbol{a}(\boldsymbol{b} \cdot \boldsymbol{c}). \tag{D.2}$$

Proof. Note that $(\boldsymbol{a} \times \boldsymbol{b}) \times \boldsymbol{c}, \boldsymbol{a}, \boldsymbol{b}, \boldsymbol{a} \times \boldsymbol{b}$ are linearly dependent since in $\mathbb{R}^3$ only three vectors can be linearly independent. But we may suppose that $\boldsymbol{a}, \boldsymbol{b}, \boldsymbol{a} \times \boldsymbol{b}$ are linearly independent (otherwise the assertion is trivial), and let $\boldsymbol{b}' = \boldsymbol{b} - \frac{(\boldsymbol{a}, \boldsymbol{b})}{|\boldsymbol{a}|^2}\boldsymbol{a}$. Then $\boldsymbol{a}, \boldsymbol{b}', \boldsymbol{a} \times \boldsymbol{b} = \boldsymbol{a} \times \boldsymbol{b}'$ are vertical vectors and

$$(\boldsymbol{a} \times \boldsymbol{b}) \times \boldsymbol{a} = |\boldsymbol{a}|^2 \boldsymbol{b}', \quad \boldsymbol{b}' \times (\boldsymbol{a} \times \boldsymbol{b}) = |\boldsymbol{b}'|^2 \boldsymbol{a}. \tag{D.3}$$

Now there exist λ, μ, ν uniquely such that $\boldsymbol{c} = \lambda \boldsymbol{a} + \mu \boldsymbol{b}' + \nu(\boldsymbol{a} \times \boldsymbol{b})$, then $\boldsymbol{a} \cdot \boldsymbol{c} = \lambda |\boldsymbol{a}|^2$, $\boldsymbol{b}' \cdot \boldsymbol{c} = \mu |\boldsymbol{b}'|^2$. By Lemma D.1,

$$\begin{aligned}
\{(\boldsymbol{a} \times \boldsymbol{b}) \times \boldsymbol{c}\} \cdot \boldsymbol{a} &= (\boldsymbol{a} \times \boldsymbol{b}) \cdot (\boldsymbol{c} \times \boldsymbol{a}) \\
&= (\boldsymbol{a} \times \boldsymbol{b}) \cdot \{-\mu(\boldsymbol{a} \times \boldsymbol{b}) + \nu \boldsymbol{b}'\} \\
&= -\mu |\boldsymbol{a} \times \boldsymbol{b}'|^2 \;=\; -\mu |\boldsymbol{a}|^2 |\boldsymbol{b}'|^2, && \text{(D.4)} \\
\{(\boldsymbol{a} \times \boldsymbol{b}) \times \boldsymbol{c}\} \cdot \boldsymbol{b}' &= (\boldsymbol{a} \times \boldsymbol{b}) \cdot (\boldsymbol{c} \times \boldsymbol{b}') \\
&= (\boldsymbol{a} \times \boldsymbol{b}) \cdot \{-\lambda(\boldsymbol{a} \times \boldsymbol{b}) - \nu \boldsymbol{a}\} \\
&= \lambda |\boldsymbol{a} \times \boldsymbol{b}'|^2 \;=\; \lambda |\boldsymbol{a}|^2 |\boldsymbol{b}'|^2, && \text{(D.5)} \\
\{(\boldsymbol{a} \times \boldsymbol{b}) \times \boldsymbol{c}\} \cdot (\boldsymbol{a} \times \boldsymbol{b}) &= 0. && \text{(D.6)}
\end{aligned}$$

Hence we know that

$$\begin{aligned}(\boldsymbol{a}\times\boldsymbol{b})\times\boldsymbol{c} &= -\mu|\boldsymbol{b}'|^2\boldsymbol{a}+\lambda|\boldsymbol{a}|^2\boldsymbol{b}' \\ &= -(\boldsymbol{b}'\cdot\boldsymbol{c})\boldsymbol{a}+(\boldsymbol{a}\cdot\boldsymbol{c})\boldsymbol{b}' \\ &= -(\boldsymbol{b}\cdot\boldsymbol{c})\boldsymbol{a}+(\boldsymbol{a}\cdot\boldsymbol{c})\boldsymbol{b}. \end{aligned} \tag{D.7}$$

□

Exercise 147. Prove that

$$\{(\boldsymbol{a}\times\boldsymbol{b})\}\cdot\{(\boldsymbol{c}\times\boldsymbol{d})\} = (\boldsymbol{a}\cdot\boldsymbol{c})(\boldsymbol{b}\cdot\boldsymbol{d})-(\boldsymbol{a}\cdot\boldsymbol{d})(\boldsymbol{b}\cdot\boldsymbol{c}), \tag{D.8}$$

whence derive the Lagrange identity.

Solution. By Lemma D.1, the left-hand side is $\{(\boldsymbol{a}\times\boldsymbol{b})\times\boldsymbol{c}\}\cdot\boldsymbol{d})$ for which we substitute (D.2) to deduce that it is

$$\{(\boldsymbol{a}\cdot\boldsymbol{c})\boldsymbol{b}-(\boldsymbol{b}\cdot\boldsymbol{c})\boldsymbol{a}\}\cdot\boldsymbol{d},$$

whence (D.8).

The Lagrange identity follows by putting $\boldsymbol{c}=\boldsymbol{a},\boldsymbol{d}=\boldsymbol{b}$. But the argument is tautology.

Exercise 148. Suppose $\boldsymbol{a},\boldsymbol{b},\boldsymbol{c}$ are linearly independent 3-dimensional real vectors (i.e. not coplanar and $[\boldsymbol{abc}]\neq 0$). Then prove that any vector $\boldsymbol{d}$ may be written as

$$\boldsymbol{d}=\frac{[\boldsymbol{bcd}]\boldsymbol{a}+[\boldsymbol{cad}]\boldsymbol{b}+[\boldsymbol{abd}]\boldsymbol{c}}{[\boldsymbol{abc}]}. \tag{D.9}$$

Solution. Applying (D.2) with $\boldsymbol{c}$ as $\boldsymbol{c}\times\boldsymbol{d}$, we have

$$(\boldsymbol{a}\times\boldsymbol{b})\times(\boldsymbol{c}\times\boldsymbol{d}) = (\boldsymbol{a}\cdot\boldsymbol{c}\times\boldsymbol{d})\boldsymbol{b}-(\boldsymbol{b}\cdot\boldsymbol{c}\times\boldsymbol{d})\boldsymbol{a} = [\boldsymbol{acd}]\boldsymbol{b}-[\boldsymbol{bcd}]\boldsymbol{a}. \tag{D.10}$$

Then viewing it as $-(\boldsymbol{c}\times\boldsymbol{d})\times(\boldsymbol{a}\times\boldsymbol{b})$, we deduce that

$$(\boldsymbol{a}\times\boldsymbol{b})\times(\boldsymbol{c}\times\boldsymbol{d}) = -(\boldsymbol{c}\cdot(\boldsymbol{a}\times\boldsymbol{b}))\boldsymbol{d}+(\boldsymbol{d}\cdot(\boldsymbol{a}\times\boldsymbol{b}))\boldsymbol{c} = -[\boldsymbol{cab}]\boldsymbol{d}+[\boldsymbol{dab}]\boldsymbol{c}. \tag{D.11}$$

Equating these and solving for $\boldsymbol{d}$, we con conclude (D.9).

Example D.1. Find the integral when $c > b^2 > 0$

$$I=\int (x^2+a)\log(b+\sqrt{x^2+c})\,\mathrm{d}x = \frac{1}{3}(x^3+3ax)\log(b+\sqrt{x^2+c}) \tag{D.12}$$

$$-\frac{1}{9}x^3+\frac{b}{6}x\sqrt{x^2+c}+\frac{b}{3}(3a+b^2-\frac{3}{2}c)\log(x+\sqrt{x^2+c})$$
$$+\frac{3a+b^2-c}{3}$$
$$\times\left(-x+\frac{1}{\sqrt{c-b^2}}\arctan\frac{x}{\sqrt{c-b^2}}-\sqrt{c-b^2}\arctan\frac{b}{\sqrt{c-b^2}}\frac{x}{\sqrt{x^2+c}}\right)+C.$$

Proof. Noting that

$$(\log(b + \sqrt{x^2+c}))' = \frac{\sqrt{x^2+c}-b}{x^2+c-b^2}\frac{x}{\sqrt{x^2+c}},$$

we write

$$x^2 + a = x^2 + \frac{c-b^2}{3} + \frac{3a+b^2-c}{3}$$

to deduce that

$$I = I_1 + \frac{3a+b^2-c}{3} I_2, \tag{D.13}$$

say, where

$$I_2 = \int \log(b + \sqrt{x^2+c})\,\mathrm{d}x. \tag{D.14}$$

By integration by parts, we have

$$I_1 = \frac{x}{3}(x^2+c-b^2)\log(b+\sqrt{x^2+c}) - \frac{1}{3}\int x(x^2+c-b^2)\frac{\sqrt{x^2+c}-b}{x^2+c-b^2}\frac{x}{\sqrt{x^2+c}}\,\mathrm{d}x.$$

The last integral is

$$-\frac{1}{3}\int x^2\left(1-\frac{b}{\sqrt{x^2+c}}\right)\mathrm{d}x = -\frac{1}{9}x^3 + \frac{b}{3}\int \frac{x^2+c-c}{\sqrt{x^2+c}}\,\mathrm{d}x$$
$$= -\frac{1}{9}x^3 + \frac{b}{3}\left(\int \sqrt{x^2+c}\,\mathrm{d}x - c\int \frac{1}{\sqrt{x^2+c}}\,\mathrm{d}x\right).$$

Using the well-known formulas

$$\int \frac{1}{\sqrt{x^2+A}}\,\mathrm{d}x = \log|x+\sqrt{x^2+A}| + C, \tag{D.15}$$
$$\int \sqrt{x^2+A}\,\mathrm{d}x = \frac{1}{2}\left(x\sqrt{x^2+A} + A\log|x+\sqrt{x^2+A}|\right) + C,$$

we find that

$$I_1 = \frac{x}{3}(x^2+c-b^2)\log(b+\sqrt{x^2+c}) - \frac{1}{9}x^3 \tag{D.16}$$
$$+ \frac{b}{6}\left(x\sqrt{x^2+c} - c\log(x+\sqrt{x^2+c})\right) + C.$$

Now by integration by parts,

$$I_2 = x\log(b+\sqrt{x^2+c}) - \int x\frac{1}{b+\sqrt{x^2+c}}\frac{x}{\sqrt{x^2+c}}\,\mathrm{d}x. \tag{D.17}$$

We rationalize the denominator to write the integrand as

$$\frac{x^2}{x^2+c-b^2}\frac{\sqrt{x^2+c}-b}{\sqrt{x^2+c}} = \left(1+\frac{b^2-c}{x^2+c-b^2}\right)\left(1-\frac{b}{\sqrt{x^2+c}}\right)$$
$$= 1-\frac{c-b^2}{x^2+c-b^2}-\frac{b}{\sqrt{x^2+c}}-\frac{b(b^2-c)}{(x^2+c-b^2)\sqrt{x^2+c}}$$

after applying Euclidean division. Hence applying (D.15) and

$$\int \frac{1}{x^2+a^2}\,\mathrm{d}x = \frac{1}{a}\arctan\frac{x}{a}+C \qquad (a\neq 0) \tag{D.18}$$

we find that

$$I_2 = x\log(b+\sqrt{x^2+c})-x+\sqrt{c-b^2}\arctan\frac{x}{\sqrt{c-b^2}}$$
$$+\,b\log(x+\sqrt{x^2+c})-b(c-b^2)I_3, \tag{D.19}$$

say, where

$$I_3 = \int \frac{1}{(x^2+c-b^2)\sqrt{x^2+c}}\,\mathrm{d}x. \tag{D.20}$$

We apply the formula whose proof is given separately:

$$\int \frac{1}{(x^2+A)\sqrt{x^2+B}}\,\mathrm{d}x = \frac{1}{\sqrt{A(B-A)}}\arctan\left(\sqrt{\frac{B-A}{A}}\frac{x}{\sqrt{x^2+B}}\right)+C, \tag{D.21}$$

where $0<A<B$. Hence

$$I_3 = \int \frac{1}{(x^2+c-b^2)\sqrt{x^2+c}}\,\mathrm{d}x = \frac{1}{b\sqrt{c-b^2}}\arctan\left(\frac{b}{\sqrt{c-b^2}}\frac{x}{\sqrt{x^2+c}}\right)+C. \tag{D.22}$$

Substituting (D.22) in (D.19), we find I_2 which we substitute in (D.13) completes the proof. □

Proof of (D.21). First rationalize the denominator and decompose into partial fractions to obtain

$$\frac{1}{(x^2+A)\sqrt{x^2+B}} = \frac{1}{B-A}\left(\frac{\sqrt{x^2+B}}{x^2+A}-\frac{1}{\sqrt{x^2+B}}\right).$$

Now rationalize the numerator of $\frac{\sqrt{x^2+B}}{x^2+A}$ and transform it in terms of the expression x^2+B. Then we obtain

$$\frac{\sqrt{x^2+B}}{x^2+A} = \frac{B}{A}\frac{1}{1+\left(\sqrt{\frac{B-A}{A}}\frac{x}{\sqrt{x^2+B}}\right)^2}\frac{1}{\sqrt{x^2+B}}.$$

From this we expect $\arctan\left(\sqrt{\frac{B-A}{A}}\frac{x}{\sqrt{x^2+B}}\right)$ and differentiate it. The result is

$$\left(\arctan\left(\sqrt{\frac{B-A}{A}}\frac{x}{\sqrt{x^2+B}}\right)\right)' = \sqrt{A(B-A)}\frac{1}{(x^2+A)\sqrt{x^2+B}}, \tag{D.23}$$

whence the result.

Fig. D.1: Weierstrass

Appendix E

Alternating algebras

It will turn out that in order to grasp the core meaning of differential forms, it is useful to introduce the notion of alternating algebras and graded modules.

Let R be a commutative ring with unity and let M be an R-module. Let $T^k(M)$ be the k**-ple tensor product**

$$T^k(M) = M \otimes \cdots \otimes M \tag{E.1}$$

which we take for granted and understand that its elements are finite linear combinations with R-coefficients, so that we may think of $x_1 \otimes \cdots \otimes x_k$ as its typical elements. In $T^k(M)$ we introduce a relation: $x \sim 0$ if they differ by elements $x_1 \otimes \cdots \otimes x_k$ in which $x_i = x_j$ for some $i \neq j$. Or letting $\mathfrak{a}_k$ be the submodule consisting of those elements $x_1 \otimes \cdots \otimes x_k$ in which at least two components coincide, we define the R-module (of k-forms)

$$\Lambda^k(M) = T^k(M)/\sim = T^k(M)/\mathfrak{a}_k; \quad \pi_\sim : T^k(M) \to \Lambda^k(M) \tag{E.2}$$

and write $\pi_\sim(x_1 \otimes \cdots \otimes x_k) = x_1 \wedge \cdots \wedge x_k$. Hence $x_1 \wedge \cdots \wedge x_k$ are alternating. We form the tensor algebra, resp. the alternating algebra by

$$T(M) = \sum_{k=0}^{\infty} T^k(M); \quad \Lambda(M) = \sum_{k=0}^{\infty} \Lambda^k(M) = \sum_{k=0}^{\infty} T^k(M)/\mathfrak{a}_k. \tag{E.3}$$

Lemma E.1.

$$\Lambda(M_1 \oplus M_2) \simeq M_1 \otimes M_2. \tag{E.4}$$

Theorem E.1. *Suppose M is a free R-module with basis $\{u_1, \cdots, u_n\}$: $M = Ru_1 \oplus \cdots \oplus Ru_n$. Then for $k > n$, $\Lambda^k(M) = 0$ and for $k \leq n$, $\Lambda^k(M)$ is a free R-module of rank $\binom{n}{k}$ with basis $\{u_{i_1}, \cdots, u_{i_k}\}$, $i_1 < \cdots < i_k$. $\Lambda(M)$ is a free R-module of rank 2^n.*

Proof. By Lemma E.1,

$$\Lambda(M) \simeq \Lambda(Ru_1) \otimes \cdots \otimes \Lambda(Ru_n).$$

Since $\Lambda(Ru_j) = R1 \oplus Ru_j$, the result follows. □

Let $R = C^r(\Omega)$ and $M = Ru_1 \oplus \cdots \oplus Ru_n$ be a free R-module of rank n. Then the k-ple tensor product $T^k(M)$ is $\{\omega^k = \sum_I a_I u_I\}$ with $I = \{i_1, \cdots, i_k\} \subset \{1, \cdots, n\}$. With $\mathfrak{a}_k$ as above and $\mathrm{d}x_j$, the space of k-forms is

$$\Lambda^k(M) = T^k(M)/\mathfrak{a}_k = \left\{ \sum_I a_I \mathrm{d}x_I \right\}, \tag{E.5}$$

where $\mathrm{d}x_j$'s have alternating properties.

Bibliography

Apostol, T. M. (1957), *Mathematical analysis*, Addison-Wesley Publishing Company, Inc., Reading, Mass.

Berndt, B. C. (1975). Character analogues of the Poisson and Euler-Maclaurin summation formula with applications, *J. Number Theory* **7**, 413–445.

Borević, Z. I. and Shafarević, I. R. (1966). *Number theory*, McGraw-Hill, Academic Press, Boston, MA.

Bourbaki, N. (1989). *General Topology: Chapters 1–4*, Springer Verl. Berlin-Heidelberg etc.

Buck, C. G. (1965). *Advanced calculus*, McGraw-Hill, New York etc. 1956, 1965.

Chakraborty, K., Kanemitsu, S., Kumagai, H. and Sato, K. (2009). Shapes of objects and the golden ratio, *J. Shangluo Univ.* **23**, 18–27.

Chakraborty, K., Kanemitsu, S. and Kuzumaki, T. (2016). *A quick introduction to complex analysis*, World Sci., Singapore etc.

Chandrasekharan, K. and Minakshisundaram, S. (1952). *Typical means*, Oxford UP, Oxford.

Comtet, L. (1974). *Advanced Combinatorics: The Art of Finite and Infinite Expansions*, Reidel, Dordrecht.

Davis, P. J. (1979). *Circulant matrices*, Wiley New York etc. 1979.

Davis, P. J. and Hersh, R. (1986) *Descartes' dream—The world according to mathematics*, Harcourt Brace Jovanovich Publ., San Diego etc.

Dieudonné, J. (1969). *Treatise on analysis* I-VIII, Pure Appl. Math., 10-I-10-8, Academic Press, New York-London, 1969–1993 (Volume I was published under the title, Foundations of modern analysis, Pure Appl. Math., 10-I (2nd ed.), 1969).

Donoghue Jr., W. J. (1969). *Distributions and Fourier transforms*, Academic Press, New York-London.

Doss, J. D. (1994). *The Shaman Sings*, Avon Books, New York, p. 115, ll. 20-21.

Engelking, R. (1977). *General topology*, PWN, Warszawa.

Flanders, H. (1963). *Differential forms with applications to the physical sciences*, Academic Press, New York-London.

Goldstein, L. J. (1971). *Analytic number theory*, Prentice Hall, New Jersey.

Hardy, G. H. (1949). *Divergent series*, Oxford UP. Oxford 1949.

Hardy, G. H. and Riesz, M. (1972). *The general theory of Dirichlet's series*, CUP. Cambridge 1915; reprint, Hafner, New York.

Hasse, H. (1930). Ein Summierungsverfahren für die Riemannsche ζ-Reihe, *Math. Z.* **32** (1930), 458–464.

Hewitt, E. and Stromberg, R. (1965). *Real and abstract analysis. A modern treatment of the theory of functions of a real variable* (GTM 25), Springer-Verl. New York.

Hille, E. (1972). *Methods in classical and functional analysis*, Addison-Wesley, New York.

Kakita, T. (1985). *Introduction to the theory of Schwarz distribution*, Nihon-hyoron-sha, Tokyo.

Kamata, Y., Ma, T.-L., Sun, Y. and Kanemitsu, S. (2019). Electric current is a continuous flow, *Advances in special functions and analysis of differential equations*, to appear.

Kanemitsu, S. and Tsukada, H. (2007). *Vistas of special functions*, World Scientific, Singapore-London-New York.

Kanemitsu, S. and Tsukada, H. (2015). *Contributons to the theory of zeta-functions: The modular relation supremacy*, World Scientific, Singapore etc.

Katznelson, Y. (2004). *Introduction to harmonic analysis*, 3rd ed., Cambridge UP, Cambridge. 2004.

Kelley, J. (1955). *General topology*, van Nostrand, New Jersey.

Kestelman, H. (1960). *Modern theories of integration*, Dover, New York.

Kitajima, H. and Kanemitsu, S. (2012). Math-Phys-Chem approaches to life, *Intern. J. Math. Math. Sci.* Volume 2012, Article ID 371825, 29 pages (doi:10.1155/2012/371825), published May 13, 2012.

J. Kotre, White gloves—How we create ourselves from our memory, The Free Press, New York etc.

Lal, B. (1954). *Mathematical theory of electromagnetism*, Asia Publishing House, London.

Lehmer, D. H. (1975). Euler constants for arithmetic progressions, *Acta Arith.* **27**, 125–142; Selected Papers of D. H. Lehmer, Vol. II, 591-6oe, Charles Babbage Res. Center, Manitoba 1981.

Levy, H. and Lessman, F. (1992). *Finite difference equations*, Dover, New York.

Li, F.-H. and Kanemitsu, S. (2019). Around boundary functions of the right half-plane and the unit disc, *Advances in Special functions and analysis of Differential Equations*, to appear.

Li, F.-H., Wang, N.-L. and Kanemitsu, S. (2913). *Number theory and its applications*, World Sci., London-Singapore-New Jersey.

Mainardi, F. and Pagnini, G. (2003). Salvatore Pincherle: the pioneer of the Mellin-Barnes integrals, *J. Comp. Appl. Math.*, 331–342.

Maurin, K. (1973). *Analiza*, Part I, PWN, Warszawa.

Mizohata, S. (1965). *The theory of partial differential equations*, Iwanami-shoten, Tokyo.

Papoulis, A. (1962). *The Fourier integral and its applications*, McGraw-Hill, New York.

Pincherle, S. (1888). Sulle funzioni ipergeometriche generalizzate, *Atti R. Acad. Lincei; Rend. Cl. Sci. Fiz. Mat. Nat.*(Ser. 4) **4**, 694–700; 792–799.

Pólya, G. and Szegö, G. (1972). Problems and theorems in analysis I, II, Springer-Verl., Berlin-Heidelberg.

Prachar, K. (1957). *Primzahlverteilng*, Springer Verl., Berlin etc. 1957, 2001.

Rademacher, H. (1973). *Topics in analytic number theory*, Springer-Verlag, Berlin.

Riemann, B. (1854). Über die Darstellbarkeit einer Function durch eine trigonometrische Reihe, *in Riemann's Collected works*, pp. 227–265.

Rudin, W. (1953). Principles of mathematical analysis, 3rd ed., 1976, MacGraw-Hill, New York 342 pp.

Rudin, W. (1966). Real and complex analysis, 3rd ed., 1987, MacGraw-Hill, New York 416 pp.

Rudin, W. (1973). Functional analysis, 2nd ed., 1991, MacGraw-Hill, New York 424 pp.

Rudin, W. (1962). *Fourier analysis on groups*, MacGraw-Hill, New York.

Serre, J. P. (1973). *A course in arithmetic*, Springer Verl., New York.

Shannon, C. E. (1948). A mathematical theory of communication, *The Bell system techn. J.* **27**, 379–423, 623–656.

Sierpinski, W. (ed. by Schinzel, A.) (1987). *Elementary number theory*, PWN Warszawa.

Sugawara, M. (1966). *Introduction to topology*, Asakura-shoten, Tokyo (in Japanese).

Sun, Y., Agarwal, P. and Kanemitsu, S. (2020). From Paley-Wiener theorem to Abel continuity theorem, to appear.

Takagi, T. (1938). Survey on analysis, Iwanami-shoten, Tokyo (in Japanese).

Temple, G. (1953). Theories and applications of generalized functions, J. London Math. Soc. **28**, 134–148.

Tsukada, H. (2015). Tychonoff's theorem as a direct application of Zorn's lemma, *Pure and Appl. Math. J.* 4(2–1): 14–17.

B. L. van der Waerden (1991). Algebra, I, II, Springer Verl., New York etc.

Wang, X.-H., Mehta, J. and Kanemitsu, S. (2020). The boundary Lerch zeta-function and short character sums a là Y. Yamamoto, to appear.

Washington, L. C. (1982). *Introduction to cyclotomic fields*, Springer Verl., New York.

Weaver, H. J. (1983). *Applications of discrete and continuous Fourier transforms*, Wiley, New York etc.

Whittaker, E. T. and Watson, G. N. (1927). *A course of modern analysis*, 4th ed., Cambridge UP, Cambridge.

Widder, D. W. (1946). *The Laplace transform*, Princeton UP, New Jersey 1946.

Widder, D. W. (1989). *Advanced calculus*, 2nd ed. Dover Publ., New York 1989.

Willard, S. (1978). *General topology*, Addison-Wesley, Mass.

Yoshida, K. (1956). *Modern analysis*, Kyoritsu Shuppan, Tokyo (in Japanese).

Zemanian, A. H. (1969). *Theory of distributions and analyis of transforms*, PWN, Warszawa.

Zygmund, A. (2003). *Trigonometrical series*, 3rd ed. Cambridge UP, Cambridge.

Index

Printed in the United States
by Baker & Taylor Publisher Services